AF808739

Reasoning with Attitude

Reasoning with Attitude

Foundations and Applications of Inferential Expressivism

LUCA INCURVATI
AND
JULIAN J. SCHLÖDER

OXFORD
UNIVERSITY PRESS

OXFORD
UNIVERSITY PRESS

Oxford University Press is a department of the University of Oxford.
It furthers the University's objective of excellence in research, scholarship,
and education by publishing worldwide. Oxford is a registered trade mark of
Oxford University Press in the UK and in certain other countries.

Published in the United States of America by Oxford University Press
198 Madison Avenue, New York, NY 10016, United States of America.

© Oxford University Press 2023

Some rights reserved. No part of this publication may be reproduced, stored in
a retrieval system, or transmitted, in any form or by any means, for commercial purposes,
without the prior permission in writing of Oxford University Press, or as expressly
permitted by law, by licence or under terms agreed with the appropriate
reprographics rights organization.

This is an open access publication, available online and distributed under the terms of a
Creative Commons Attribution – Non Commercial – No Derivatives 4.0
International licence (CC BY-NC-ND 4.0), a copy of which is available at
http://creativecommons.org/licenses/by-nc-nd/4.0/.

You must not circulate this work in any other form and you must impose this same condition on any acquirer

CIP data is on file at the Library of Congress

DOI: 10.1093/oso/9780197620984.001.0001

ISBN 978–0–19–762098–4

Integrated Books International, United States of America

MIX
Paper
FSC
FSC® C183721

To our parents

Contents

Preface

Words, sentences, and languages are endowed with meaning. Language allows us to convey thoughts about events occurring in distant places, and to coordinate on actions to be undertaken by our future selves. But what does the meaning of linguistic expressions consist in? A longstanding view in the history of philosophy, tracing at least as far back as Plato's *Cratylus*, identifies the meaning of an expression with what it stands for—its reference. This view is known as referentialism and has become something of a paradigm within the study of meaning. This is no doubt due, at least in part, to the success of formal semantics, which has over the years provided increasingly sophisticated referential analyses of the contribution of a large number of expressions to the meaning of the sentences in which they occur.

This is not to say that referentialism has gone unchallenged. One strand of opposition comes from the expressivist analysis of moral vocabulary. Expressivism emerged in the heyday of logical positivism as an attempt to reconcile moral discourse with an empiricist metaphysics and has since then had many reincarnations. These reincarnations share the idea that meaning is to be explained in terms of speech acts, which provide linguistic expression to our inner lives. Another strand of opposition to referentialism comes from the inferentialist analysis of logical vocabulary. Inferentialism has its roots in Wittgenstein's remarks in the *Philosophical Investigations* that meaning is use, and takes the meaning of expressions to be explained in terms of their role in inferences.

A central contention of this book is that, their differences notwithstanding, expressivism and inferentialism are best seen as opposing referentialism on the basis of the same pragmatist insight: that semantic explanations should not go beyond what is needed to explain the role of words in our practices. Expressivists focus on the attitudes that words are used to express; inferentialists focus on the inferences that words are used to draw. In this book, we lay the foundations for *inferential expressivism*, a theory of meaning which countenances both aspects of our linguistic practice and explains meaning in terms of the inferences we draw involving the attitudes we express.

Inferentialism and expressivism face challenges of their own. The history of ethical expressivism and its many reincarnations is, at least in part, the history of the struggle to account for the behaviour of expressions in embedded contexts. This is the core of what is known as the Frege–Geach Problem. Inferentialism, for its part, has remained limited in scope, so much so that doubts have been rightly raised about its ability to provide a theory of meaning for expressions other than simple logical constants such as *and* and *or*.

We argue that, by joining forces, expressivism and inferentialism can meet their key challenges. By explaining meaning inferentially, we contend, expressivists can solve the Frege–Geach Problem. The problem, as we show, arises not only in meta-ethics, but in several domains in which an expressivist analysis is called for. The resources of inferentialism afford expressivists the means of solving the problem across the board. By enriching their repertoire with attitude expressions of various kinds, in turn, inferentialists can provide semantic analyses of a variety of linguistic items going well beyond the simple logical constants. We demonstrate this by providing applications of the inferential expressivist approach to a diverse range of linguistic items, including epistemic modals, probability operators, conditionals, moral predicates, the truth predicate, and propositional attitude predicates.

Referential semantics is typically carried out in a Tarskian model-theoretic framework. We argue that inferentialist expressivist semantics is best carried out in what we call a *multilateral* proof-theoretic framework. This framework makes use of inference rules involving formulae decorated with signs standing for speech acts expressing attitudes. We obtain this framework by extending standard *bilateral* frameworks, which include signs for assertion and rejection, with signs for speech acts such as weak and strong assertion, approval and disapproval, and supposition. Adopting this framework opens up the possibility of distinguishing between two notions of legitimate inference: legitimate inference as preserving commitment and legitimate inference as preserving evidence. Once properly incorporated into the framework, this distinction motivates principled restrictions on the meta-rules of classical logic and leads to a uniform solution to puzzles and paradoxes of, among others, truth, epistemic modality, and conditionals. At the same time, the theories that we develop can account for the linguistic data in the relevant domains that have preoccupied semanticists whilst validating all classically valid arguments. The prospects for a research programme combining expressivism and inferentialism are bright indeed.

As should be clear, then, there are different themes running through this book: the pragmatist theme concerning the shape our semantic explanations should take; the foundational theme concerning how to combine expressivism and inferentialism into a novel theory of meaning; the expressivist theme concerning how the resulting approach has the resources to solve the Frege–Geach Problem; the formal semantics theme concerning how the approach can be applied to provide rich semantic theories of notions from a variety of domains; the logical theme concerning how to solve semantic and epistemic paradoxes of various kinds. Although we deem these themes to be intertwined and complementary, we hope that each of them will be of interest in its own right.

* * *

We now provide an outline of the structure of the book and of the content of its ten chapters. In the first two chapters, we provide opinionated introductions to the semantic frameworks of expressivism and inferentialism, their appealing features, and the distinctive problems they face. We argue that both expressivism and inferentialism are best motivated on the basis of their adherence to a pragmatist approach to semantic theorizing. Expressivism, in particular, complies with the meta-theoretical principle we dub the *Pragmatist Razor*, which underwrites our challenges to various forms of referentialism throughout the book. We also make a start on solving some of the problems expressivism and inferentialism are confronted with but argue in Chapter 1 that to solve the Frege–Geach Problem expressivists had better become inferentialists, and in Chapter 2 that to solve the Problem of Limited Applicability inferentialists had better become expressivists.

In Chapter 3, we show how to combine expressivism and inferentialism into inferential expressivism. Our point of departure is the bilateralist approach to meaning which, we argue, is best understood as a form of inferential expressivism about negation. We go on to present the inferential expressivist solution to the Frege–Geach Problem as applied to the case of embeddings of negation in conditional antecedents. We conclude the chapter by arguing that bilateralism faces the problem of weak rejections and that to solve this problem, bilateralists should become multilateralists.

In Chapter 4 we show how the extension from bilateralism to multilateralism opens up the possibility of solving the Problem of Limited Applicability. In particular, we extend the basic multilateral logic of the previous chapter into an epistemic multilateral logic and use this logic as the basis for an inferential expressivist explanation of the meaning of the epistemic modal *might*. We note that the strategy used to explain the meaning of *might* is an application of a general *multilateral methodology* for providing inferential expressivist explanations of the meaning of linguistic items.

The remaining chapters apart from the concluding one detail applications of the multilateral methodology to arrive at inferential expressivist accounts of further pieces of vocabulary. We begin by providing an inferential expressivist account of moral vocabulary in Chapter 5. We show that this account has the resources to address a particularly strong version of the Frege–Geach Problem, to solve the Wishful Thinking Problem for ethical expressivism, and to account for moral motivation.

We then turn to attitude ascriptions in Chapter 6. We give an inferential expressivist semantics for ascriptions of various kinds of attitudes and show that this semantics is compatible with what we deem to be the correct shape of a solution to the Many Attitudes Problem for expressivism. We go on to discuss the question of realism about the mind from the point of view of inferential expressivism.

In Chapter 7, we provide an inferential expressivist account of the truth predicate. We prove that the resulting theory of truth has the resources to provide a principled solution to the semantic paradoxes and their revenge versions. The theory, moreover, can be provably shown to share profound similarities with supervaluational approaches to truth.

In Chapter 8, we tackle conditionals. We extend the multilateral framework so as to include not only signs for categorical speech acts, but also for conditional ones. We then explain the meaning of conditionals in terms of such binary speech acts. We show that the resulting inferential expressivist account of conditionals has several attractive features. Notably, it avoids Gibbard's collapse problem and Curry's Paradox in a natural way.

As a final application of the multilateral methodology, we present in Chapter 9 an inferential expressivist account of epistemic uses of probabilistic vocabulary. We give an inferential expressivist analysis of *it is probable that* which accounts for several linguistic data concerning probability expressions and avoids the Frege–Geach Problem. We end the chapter by explaining how the account can be extended to expressions of comparative probability and to quantitative probability talk.

We conclude the book in Chapter 10 by illustrating the road ahead for the inferential expressivist approach to meaning. We present a number of further possible applications of the approach and outline how one may go about carrying out these applications.

Some of the chapters are based on material that has appeared elsewhere, though often in a very different form. Chapter 3 draws on 'Weak rejection' (*Australasian Journal of Philosophy*, 2017), Chapter 4 draws on 'Weak assertion' (*Philosophical Quarterly*, 2019) and 'Epistemic multilateral logic' (*Review of Symbolic Logic*, 2022a), Chapter 5 draws on 'Inferential expressivism and the negation problem' (*Oxford Studies in Metaethics*, 2021), and Chapter 7 draws on 'Inferential deflationism' (*Philosophical Review*, Forthcoming). We are grateful to the editors and publishers for permission to use material from these articles.

* * *

On 13 June 2014 the *LoLa Day* took place at the University of Amsterdam. During this meeting, several members of the Logic and Language (LoLa) group of the Institute for Logic, Language and Computation gave talks on aspects of their current research. We both presented material on the speech act of rejection. Having discovered that we were both working on what we considered an unjustly neglected topic, we decided to exchange notes. So began the collaboration that led to this book. We would like to thank Franz Berto and Robert van Rooij for organizing that meeting as heads of LoLa at the time.

Over the years, the material in this book has benefitted from conversations and feedback from many friends and colleagues. We are especially grateful to Maria Aloni, Daniel Altshuler, Nicholas Asher, Bahram Assadian, Dorit Bar-On, Franz Berto, Arianna Betti, Justin Bledin, Paul Bloomfield, Lwenn Bussière, Tim Button, Matthew Chrisman, Mariangela Cocchiaro, Paul Dekker, Imogen Dickie, Cian Dorr, Raquel Fernández, Filippo Ferrari, Stephen Finlay, Salvatore Florio, Melissa Fusco, Manuel García-Carpintero, Lewis Gordon, Magdalena Kaufmann, Angelika Kratzer, Manfred Krifka, Alex Lascarides, Hannes Leitgeb, Bill Lycan, Michael Lynch, Matthew Mandelkern, Teresa Marques, Thomas Müller, Julien Murzi, Carlo Nicolai, Alejandro Pérez Carballo, Francesca Poggiolesi, Huw Price, Greg Restall, Dave Ripley, Lucas Rosenblatt, Marcus Rossberg, Lorenzo Rossi, Giorgio Sbardolini, Thomas Schindler, Lionel Shapiro, Stewart Shapiro, Sebastian Speitel, Una Stojnić, Martin Stokhof, Christine Tiefensee, Robert van Rooij, Frank Veltman, Jack Woods, and Hedde Zeijlstra. We would also like to thank Peter Ohlin and the team at Oxford University Press for their help and support throughout the writing and production process.

We have presented some of the material in this book in a number of forums. We are grateful to the organizers and audiences of the Mathematical Logic Colloquium at the University of Bonn, the Theoretical Philosophy Colloquium at Utrecht University, the PLM Masterclass with Angelika Kratzer at the University of Salzburg, the Philosophy Colloquium at the University of Gothenburg, the New York Philosophy of Language Workshop, the XIV Annual Madison Metaethics Workshop, the Semantics and Philosophy in Europe 11 Conference at the University of Warsaw, the Higher Seminar in Theoretical Philosophy at Uppsala University, the Monday Colloquium at the University of Bonn, the Workshop on Propositions, Properties, Sets, and Other Abstract Objects at the University of Amsterdam, the SPAGAD2 Workshop and the Mini-Workshop on Discourse of the ZAS (Berlin), the Logic Colloquium at the Munich Center for Mathematical Philosophy, the Logic Colloquium and Philosophy Colloquium at the University of Konstanz, the London Group for Formal Philosophy, the Workshop on Classical and Non-Classical Approaches to Paradoxes at the University of Buenos Aires, the von Weizsäcker Colloquium at the University of Tübingen, the Philosophy of Hybrid Representations reading group at the University of Barcelona, the Logic Colloquium at the University of Connecticut, the Varieties of Philosophical Expressivism Workshop at the University of Mannheim, and the Rhineland Expressivism Workshop at the University of Bonn. We are also grateful to the audiences of the EXPRESS Workshops on Bilateral Approaches to Meaning and on Expressivist Approaches to Meaning, which we organized at the University of Amsterdam together with Lwenn Bussière and Giorgio Sbardolini.

This work has received funding from the European Research Council (ERC) under the European Union's Horizon 2020 research and innovation programme (grant agreement No. 758540) within the project *From the Expression of Disagreement to New Foundations for Expressivist Semantics* (EXPRESS). The ERC funding made it possible to devote ourselves full time to this project and to publish the book open access. We are deeply grateful.

Finally, before we begin, some words from each of us individually.

From Luca: Thank you to Julian for being such a wonderful collaborator and for being so much fun to work with; to Tim Smiley for introducing me to rejection when I was a graduate student; to the Cambridge pragmatists for influencing me in ways that I am only now fully realizing; to Sarah for her continuous love and not minding too much me testing her acceptability judgements about embedded epistemic modals; to Tommaso for bringing so much joy to my life; and to my parents for always believing in me.

From Julian: I am grateful to Luca for being, over the years, an outstanding advisor, supervisor, colleague, and co-author; to Raquel Fernández, Alex Lascarides, and Paul Dekker for their teachings which have stayed with me more than they might know; to Yu'an for her love and toleration of late nights working; and to my parents for their unwavering support.

List of Figures

1
Expressivism

This book presents and develops *inferential expressivism*, a novel approach to the study of meaning which combines elements of the inferentialist and expressivist programmes. This chapter introduces expressivism and the challenges it faces. The next one introduces inferentialism.

Both expressivism and inferentialism come in local and global varieties. Global approaches to meaning take their central tenets to apply across the board; local approaches restrict attention to some specific set of expressions or some particular area of discourse. Historically, expressivism originated as an approach to the meaning and function of moral vocabulary, so we focus in this chapter on ethical expressivism, but much of what we say can be easily applied to expressivist treatments of other areas. We shall indicate when this is not the case.

We begin by introducing some key distinctions within the theory of meaning which will be useful in the remainder of the book. We then present what we see (and are often seen) as the two main concerns that have been particularly influential in the emergence and development of ethical expressivism. We explain how the related view of speaker subjectivism can be seen as an early attempt to deal with these concerns. According to ethical subjectivism, the meaning of moral vocabulary consists in *reports* of attitudes, and the constitutive function of moral statements is to report these attitudes. Speaker subjectivism faces the problem of accounting for the phenomenon of moral disagreement. This problem motivates the focus on *expressions* of attitudes, which is the central innovation of expressivism. Traditional ethical expressivism takes the meaning of moral vocabulary to consist in expressions of attitudes, and the function of moral judgements to be that of expressing these attitudes.

Traditional expressivism faces a number of problems, which we also introduce in this chapter. Of these, the so-called *Frege–Geach Problem* is perhaps the best known, and the one that has historically been regarded as a thorn in the side of expressivism. Indeed, sophisticated and hybrid forms of expressivism have been developed, at least in part, in response to

Reasoning with Attitude: Foundations and Applications of Inferential Expressivism. Luca Incurvati and Julian J. Schlöder, Oxford University Press. © Oxford University Press 2023.
DOI: 10.1093/oso/9780197620984.003.0001

this problem. One of the central contentions of the book is that inferential expressivism offers the best available solution to this problem, as it arises not only in the ethical domain, but in other domains as well.

1.1 Semantics, postsemantics, and meta-semantics

This book is concerned with the theory of meaning. Philosophers have historically assigned different meanings to 'theory of meaning'. The central distinction here, which has become customary in recent years, is that between semantics and meta-semantics (Kaplan 1989a).

Semantics is concerned with the assignment of semantic values to the expressions of a language. Semantic theories are typically *compositional*: they assign semantic values to the simple expressions of the language, and determine the semantic value of complex expressions as a function of the semantic value of the simpler ones. The compositionality of semantic theories is intended to account for the fact that speakers can understand and produce complex expressions which they have never encountered before.

In standard, *referential* semantic theories, expressions are assigned denotations as their semantic values. That is, a referential semantic theory specifies a mapping $[\![\cdot]\!]$ from expressions to their denotations. Thus, for instance, the semantic theory computes the semantic value of *Andrea rests* by assigning Andrea as the denotation of *Andrea*, the RESTING function—that is the function mapping the things that rest to the value 1 and all other things to the value 0—as the denotation of *rests*, and determining how these semantic values combine to determine the semantic value of the sentence. In symbols, and using the lambda notation to represent functions:

$$[\![\text{Andrea}]\!] = \text{Andrea}$$
$$[\![\text{rests}]\!] = \lambda x.x \text{ rests}$$
$$[\![\text{Andrea rests}]\!] = 1 \text{ iff Andrea rests}$$

Typically, the RESTING function is taken to represent the property of resting, and the values 1 and 0 to represent truth and falsity. Under this interpretation, the semantic theory specifies the truth conditions of *Andrea rests* on the basis of the meaning of its components.

Standard semantic theories in the tradition of Richard Montague (1970) or David Lewis (1970) typically assign semantic values to the simple expressions relative to an *index i*, an *n*-tuple whose elements are typically taken to represent the circumstances of evaluation. Accordingly, the semantic value of a sentence would be a function from indices to 0 or 1. Suppose, for instance, that the index includes elements *w* and *t*, standing, respectively, for the world and time of evaluation. Then the semantic clauses of our toy example look as follows.

$$[\![\text{Andrea}]\!]^{t,w} = \text{Andrea}$$
$$[\![\text{rests}]\!]^{t,w} = \lambda x.x \text{ rests at } t \text{ and } w$$
$$[\![\text{Andrea rests}]\!]^{t,w} = 1 \text{ iff Andrea rests at } t \text{ and } w$$

To deal with indexical phenomena in language, it is customary to also take the *context of utterance* into account. On the standard picture going back to David Kaplan (1989b), this is done by relativizing semantic values to a parameter *c*, representing the context. This is what Lewis (1980) calls the *variable-but-simple-semantic-value* approach to semantics. On this approach, semantic values are variable because they change from context to context, at least for some sentences, such as those containing indexical vocabulary; they are simple because the semantic values of sentences are mere functions from indices to truth values. On the *constant-but-complicated-semantic-value* approach to semantics, by contrast, semantic values are constant across contexts; they are complicated because they are functions from indices and contexts to truth values. The constant-but-complicated-semantic-value approach has recently been revived in the work of Brian Rabern (2012), Seth Yalcin (2014), and Paolo Santorio (2019) among others.

The referential semantic theory computes a semantic value for each sentence of the language. Supposing, for simplicity, that we have settled for the variable-but-simple-semantic-value approach, the semantic theory specifies, for each sentence, the conditions under which the sentence receives value 1 (representing truth) at a context and at an index. This is what John MacFarlane (2003) calls the *semantics proper*. The next stage is to define suitable notions of truth and validity for the language, the stage MacFarlane calls the *postsemantics*. As MacFarlane puts it, at the end of the day we are interested in truth *simpliciter* (perhaps relativized to a context), rather than truth at an index. But the semantics proper affords the means to define truth *simpliciter* in terms of truth at an index: we can take a sentence to be true *simpliciter* just in case it is true at the index i_c of the context of utterance.

In case the index *i* includes a world and a time, for instance, i_c will include the world and time of the context of utterance. As for validity, according to the standard definition (and suppressing context sensitivity for simplicity) an argument is valid just in case, for every index *i*, if the premisses of the argument are true at *i*, then the conclusion is true at *i*. Logical truth is then defined as the special case in which the set of premisses is empty, that is by saying that a sentence is logically true just in case it is true at every index.

The formal semantic theory just described assigns Andrea to *Andrea* and the RESTING function to *rests* (relative to an index *i*). Moreover, it assigns *Andrea rests* a function from *i* to the values 0 and 1 (relative to the parameter *c*) as its semantic value. As noted, it is customary to interpret the functions assigned to predicates as representing the corresponding properties, so that the RESTING function represents the property of resting. Similarly, it is customary to interpret the members of the index as representing the circumstances of evaluation and the parameter *c* as representing the context of utterance. Finally, the values 1 and 0 are typically interpreted as standing, respectively, for truth and falsity. However, all of this is not part of the semantic theory, but rather of the *interpreted semantic theory*: the semantic theory under a particular interpretation. One way of looking at the situation is to consider the formal semantics as providing a model, and the interpretation as specifying what the elements of the model are taken to represent. What the elements of the model are taken to represent is the answer to the Interpretation Question.

Interpretation Question. What does the semantic value of an expression represent?

Of course, the interest of a semantic theory typically lies in its capacity to explain facts that belong to the domain being modelled, so it is typically interpreted semantic theories that are the objects of interests of semanticists and philosophers. Thus, *referentialists*, who hold that the meaning of an expression is given by its referent, will typically take, for instance, the semantic value of a predicate to be a function, and this function to represent a property. Nonetheless, it is important to always keep track of the distinction between semantic theory as an abstract structure and its interpretation.

So far, we have presented semantics (which includes the semantics proper, the postsemantics, and their interpretation). This is one sense of 'theory of meaning'. Another sense is provided by *meta-semantics*, the study of foundational questions about semantics. Meta-semantic questions abound. Here we highlight a number of meta-semantic questions that will be particularly relevant in the remainder of the book. We use our toy example from standard referential semantics as a running example.

The semantic theory tells us that Andrea is the semantic value of *Andrea* and the RESTING function is the semantic value of *rest*. But what makes it the case that this is so? One central meta-semantic question is the *metaphysical* question of what makes it the case that an expression has the semantic value it does.

Meaning Determination Question. In virtue of what does an expression have the semantic value it does?

A semantic theory does not provide an answer to the Meaning Determination Question—nor does it intend to. A theory whose aim is to provide an answer to the Meaning Determination Question is known as a *foundational semantics* (Speaks 2021)—in effect, a branch of meta-semantics.[1] One influential answer to the Meaning Determination Question is provided by representationalism. According to *representationalism*, an expression has the semantic value it does in virtue of what it represents. Thus, for instance, *Andrea* has the semantic value it does because it purports to represent Andrea, and *rests* has the semantic value it does because it purports to represent the property of resting.

The Meaning Determination Question is to be distinguished from the question of what *led* to an expression having the semantic value it does. This question concerns the etiology of meaning. Of course, it is in principle possible to think that the natural history of an expression *is* what makes it the case that it has the semantic value it does. But this point of view is not forced upon us. For instance, one may accept a game-theoretic account of the origins of meaning, whereby expressions come to have the semantic values they do as a result of some coordination game. This is compatible with thinking, for instance, that what makes it the case that expressions have the

[1] Following Stalnaker (1997), it has become relatively common to contrast foundational semantics with *descriptive semantics* (García-Carpintero 2012). This is unfortunate, because it makes it sound as if (first-order) semantics is a merely descriptive enterprise without any explanatory ambitions, a description which most working semanticists would likely resist, as noted by Szabo (2019). As Szabo points out (and Stalnaker himself stresses), semantic theories crucially purport to explain speaker productivity. The fact that semantic theories are compositional is, as noted, intended to account for this.

semantic value they do is the existence of a representation relation between the expressions and their semantic values.

Speakers possess linguistic knowledge and part of this knowledge consists in knowledge of the meaning of simple expressions of the language and how their meaning contributes to the meanings of more complex expressions. The second meta-semantic question is the *epistemological* question about what makes it the case that speakers have the semantic knowledge they do.[2]

Meaning Knowledge Question. In virtue of what does a speaker knowing the semantic value of an expression have that knowledge?

The third meta-semantic question concerns the constitutive function of a certain class of expressions of a given language, and of statements involving these expressions.

Functional Role Question. What is the constitutive function of an expression (or statements involving this expression) belonging to a certain area of discourse?

Consider the sentence *Andrea rests*. It seems natural to say that its constitutive function is to *describe* the world as being a certain way. It is commonly assumed that in describing the world a certain way, we give voice to our ways of representing the world as being a certain way—we express *beliefs* that represent the world as being a certain way. It follows that in uttering *Andrea rests*, a speaker expresses a belief representing the world as being a certain way.

The Functional Role Question is a question about the *raison d'être* of having certain expressions in our language. As such, it is a meta-semantic question. However, it is intimately connected with semantic issues. In particular, a semantics should assign meanings to expressions in such a way that it is possible to explain how those expressions can fulfil the functional role they have. When a sentence has a descriptive function, as it would seem to do in the case of *Andrea rests*, referential semantics (under the standard answer to the Interpretation Question) would seem to allow one to provide

[2] Yalcin (2014) argues that the metaphysical meta-semantic question should be replaced by the epistemological meta-semantic question.

a natural story about how it can serve that function: in uttering *Andrea rests*, one communicates that its truth conditions are satisfied.

Ethical expressivism, as we shall now see, challenges the idea that the constitutive function of moral statements can be regarded as a descriptive one.[3]

1.2 Why expressivism?

It appears straightforward to give a referential semantics for simple moral sentences along the lines sketched in the previous section. Consider the sentence *Lying is wrong*. A simple referential semantics will assign the act of lying to *lying*, assign the WRONG function to *wrong*, and compute the conditions under which *Lying is wrong* receives the value 1 on the basis of these assignments. Given the standard, referentialist answer to the Interpretation Question, the referential semantics therefore gives the truth conditions of *Lying is wrong*, which are determined by the fact that *lying* stands for the act of lying and *wrong* stands for the property of being wrong.

A semantics must make it possible to explain how moral statements, just like other statements, can perform their constitutive function. Which function is this? *Moral descriptivism* holds that it is the same function as that played by statements such as *Andrea rests*. That is, the function of moral talk too is to describe how things are in the world. Again, assuming that by describing the world as being a certain way we express beliefs that represent the world as being in a certain way, it follows that in uttering, say, *Lying is wrong*, a speaker expresses a belief representing the world as being a certain way. This view is known as moral cognitivism.

Moral Cognitivism. Moral statements express beliefs.

The answer to the Functional Role Question concerning moral vocabulary provided by moral descriptivism allows one to extend to moral sentences the explanation we gave above of how sentences such as *Andrea rests* perform

[3] In Chapter 7, we will see that deflationism about truth challenges the idea that the constitutive function of truth talk is descriptive. Instead, according to deflationists, the constitutive function of the truth predicate is to allow speakers to make compendious and indirect endorsements.

their constitutive function: uttering a moral sentence such as *Lying is wrong* serves to communicate that the truth conditions of this sentence are satisfied.

Moral descriptivism extends to moral discourse what appears to be the constitutive function of common-or-garden talk about middle-sized objects. This makes it possible to provide a natural story of how moral sentences can perform their constitutive function on the basis of referential semantics. This is certainly an attraction of moral descriptivism. But it carries, prima facie, certain familiar costs.

Moral descriptivism holds that moral sentences serve to describe the world, so that in uttering *Lying is wrong* one communicates that its truth conditions are satisfied. But according to referential semantics on its standard interpretation, these truth conditions are satisfied just in case the act of lying has the property of wrongness. But what kind of property is this? If this is a property belonging to the natural furniture of the world, then it is not clear what property that might be. If it is a non-natural, *sui generis* property, then the interpreted referential semantics would seem incompatible with a scientific conception of the world.

Naturalistic scruples have been high on the anti-descriptivist agenda. Reliance on such scruples, however, has often obscured what we take to be the motivation for adopting the expressivist approach that we will defend in this book, namely that whatever answers one gives to questions concerning the nature of moral properties appears to make no difference to our understanding of moral discourse. This suggests that the functional role of moral sentences is not that of describing the world.

It is helpful here to contrast ethical expressivism with moral error theories (Mackie 1977; Olson 2014; Streumer 2017). Error theories hold that albeit purporting to describe the world, moral judgements systematically fail to do so: neither the natural nor the non-natural furniture of the world includes moral properties, and hence moral statements are typically false or truth-valueless (depending on one's favourite view of reference failure). Naturalistic scruples can be assuaged by embracing the moral error theory variety of descriptivism. But from the expressivist viewpoint which we shall introduce in the next section, error theorists make the same mistake as the standard descriptivist who posits a realm of natural or non-natural properties to serve as the entities that the semantic values of predicates are supposed to model. Both types of descriptivists appeal to features that appear to be unnecessary to explain the *raison d'être* of moral discourse. Indeed, moral error theorists typically take this *raison d'être* to consist in the social function of moral discourse. But then it seems more straightforward to directly explain the meaning of moral vocabulary in terms of this function.

The broadly pragmatist insight at work here is that the meaning attributed to expressions or sentences should not go beyond what is needed to explain the *raison d'être* of those expressions or sentences.[4]

The Pragmatist Razor. Avoid semantic explanations that are not needed to account for the functional role of an expression or sentence.

What does the Pragmatist Razor have to recommend it? On the face of it, it may seem motivated by ontological parsimony. But the example of moral error theories shows that the Pragmatist Razor cuts both ways: in the case of moral vocabulary, it cautions against semantic explanations that are committed to the existence of moral properties as well as ones that are committed to their non-existence. Rather, the razor prevents a semantic theory from being hostage to the outcome of metaphysical disputes whose resolution would seem to have no effect on moral discourse.

A different source of concerns for moral descriptivism has traditionally come from moral motivation. Someone judging that helping others is good appears, at least in standard circumstances, to be motivated to help others. *Motivational internalism* takes this appearance at face value: if someone sincerely judges that helping others is good, they are, at least to some degree, motivated to help others.[5] The traditional expressivist approach we will introduce in the next section *implies* motivational internalism and is therefore particularly well equipped to account for the intimate connection it postulates between moral judgement and motivation. Standard moral descriptivism, by contrast, appears compatible with the possibility of someone judging that helping others is good without being thereby motivated to

[4] Brandom (1994: 144) appears to invoke something akin to the Pragmatist Razor when he writes that 'it is pointless to attribute semantic structure or content that does no pragmatic explanatory work', where the pragmatic explanatory work is to explain 'the circumstances under which judgements and inferences are properly made and the proper consequences of doing so'. The proper circumstances are the domain of discourse to which an expression belongs and the proper consequences are what the expression is used for. That is, according to Brandom, attributions of semantic content should not exceed what is required to explain an expression's function in its domain.

[5] The distinction between internal and external motivation is explicitly made by Frankena (1958). The label *motivational internalism* (and that for the opposing view, *motivational externalism*) were explicitly introduced by Svavarsdottir (1999). Classic discussions of the distinction include Darwall 1983: ch. 5 and Brink 1989: ch. 3.

help others: to ascribe the natural or non-natural property of goodness to helping others does not seem to imply being motivated to do so.

Motivational internalism is controversial. It makes the existence of amoralists—subjects who are prepared to judge an act as right or wrong without thereby having the associated moral motivation—a conceptual impossibility. If there are amoralists—or, at least, if they are not conceptually impossible—then a satisfactory account of moral discourse and practices ought not to vindicate motivational internalism (see, e.g., Brink 1989: ch. 3; Mele 1996; Svavarsdottir 1999). Nonetheless, such an account ought to explain why there appears to be an intimate connection between moral judgement and moral motivation. As we shall see in Chapter 5, the inferential expressivist account of moral vocabulary allows one to vindicate the motivating power of moral judgement without precluding the possibility of amoralists.

A third supposed advantage of expressivism is that it appears to have the resources to account for the phenomenon of moral disagreement. To explain this aspect of the view, it will be helpful to first consider a related but importantly different view, speaker subjectivism.

1.3 Speaker subjectivism

Rather than leading one to suppose that moral discourse is not in the business of describing moral states of affairs, the Pragmatist Razor and the phenomenon of moral motivation might lead one to hold that these states of affairs are of a different nature than standard moral descriptivism supposes.

The Pragmatist Razor recommends avoiding semantic explanations that are not needed to account for the functional role of an expression or sentence. Moral discourse seems intimately connected with our practice of approving, disapproving, praising, and blaming—with certain non-cognitive attitudes. To comply with the Razor one might therefore take moral sentences to be in the business of describing the states of affairs that obtain whenever the speaker has certain non-cognitive attitudes. Similarly, it is true that, on the face of it, ascribing the natural or non-natural property of goodness to helping others would not seem to have anything to do with moral motivation. But if the property that is being ascribed to helping others is that of being approved by the speaker, then moral judgement is in fact intimately connected with moral motivation—indeed, a form of motivational internalism would seem to hold.

These considerations lead to *speaker subjectivism*. Semantically, this is the view that an utterance of, say, *Helping others is good* is true just in case the speaker approves of helping others. These truth conditions are obtained by taking the satisfaction conditions of *good* to be satisfied just in case the speaker has a certain non-cognitive attitude such as approval towards helping others.

Speaker Subjectivism (Semantic). Terms receiving subjectivist treatment denote the property an object has just in case the speaker has a certain non-cognitive attitude towards it.

At the meta-semantic level, speaker subjectivism takes the constitutive role of moral statements to be that of describing the non-cognitive attitudes of the speaker.

Speaker Subjectivism (Meta-semantic). The function of moral statements is to describe the non-cognitive attitudes of the speaker.

Speaker subjectivism appears to have the resources to deal with the problems for standard forms of descriptivism. However, it has difficulties in accounting for the phenomenon of moral disagreement. When someone asserts that lying is wrong and someone denies that it is, they appear to be disagreeing. Yet, according to speaker subjectivism, they are not: the first speaker is reporting the fact that they disapprove of lying, and the second speaker is reporting the fact that they do not. And someone disapproving of lying is of course perfectly compatible with someone else not disapproving of it. An utterance of *Lying is wrong* and an utterance of its negation performed by different speakers are no more incompatible than an utterance of *It is raining here* and its negation performed in two different locations.

1.4 Traditional expressivism

According to traditional expressivism, terms receiving expressivist treatment do not have a semantic value, but serve to indicate the expression of an attitude. Thus, for instance, traditional ethical expressivism (Ayer 1936; Stevenson 1937) holds that *is wrong* does not denote a function mapping the wrong things to the truth. Rather, *is wrong* indicates the expression of the attitude of disapproval. This means that *Lying is wrong* does not *say* anything but expresses disapproval of lying. As a result, *Lying is wrong* is neither true nor false. As A. J. Ayer puts it:

> sentences which simply express moral judgements do not say anything. They are pure expressions of feeling and as such do not come under the category of truth and falsehood. (Ayer 1936: 108)

So understood, traditional ethical expressivism combines semantic and meta-semantic ingredients. At the semantic level, traditional ethical expressivism combines a negative thesis, namely the rejection of a referential semantics for moral vocabulary, and a positive thesis about what the appropriate semantics for that vocabulary is: that moral terms are to be analysed not as having a denotation but as indicating the expression of an attitude.

Thus, for instance, *Lying is wrong* is to be analysed as ☹*m*, where ☹ indicates the expression of the attitude of disapproval and *m* stands for the act of lying. Disapproval here is to be understood in very general terms: *Lying is wrong* expresses a desire-like attitude against lying. For the traditional ethical expressivist, an utterance of *Lying is wrong* is an outburst of emotion, akin to shouting *Boo!* against all acts of lying. One can then go on to analyse *Lying is right* as ☺*m*, where ☺ indicates the expression of approval, akin to shouting *Hurrah!* for all acts of lying. This is why Ayer's original account is also known, tongue in cheek, as the *boo/hurrah theory* of moral language. Traditional ethical expressivism starts from the fact that we use moral statements to express approval, disapproval, praise, and blame. It then explains the meaning of moral vocabulary precisely in terms of this, and so complies with the Pragmatist Razor.

Abstracting from the specificities of the ethical case, we can therefore lay down the two following semantic theses characterizing traditional expressivism.[6]

No Referential Semantics. Terms receiving expressivist treatment do not have a referential semantic value.

[6] To endorse the No Referential Semantics thesis is to reject the referentialist style of assigning semantic values altogether. As we will see below, certain more recent developments of expressivism, notably Gibbard 2003, retain a referential semantics, but reject the referentialist interpretation of the values provided by such a semantics. In particular, although the formal semantics assigns functions as the referents of moral predicates, these functions need not be interpreted as representing normative properties.

Traditional Expressivism (Semantic). Terms receiving expressivist treatment indicate the expression of an attitude.

At the meta-semantic level, traditional ethical expressivism combines again a negative thesis and a positive one. The negative thesis is the rejection of moral cognitivism and, as a result, of moral descriptivism: moral statements do not express beliefs and so their constitutive function is not to describe how things are in the world. The positive thesis, which distinguishes it from other forms of non-cognitivism, is that the function of moral statements is to express non-cognitive attitudes. A somewhat different suggestion was made by Charles Stevenson (1937). According to his account, the function of moral statements is not simply to *express* non-cognitive attitudes, but to *change* other people's attitudes. Ayer's and Stevenson's views are often taken to be in opposition, but in Chapter 5 we will show how they can in fact be seen as complementary. For now, we can again abstract from the ethical case and lay down the two following meta-semantic theses characterizing traditional expressivism.

Anti-Descriptivism. The function of terms or sentences receiving expressivist treatment is not to describe how things are in the world.

Traditional Expressivism (Meta-semantic). The function of terms or sentences receiving expressivist treatment is to express attitudes.

Traditional ethical expressivism deals with standard descriptivism's problems whilst avoiding the pitfalls of speaker subjectivism. Traditional ethical expressivism agrees with speaker subjectivism that the constitutive function of moral statements is not that of describing moral states of affairs. But this is not because their function is to describe our psychology. It is because their function is not to describe at all. The key distinction here is that between *reporting* an attitude and *expressing* it. To report an attitude is to describe oneself as having that attitude. To express an attitude is not to report an attitude. If one says that Andrea rests, one is not describing the world as being such that one believes that Andrea rests. This is what is described by saying that one believes that Andrea rests. Instead, in saying that Andrea rests, one expresses one's belief that Andrea rests. According to the traditional expressivist, something similar happens in the case of moral statements, except that what one expresses is not the cognitive state of belief but some non-cognitive state such as approval or disapproval. Thus, in saying that lying is wrong, one is not describing oneself as disapproving of lying. This is what is described by saying that one believes that lying is wrong. Instead, in saying that lying is wrong, one expresses one's disapproval of lying.

Traditional ethical expressivism, just like speaker subjectivism, has a natural story about moral motivation. When a speaker utters *Helping others is right*, they are expressing approval of helping others. On the assumption that their utterance was sincere, this means that they approve of helping others. If non-cognitive states such as approval and disapproval are intrinsically motivating, it follows that they are motivated to help others.

Finally, traditional ethical expressivism appears to have a plausible explanation of the phenomenon of moral disagreement. If someone utters *Lying is wrong*, they express disapproval of lying. And if someone else utters *Lying is not wrong*, they express another non-cognitive attitude, incompatible with disapproval of lying. The expression of these two incompatible attitudes constitutes a disagreement, just as uttering *Andrea is in Rome* and *Andrea is not in Rome* constitutes a disagreement about Andrea's whereabouts. To say exactly what the attitude expressed by *Lying is not wrong* is and whether talk of incompatible attitudes can be made sense of are difficult questions (Schroeder 2008a), which we will address in Chapter 5.

1.5 Attitude expression

In the previous section, we introduced traditional expressivism as involving both a semantic and a meta-semantic component. Both components are formulated in terms of the notion of attitude expression. We would now like to say more about this notion.

It is customary to distinguish between an achievement and a non-achievement reading of attitude expression (Vendler 1967). According to the achievement reading, attitude expression implies attitude possession: one can only express attitudes one does in fact have. Expressivists are not concerned with the achievement reading. For they should allow for the possibility of someone saying that lying is wrong even if in fact they do not disapprove of lying. But by expressivist lights, in having said that lying is wrong, they have expressed disapproval of lying.

According to the non-achievement reading, attitude expression does not imply attitude possession. How is attitude expression on the non-achievement reading to be understood? John Searle (1969) lays down a number of conditions a speech act must satisfy to be felicitous. One of these conditions is a *sincerity* condition: the felicitous performance of a speech act requires the speaker to have the attitude the speech act expresses. The speech act then counts as an expression of a certain attitude whenever having that attitude is required to satisfy the sincerity condition of the speech act (Searle 1969: 65). This is compatible with the sincerity condition not being satisfied: assertion expresses belief because its sincerity condition requires the speaker to believe the content of the assertion; but one may assert things one doesn't believe. Searle's view, further developed in joint work with Daniel Vanderverken (Searle and Vanderveken 1985), takes attitude expression to be a primitive, not to be analysed in terms of further notions.

An analysis of attitude expression along Gricean lines is offered by Kent Bach and Robert Harnish (1979: 15). According to their account, a speaker *S* expresses an attitude just in case *S* has the intention that, by means of recognizing this intention, the addressee takes *S*'s utterance as reason to think that *S* has that attitude. On this account, expressing an attitude involves communicative *reflexive* intentions (*R-intentions* for short): intentions to produce an effect in the audience by means (at least in part) of the recognition of this intention. Thus, for instance, a speaker expresses a certain belief by means of an utterance just in case they *R*-intend the addressee to take their utterance as reason to think that they have that belief.

Communicative reflexive intentions were introduced by Paul Grice (1957) to provide an analysis of speaker meaning. Grice argued that speaker meaning that *A* cannot be reduced to performing an action which may be taken as reason to form a belief that *A*: Ann's taking a seat at the table is reason to believe that she is about to eat, but Ann is not communicating that she is about to eat. Furthermore, Grice continued, speaker meaning cannot be reduced to performing an action with the intention that it be taken as reason to form the relevant belief: in intentionally leaving some footprints leading to a tree, the child who plays hide-and-seek and hides behind the bush is not communicating that they are hiding behind the tree. Ultimately, Grice concluded, reflexive intentions are needed: speaker meaning is analysed in terms of intentions to produce an effect in the audience by means (at least in part) of the recognition of this intention. Similar examples may be used to motivate Bach and Harnish's use of *R*-intentions: attitude expression can be reduced to neither performing an action providing evidence that one has that attitude nor to performing an action with the intention of providing such evidence.[7]

It is easy to see that Bach and Harnish's account allows for the possibility to express attitudes one does not have: one can have the *R*-intention that the addressee takes one's utterance as a reason to think that one has a certain attitude even though one does not, in fact, have that attitude. To this extent, the account is congenial to the expressivist. In appealing to reflexive intentions, however, the account faces similar challenges to those faced by Grice's account of speaker meaning, namely that reflexive intentions appear too complex to be reasonably ascribed to ordinary speakers, at least with the frequency that the account would demand (Davis 2003: 88; Siebel 2003: 356).

Grice (1969) replaces reflexive intentions with iterative ones, and it is straightforward to reformulate Bach and Harnish's account in those terms (McGlynn 2010: 24): a speaker *S* expresses an attitude just in case *S* has the intention that (i) the addressee takes *S*'s utterance as reason to think that *S* has that attitude and that (ii) the addressee recognizes that *S* intends (i) on the basis of *S*'s utterance. However, Gilbert Harman (1974) argued that in the absence of reflexive intentions, a satisfying Gricean analysis of the openness of communicative intentions will need to ascribe to speakers an infinite list of intentions, thereby making the account implausible as a realistic description of the psychological phenomena. Following a suggestion made but not endorsed by Grice (1969: 104–105) for speaker meaning, Aidan McGlynn (2010: 25) suggests that the Gricean account of attitude expression should include a proviso that the speaker *S* does not have *sneaky* intentions—intentions that the addressee be mistaken about *S*'s intention that condition (ii) be fulfilled.

The jury is out on whether a reductive account of attitude expression in Gricean terms can be given and, if so, whether it is psychologically plausible. However, it should be stressed that the possibility of providing such a reductive account is not necessary for the success of the expressivist programme. For several key philosophical notions it is in any case doubtful whether a reductive account can be given. From the expressivist perspective, it would seem that attitude expression is a natural candidate for a notion that should be taken as primitive. However, this does not prevent the expressivist

[7] Performing an action providing evidence that one has that attitude is a case of what Davis (2003: §3.1) calls *evidential* expression, as opposed to *speaker* expression, which is what the expressivist is concerned with.

from saying informative things about this notion. Compare with Timothy Williamson (2000) on knowledge: although on his view knowledge is a primitive notion that cannot be analysed in simpler terms, it is nonetheless possible to say informative things about it, such as that knowledge entails truth and that it entails belief.

One key aspect of attitude expression is its inherently social character. A sincere attitude expression makes one's attitude manifest: a sincere assertion makes one's belief manifest; a sincere expression of disapproval makes that disapproval manifest (Green 2007, 2009).[8] As we noted in the previous section and we will emphasize in this book, an important aspect of the expressivist story about moral vocabulary is that it serves to coordinate attitudes. Cooperative life requires us to be able to influence others' attitudes and be influenced by them. Through attitude expression, one puts one's attitude on the table.

This does not imply that one must have that attitude. It is possible to put an attitude on the table for the purposes of coordination even though one does not in fact have that attitude. Assertions, promises, and other speech acts need not be sincere. However, in putting an attitude on the table one is explicitly committing to having that attitude. Thus, for instance, by expressing a belief that it will rain tomorrow, one is explicitly committing to having that belief. Not all commitments are explicit. In Chapter 3, we will see how inference can be understood as unpacking the commitments—implicit or explicit—one has in virtue of having explicitly undertaken certain commitments through attitude expression.

1.6 The Frege–Geach Problem

Traditional expressivism faces what is perhaps the best known challenge for expressivism, the *Frege–Geach Problem*. Recall that according to traditional expressivism, terms receiving expressivist treatment indicate the expression of an attitude. Thus, for instance, traditional ethical expressivism holds that *wrong* indicates the expression of the attitude of disapproval.

[8] Green (2009) further argues that signalling theory may be used to make sense of how speech acts succeed in expressing attitudes. See Graham 2020 for discussion.

The Frege–Geach argument against this view now goes as follows. Consider the following seemingly valid inference.

(1) a. If lying is wrong, it should be punished.
    b. Lying is wrong.
    ___
    c. Lying should be punished.

The phrase *lying is wrong* in the first premiss cannot indicate the expression of disapproval of lying, since somebody uttering the first premiss might *approve* of lying. Thus, the Frege–Geach argument goes, the semantic function of *wrong* in the first premiss cannot be that of expressing the attitude of disapproval towards lying.

Suppose we insist that in unembedded contexts such as (1b), the second premiss of the inference, *Lying is wrong*, expresses disapproval of lying, as the traditional expressivist would have it. Then, the content of the second premiss does not coincide with that of the antecedent of the first premiss, and the inference (1) cannot be validated by *modus ponens*. For the inference to be an instance of *modus ponens*, the semantic function of *wrong* in the second premiss of the inference must be the same as that in the first premiss. But then *wrong* cannot express disapproval.

In 'Die Verneinung' (1919), Gottlob Frege originally presented the problem as applied to a form of traditional expressivism about negation. It was Peter Geach (1965) who subsequently observed that the problem affects traditional ethical expressivism and indeed any putative expressivist treatment of a term which can embed in contexts such as conditional antecedents. We shall return to traditional expressivism about negation in Chapter 3. As we will then see, understanding how the problem can be solved for the case of negation paves the way for a solution to the problem which can be successfully applied across the linguistic board.

A clarification is in order. What we have presented in this section is the original, *conditional* version of the Frege–Geach Problem. In the last twenty years or so, however, the debate in meta-ethics has centred around the *negation* version of the problem (Unwin 1999; Schroeder 2008b). We will present this version of the problem in Chapter 5, where we shall show that inferential expressivism has the resources to address this problem too.

Historically, expressivists have followed R. M. Hare (1952) in taking the Frege–Geach argument to be, at its core, a special instance of the problem of providing a compositional semantics for terms receiving expressivist treatment. The ethical expressivist, for instance, must specify how the meaning of complex sentences containing moral terms is determined by its constituents. The conditional version of the Frege–Geach argument, on this reading, challenges the expressivist to explain the meaning of conditionals with moral vocabulary in their antecedents. In so doing, the expressivist must make sure that seemingly valid inferences involving these conditionals are validated.

Early attempts to solve the Frege–Geach Problem proceed by trying to specify the meaning of embedding-forming operations in such a way that terms occurring in the contexts these operations form can still indicate the expression of an attitude. That is to say, early attempts to tackle the Frege–Geach Problem try to salvage traditional expressivism. Since the focus was on the conditional version of the Frege–Geach Problem, these attempts concentrate on the meaning of the conditional.

A notable example is Simon Blackburn's (1984) proposal. Blackburn's idea is that the conditional is used to express *higher-order attitudes*. According to Blackburn, a conditional expresses the attitude of disapproval of holding a particular combination of attitudes, namely having the attitude in the antecedent but not having the attitude in the consequent. Thus, for instance, the conditional premiss of the Frege–Geach argument, *If lying is wrong, it should be punished*, expresses disapproval of, at the same time, disapproving of lying and not believing that lying should be punished.

This is supposed to account for the validity of the inference from *Lying is wrong* and *If lying is wrong, it should be punished* to *Lying should be punished* as follows. Suppose someone utters the premisses of the inference. In uttering the first premiss, they express disapproval of lying; in uttering the second, they express disapproval of, at the same time, disapproving of lying and not believing that lying should be punished. Should they refuse to express belief that lying should be punished, they would be, at the same time, expressing a certain combination of attitudes and expressing disapproval of holding that combination of attitudes. A clash of attitudes occurs.

Blackburn's solution does not do justice to the problem. On the proposed solution, someone who accepts the premisses but not the conclusion of the inference from *Lying is wrong* and *If lying is wrong, it should be punished* to *Lying should be punished* is making a moral or practical mistake: it is the same mistake made by someone who accepts *Lying is wrong* and *Disapproving of lying and, at the same time, refusing to think that lying should be punished is wrong* but refuses to think that lying should be punished. But, as noted by

Bob Hale (1993), the mistake made in accepting the premisses but not the conclusion of the Frege–Geach inference is, on the face of it, a logical one. Indeed, on pain of making conditionals ambiguous depending on whether they involve moral vocabulary, Blackburn must hold that someone who accepts the premisses but refuses to accept the conclusion of any *modus ponens* argument is making a moral or practical mistake, as opposed to a logical one.

Hale concludes that the prospects for solving the Frege–Geach Problem by appealing to a logic of attitudes are bleak. As we will see in the remainder of the book, Hale's pessimism is unwarranted. A logic of attitudes can be developed in such a way that someone who accepts the premisses but refuses to accept the conclusion of the Frege–Geach inference is making a logical mistake.

1.7 Sophisticated expressivism

The Frege–Geach argument appears to sink traditional expressivism about linguistic items such as moral terms. Expressivists have responded by adding sophistication to their view. *Sophisticated expressivists* reject the idea that terms receiving expressivist treatment *merely* indicate the expression of an attitude and that the function of statements belonging to the relevant class is *merely* that of expressing non-cognitive attitudes.[9]

A notable proponent of a form of sophisticated expressivism is Allan Gibbard (2003). Formally, Gibbard takes a standard formal referential semantics in which semantic values are assigned relative to a customary, factual world parameter but adds an additional *practical* world parameter representing a plan about what to do in the circumstances of the factual world parameter.[10] Just like standard possible worlds, factual worlds are complete in that they specify every circumstance of the (factual) world; additionally, practical worlds are complete too: plans are *hyperplans* in that they specify what to do for any given circumstance of the factual world. As in the standard case, the semantics then specifies, for each sentence, the conditions

[9] The label 'sophisticated expressivism' is Tiefensee's (2021). The distinction between traditional expressivism and sophisticated expressivism seems to be the same distinction drawn by Gibbard (1993) between narrow and broad expressivism.

[10] Gibbard (1990) interprets the additional parameter as representing the set of norms that are accepted in the circumstances of the world parameter. We focus on the more recent version of Gibbard's view here.

under which the sentence receives value 1 at a factual and practical world pair (a *fact-prac world* for short). In the case of complex sentences, in particular, the semantics assigns semantic values to them in the usual way via set-theoretic operations, for instance by taking conjunction to denote intersection and negation to denote complementation. Thus, the judgement expressed by *If lying is wrong, it should be punished* is the judgement which rules out exactly those fact-prac worlds in which one is planning to blame for lying but not punish it. An inference is then valid if the denial of the conclusion is inconsistent with the premisses. Now, *Lying is wrong* is compatible with those fact-prac worlds in which one is planning to blame for lying. Moreover, *Lying should not be punished* is compatible with those fact-prac worlds in which one is planning not to punish lying. Hence, the premisses and the conclusion are inconsistent, which validates the Frege–Geach inference.

Gibbard is clearly a sophisticated expressivist: he denies that normative statements directly express conative attitudes. Nonetheless, he holds that his formal semantics can be understood in expressivist terms. For the meaning of a purely normative judgement is explained solely by reference to the practical worlds (representing plans) which are ruled out by the judgement. By contrast, the meaning of a purely descriptive judgement is explained solely by reference to the factual component of fact-prac worlds, that is which factual worlds are ruled out by the judgement. As for mixed judgements such as the one expressed by *Lying is wrong and grass is green*, their meaning is explained by reference to both components of fact-prac worlds. What notion of explanation is involved here is not entirely clear, and we will return to the issue when discussing the Problem of Creeping Minimalism in Section 1.10.

Gibbard's proposal is one of the best worked out attempts to deal with the Frege–Geach Problem. As we shall see, the inferential expressivist approach agrees with Gibbard that, in order to address the Frege–Geach Problem, we must regard normative statements as expressing beliefs. However, Gibbard's proposal faces issues having to do with the negation version of the Frege–Geach Problem, which we shall discuss in Chapter 5.

An early form of sophisticated expressivism was advocated by Blackburn (1988, 1998) in response to the difficulties raised by Hale. Blackburn develops a version of commitment semantics, according to which the meaning of a sentence is given by the attitudes to which its assertion commits the speaker. Assertions of descriptive sentences commit the speaker to beliefs; assertions of moral sentences commit the speaker to non-cognitive attitudes such as disapproval. Blackburn then argues that in asserting a disjunction, one 'ties oneself to a tree' with two branches: in asserting *A or B*, the speaker undertakes a pair of conditional commitments, namely the commitment to accepting *A* should *B* turn out to be untenable, and to accepting *B* should *A* turn out to be untenable. As Gibbard (1993) notes, Blackburn's proposal shares many structural similarities with his. For the same reason, it faces a negation version of the Frege–Geach Problem.[11]

Hybrid ethical expressivism holds that moral statements have a cognitive component and a non-cognitive one: they express beliefs as well as conative attitudes. Hence, at the meta-semantic level, hybrid expressivists reject the negative thesis of traditional expressivism, namely moral non-cognitivism. At the semantic level too, hybrid expressivists reject the negative thesis of traditional expressivism: moral terms do have a standard referential semantics. In particular, moral terms denote an extension, representing a property.

On one type of hybrid theory, this property is determined by the non-cognitive attitudes of the speaker, which can vary across speakers and even across time for the same speaker (Barker 2000; Ridge 2006, 2014). This variability is inherited by the property denoted by moral terms. Thus, for instance, someone uttering *Lying is wrong* expresses both disapproval of things that are *G* and belief that lying is *G*, where *G* can vary from speaker to speaker—it can be, for instance, the property of non-maximizing utility or that of violating the rights of others. The Frege–Geach Problem is then solved by offloading logical complexity into the content of the cognitive component of the mental state expressed by moral statements. For instance, *If lying is wrong, it should be punished* expresses disapproval of things that are *G* (where *G* is the property the speaker takes to be denoted by *wrong*) and a belief that if lying is *G*, it should be punished.

On another type of hybrid theory, both the property denoted by moral terms and the attitude associated with them are context-invariant (Copp 2001; Boisvert 2008; Hay 2013). *Lying is wrong* expresses both disapproval of things that have the property invariably denoted by *wrong* and belief that lying has that property. The model here is a standard understanding of slurs, which are taken to involve both a descriptive component picking out a property and a constant negative attitude towards the things

[11] This was noted by Hale (2002), who discovered the Negation Problem independently of Unwin (1999). For discussion of the Negation Problem in connection with Blackburn's commitment semantics, see Sinclair 2011.

having that property (Potts 2007; McCready 2010; Jeshion 2013). Again, the Frege–Geach Problem is solved by offloading logical complexity into the cognitive component of the mental state expressed by moral statements.

The hybrid solution does not get to the heart of the Frege–Geach Problem. The Frege–Geach insight is that an utterance of *If lying is wrong, it should be punished* need not express disapproval, not only of lying, but of anything at all. However, on hybrid theories, the utterance expresses disapproval of things that are *G*, whether *G* is held constant across speakers or not.[12] Our solution to the Frege–Geach Problem will be such that an utterance of *If lying is wrong, it should be punished* does not express disapproval of anything at all. However, the solution will retain an important aspect of hybrid theories, namely that the key to solving the Frege–Geach Problem is to take moral statements to express beliefs.

1.8 The Many Attitudes Problem

The Frege–Geach Problem purports to show that ethical expressivism cannot account for the semantic behaviour of moral vocabulary when this vocabulary is *embedded.* The conditional version of the problem focuses on embeddings in conditional antecedents. The negation version of the problem, as we shall see, focuses on embeddings under negation.

These embeddings arise from the use of logical vocabulary. But other embeddings arise because of the presence of propositional attitude verbs. The expressivist would seem to have a natural story to tell about what it is to *believe* that lying is wrong—it is just to disapprove of lying.[13] Thus, an utterance of *Ann believes that lying is wrong* will be true just in case Ann disapproves of lying. However, besides believing that lying is wrong, one can

[12] We take it that this is the reason why Ridge's most recent version of his view (Ridge 2014: 113) adopts the notion of a normative perspective, which is what is expressed by the utterance of a statement containing moral vocabulary. Normative perspectives are characterized in negative terms. Thus, Ridge says that someone uttering *If eating meat is wrong, then eating beef is wrong* is committing to not accepting moral standards which simultaneously permit eating meat but do not condemn eating beef. The problem, however, is that the utterance commits the speaker to *Eating beef is wrong* should they also utter *Eating meat is wrong*: the speaker is not only refraining from endorsing the negation of the antecedent and the negation of the consequent; they are denying them. In particular, it seems to us, taking normative perspectives to be characterized in purely negative terms prevents Ridge from using them to account for moral motivation, which was one of the initial attractions of expressivism.

[13] Later on, we will challenge the idea that the expressivist is committed to this account of what it is to believe normative contents. And, as we have seen, a number of sophisticated expressivists would take issue with this idea too.

also *fear* that lying is wrong, *hope* that it is, *wonder* whether it is. But what is it, for the expressivist, to have such attitudes towards lying being wrong? This is known as the *Many Attitudes Problem* for expressivism (Schroeder 2008a, 2013; Shiller 2017).

Bob Beddor (2020) presents a solution to the Many Attitudes Problem. He begins by providing a reductive account of desire in terms of dispositions to act in a certain way and to have certain experiences given that one has certain beliefs. In the case of *normative* desire, some of these beliefs will themselves be normative; but the expressivist, he notes, already has an account of normative belief. Thus, the expressivist can provide an account of normative desire. Beddor then argues that the many attitudes the Many Attitudes Problem is concerned with can be reduced to suitable combinations of belief and desire. Thus, he concludes, the expressivist can provide an account of what it is to *think* about normative matters: what it is, for instance, to fear or hope that lying is wrong, or what it is to wonder whether donating to charity is right.

The dispositional account of desire is controversial, and so is the idea that mental attitudes can be reduced to suitable combinations of belief and desire. As Derek Baker (2022) notes, this means that Beddor's proposal does not improve expressivists' dialectical position. On the contrary, it burdens them with commitments that might be difficult to defend.[14] In addition, we submit, reductionism and expressivism make for strange bedfellows. Expressivism aims to give its due to the multiplicity of attitudes one can take towards contents and objects: belief but also approval, disapproval, blame, and praise. Reductionism reduces this multiplicity to a combination of belief and desire.

Can a non-reductionist solution to the Many Attitudes Problem be given? Baker argues that solving the Many Attitudes Problem does not require one to provide a detailed account of the many attitudes themselves, but only of what they are attitudes towards. Expressivists do not need to say in any great detail what it is, say, to *hope* that lying is wrong; they need to say what it is to hope that *lying is wrong* (as opposed, to say, hope that it is not raining outside). If the Many Attitudes Problem is best understood as the attitude verb version of the Frege–Geach Problem, this is certainly correct. The conditional version of the Frege–Geach Problem challenges the expressivist to give an account of the meaning of, say, *wrong* in such a way

[14] Baker argues that Köhler's (2013) proposed solution to the Many Attitudes Problem faces similar difficulties.

that it remains possible to explain its behaviour in conditional antecedents. The Many Attitudes Problem challenges the expressivist to give an account of *wrong* in such a way that it remains possible to explain its behaviour under propositional attitude verbs.

Baker argues that if a solution to the Frege–Geach Problem is available, then a solution to the Many Attitudes Problem is within reach. In particular, he thinks (and we agree) that the shape of a good solution to the Frege–Geach Problem consists in explaining the contribution of *wrong* in conditional antecedents by taking the state expressed by *If lying is wrong, then it should be punished* to be a belief bearing certain appropriate relations to disapproval of lying. One can then tell a similar story about embedding under propositional attitudes. To hope that lying is wrong is to bear the right kind of relation to the belief that lying is wrong or to other attitudes in turn suitably related to that belief. This explanation does not require hope to be reducible to a combination of belief and desire. While we agree that this is the right shape of a solution to the Many Attitudes Problem, we believe that, just as in the case of the ordinary Frege–Geach Problem, the Many Attitudes Problem does place further constraints on a solution to the Frege–Geach Problem: a satisfying story about the relation between uttering *Lying is wrong* and disapproval of lying must make it possible not only to validate the Frege–Geach inference but also to provide a satisfying account of propositional attitude verbs. We will tackle the issue in Chapter 6.

1.9 The Wishful Thinking Problem

Suppose expressivists succeeded in providing a successful solution to the Frege–Geach Problem, at least in its conditional version. That is, suppose they succeeded in providing a compositional semantics for *wrong* in such a way that the following inference is validated.

(2) a. If lying is wrong, the souls of liars will be punished in the afterlife.
b. Lying is wrong.
―――――――――――――――――――――――――
c. The souls of liars will be punished in the afterlife.

Cian Dorr (2002) has shown that expressivists still face a different but related problem. Even if the conclusion of the Frege–Geach inference follows from the premisses, Dorr argues, expressivism implies that one cannot justifiably come to accept the conclusion on the basis of their acceptance of the premisses. But, Dorr continues, it is plainly the case that one can, so expressivism should be rejected.

Specifically, Dorr describes the case of Edgar, who accepts (2a) on the basis of the fact that he was told to do so by his teachers. Initially, Edgar does not accept (2b): his teachers never mentioned lying, and it never occurred to him that there might be anything wrong with it, provided that you can get away with it. Furthermore, Edgar is agnostic about (2c): 'No-one whose opinion he trusts has told him anything one way or the other; nor does he have evidence of his own that bears on the specific nature of the afterlife' (Dorr 2002: 98). Subsequently, however, Edgar comes to accept (2b) by reading a philosophy book. On the basis of his acceptance of (2b), Edgar comes to accept (2c). Edgar, says Dorr, appears to be justified in doing so.

But how can expressivists account for this fact, asks Dorr? Before coming to accept (2b), Edgar's evidence does not seem to justify his acceptance of (2c): it would be irrational for him to believe that the souls of liars will be punished in the afterlife. However, says Dorr, according to expressivists when Edgar comes to accept (2b) there is a change in his non-cognitive attitudes, but Edgar 'acquired no new evidence or other beliefs relevant to the question of the fate of liars in the afterlife' (Dorr 2002: 98). So if it was irrational for Edgar to believe that the souls of liars will be punished in the afterlife *before* accepting (2b), it would be irrational for him to do so *after* that. To come to form new descriptive beliefs not on the basis of a change in one's evidence, but on the basis of a change in one's non-cognitive attitudes, is wishful thinking.

Hybrid expressivists have an easy time with the Wishful Thinking Problem. Upon accepting (2b), Edgar justifiably acquires a new descriptive belief. So he acquires a belief 'relevant to the question of the fate of liars in the afterlife'. To come to justifiably form new descriptive beliefs on the basis of acquiring certain descriptive beliefs is not wishful thinking. However, the hybrid solution is reached at the price of making the non-cognitive component expressed by moral statements completely irrelevant to the use of moral vocabulary in inferences. We think expressivists can do better and we will discuss the issue in Chapter 5.

1.10 The Problem of Creeping Minimalism

In the beginning, there was traditional expressivism. Traditional expressivism denies that *wrong* makes any contribution to the truth conditions of the sentences in which it occurs. Rather, it indicates the expression of a non-cognitive attitude. *Lying is wrong* does not express a proposition and it is neither true nor false. Moreover, traditional expressivism denies that moral statements express beliefs. Traditional expressivism rejects several of the central tenets of standard realist approaches to morality and so it is clearly distinct from them.

But then expressivism became sophisticated. Expressivists began saying that moral statements do express beliefs. This allowed them to provide a solution to the Frege–Geach Problem but was also independently motivated. According to sophisticated expressivists, assertions express beliefs. Indeed, many sophisticated expressivists would agree with Crispin Wright that there is a tight conceptual connection between assertion and belief.

> Assertion has the following analytic tie to belief: if someone makes an assertion, and is supposed sincere, it follows that she has a belief whose content can be captured by means of the sentences used.
>
> (Wright 1992: 14)

Moreover, for an utterance to count as an assertion it suffices for it to be made by means of a *syntactically disciplined* sentence: by means of a sentence that is syntactically sophisticated in that it is capable of featuring in various types of embeddings and subject to standards of appropriate and inappropriate use (Sinclair 2006). In short, an utterance of an embeddable declarative sentence serves to perform an assertion.[15] Thus, if a competent speaker sincerely utters a syntactically disciplined sentence, then they have a belief whose content can be captured by means of that sentence. This *minimalist* conception of belief (Wright 1992; Divers and Miller 1995) allows expressivists to vindicate everyday ascriptions of moral beliefs. Since moral sentences are syntactically disciplined, it follows that moral statements express beliefs. Moreover, when someone utters a moral sentence sincerely, they have the belief expressed by that sentence. Thus, when folk attribute to the Pope the belief that lying is wrong, we can take them at face value. The expressivist can say, with the realist, that in uttering *Lying is wrong*, the Pope expresses a belief.

Folk also ascribe truth to moral statements. *It is true that lying is wrong* is as good a piece of English as any. To accommodate this aspect of the use

[15] It is worth stressing that the requirement that the sentence be capable of featuring in various types of embeddings is crucial. As we will argue in Chapters 4 and 9, there are declarative sentences—performed using adverbs such as *perhaps* and *probably*—whose utterance does not result in an assertion. However, such sentences resist embedding in contexts such as conditional antecedents.

of moral language, sophisticated expressivists appeal to a more familiar kind of minimalism, one about truth.[16] Truth minimalism is a broad church, and we will defend our take on it in Chapter 7. For now, we can take the view to consist in the idea that the meaning of the truth predicate is exhausted by all instances of the *T-Schema*:

$$\ulcorner A \urcorner \text{ is true if and only if } A,$$

where $\ulcorner A \urcorner$ is a name for *A*.[17] Assuming that expressivists have explained the meaning of *Lying is wrong*, they can use the T-Schema to explain the meaning of *It is true that lying is wrong*. The expressivist can say, with the realist, that it is true that lying is wrong.

But then, asks James Dreier (2004), what distinguishes the expressivist from the realist? This is the *Problem of Creeping Minimalism*. One might have thought that unlike the realist, the expressivist holds that moral sentences do not express propositions or that moral terms do not denote properties. But not so: sophisticated expressivists tend to subscribe to minimalist views concerning propositions and properties as well. Thus, what propositions are is exhausted by the schema

$$\ulcorner A \urcorner \text{ expresses the proposition that } A,$$

and what properties are is exhausted by the schema

$$a \text{ has the property of being } F \text{ if and only if } a \text{ is } F.$$

The expressivist can say, with the realist, that *Lying is wrong* expresses a proposition and that *wrong* denotes a property. To address the Problem of Creeping Minimalism, expressivists must explain what distinguishes their view from other meta-ethical approaches in a way that retains the original motivation of their view.

Dreier (2004) offered what he called the *explanation explanation* of what makes expressivism a distinctive meta-ethical approach. According to the

[16] Some (e.g., Horwich 1993) have argued that minimalism about truth allows the expressivist to solve the Frege–Geach Problem for cheap. We agree with Dreier (1996: 34–38) that this is a mistake.

[17] To be precise: the expression '*A*' is a meta-language name for an object-language sentence and '$\ulcorner A \urcorner$' is a meta-language name for an object-language name for the object-language sentence denoted by '*A*'.

explanation explanation, expressivists are distinguished by their claim that 'to explain what it is to make a moral judgement, we need not mention any normative properties' (Dreier 2004: 39). Although there are normative properties in an appropriately minimal sense of *property*, they are not needed to explain what it is to have a moral belief. One can explain what it is to believe that lying is wrong without appealing to the property of wrongness. The expressivist can agree with the realist that *Lying is wrong* expresses a true moral proposition and that the Pope really believes that lying is wrong. But the Pope believes that lying is wrong in virtue of the fact that he disapproves of lying. Normative properties are dispensable when giving an answer to the metaphysical question of what makes it the case that moral beliefs have the semantic content that they do. In other words, the difference between the expressivist and the realist lies in the kind of answer they admit to the Meaning Determination Question concerning the content of moral beliefs. The explanation explanation is naturally extended to the meaning of moral sentences: unlike the realist, the expressivist thinks that it is possible to explain what it is for *Lying is wrong* to have the meaning that it does without appealing to normative properties.

Dreier's explanation explanation is appealing. However, it has been criticized for classifying moral error theories together with expressivist theories (Simpson 2018). For according to moral error theories, one can explain what it is for *Lying is wrong* to have the meaning it does without recourse to normative properties, since, after all, according to these theories there are no moral properties. Similar things can be said about explaining the content of the belief that lying is wrong. Dreier (2018: 534–535) notes that his explanation explanation appears nonetheless to capture an important distinction: there is an important sense in which moral error theories are not realist theories. However, he concedes that his earlier account has failed to capture *another* distinction, namely that between traditional realist theories and error theories on the one hand and expressivism on the other.

How to draw this distinction? Matthew Simpson (2020) suggests that the distinction is one between representationalists and non-representationalists and that we can provide an explanation explanation of this distinction if we take what is dispensable according to the expressivist to be not simply normative properties, but subject matter. The subject matter of a sentence or belief is what that sentence or belief is about. Crucially, the subject matter of a sentence need not exist: *Superman can fly* is about Superman much in the way in which *Andrea rests* is about Andrea. By reformulating the explanation

explanation in terms of subject matter, Simpson argues, the moral error theory need no longer be wrongly classified with expressivism: there being no normative properties is compatible with explaining why *Lying is wrong* has the meaning that it does in terms of its subject matter, namely wrongness. Simpson considers the objection that his view would classify with expressivism the reductive naturalist view which says that *good* means good because it causally tracks the property of pleasure maximization. This view does not mention goodness but is clearly representationalist. Simpson replies that his version of the explanation explanation does distinguish reductive naturalism from expressivism because according to reductive naturalism pleasure maximization *is* part of the subject matter of ethics. This is correct, but then, by the same token, plan coordination is part of the subject matter of ethics for Gibbard. Since an account of why *Lying is wrong* has the meaning that it does must mention planning, Gibbard's view would then be wrongly classified as not being in the expressivist camp.

Simpson wanted to draw the distinction between traditional realist theories and error theories on the one hand and expressivist theories on the other as the meta-semantic distinction between representationalism and non-representationalism. However, the distinction, as we see it, is primarily the semantic one between referentialism and its opposition, non-referentialism. Recall that from the point of view of expressivism as we understand it, moral error theories are making the same kind of mistake that traditional realist theories make: they both fail to comply with the Pragmatist Razor. The reason for this is that they are both referentialist theories, so it makes sense to distinguish these two theories from expressivism on the basis of their referentialist underpinnings.

The distinction between referentialism and non-referentialism may seem to be easily drawn: referentialists take the meaning of an expression to be given by its referent, whereas non-referentialists do not. The question, however, is what it means for the meaning of an expression to be given by its referent. Presumably, sophisticated expressivists endorse some form of minimalism about reference too, so that they can say, with the referentialist, that *wrong* refers to the properties of wrongness and use this fact to give the meaning of *wrong*. Simpson is therefore correct that it is useful to resort to an explanation explanation of the distinction between traditional realist theories and error theories on the one hand and expressivist theories on the other. However, the explanation explanation of this distinction too should be formulated focusing on normative properties rather than subject matter, following Dreier. The difference with Dreier's original explanation explanation is that we should not focus on the *meta-semantic* explanation of what makes it the case that moral vocabulary has the meaning that it does—on how these theories answer the Meaning Determination Question in meta-semantics. Rather, as Baker (2021) has recently argued, we should focus on their *semantic* explanation of what the meaning of moral vocabulary is. This, to stress, is in keeping with the pragmatist motivation for the expressivist project that we have been advocating, as captured by the Pragmatist Razor. According to the Razor, one should avoid semantic explanations that are not needed to account for an expression's or sentence's functional role. The focus here is on semantic explanations rather than meta-semantic ones. Accordingly, this is what the characterization of the expressivist approach should focus on.

Thus, what distinguishes non-referentialists is that they hold that the meaning of, say, *wrong* can be completely explained in the meta-language without using expressions ascribing moral properties. Expressivists hold, in addition, that this meaning can be completely explained in terms of desire-like attitudes. Similarly, non-referentialists hold that the content of the belief that lying is wrong can be completely explained without reference to moral properties.[18] Expressivists hold, in addition, that this content can be completely explained in terms of desire-like attitudes.[19]

As Baker notes, this version of the explanation explanation correctly classifies moral error theories in the same camp as traditional realist theories. Moral error theorists must think that one can explain why moral sentences have the meaning that they do without appealing to moral properties because they think there are no moral properties. But they also think that one cannot explain what the meanings of moral sentences *are* without appealing to moral properties: that is why moral sentences are typically false. Moreover, as Baker also explains, the semantic version of the explanation explanation can

[18] It should be stressed, however, that on our proposal it does not suffice for someone to count as a non-referentialist that the meaning of moral vocabulary is such that moral assertions bear suitable inferential connections to attitudes. Chrisman (2008) has suggested a characterization of non-representationalism (thus, at the meta-semantic rather than semantic level) along similar lines but, as Tiefensee (2016) and Dreier (2018) have noted, a representationalist can agree that there are such connections. In a similar fashion, a referentialist can agree that moral assertions bear suitable inferential connections to attitudes, but insist that to completely explain the meaning of moral vocabulary one must invoke moral properties.

[19] Baker (2021: 2323) says that expressivists must explain an assertion of *Lying is wrong* by indicating which desire-like attitude it expresses, and that the belief that lying is wrong must be explained by describing the belief as a desire-like attitude. This seems to be too strong, and it would classify Gibbard as well as our own view as non-expressivist.

be generalized to domains other than ethics: the expressivist about a certain expression *E* holds that one can explain the meaning of assertions involving *E* without appealing to what *E* is traditionally taken to stand for but rather in terms of some non-assertoric speech act expressing some attitude other than belief. Similarly, the content of beliefs that . . . *E* . . . can be explained without reference to what *E* is traditionally taken to stand for but rather in terms of some attitude other than belief. We agree with Baker, but with a caveat: we think that it is not necessary for an expressivist about *E* to hold that the meaning of assertions involving *E* is to be explained in terms of some *non-assertoric* speech act: it suffices that the relevant explanation of the meaning of *E* be given not in terms of what *E* is traditionally taken to stand for, but in terms of speech acts expressing attitudes. These attitudes may well include belief, which is expressed by assertion. Similarly, the content of beliefs that . . . *E* . . . can be explained without reference to what *E* is traditionally taken to stand for, but in terms of some attitudes (including belief) whose content can be described without invoking what *E* is traditionally taken to stand for. What matters, it seems to us, is whether the explanation features the traditional reference of *E*, not whether it features assertion or belief. This will become relevant in Chapter 7.

Following Baker (2021), we have argued that the distinction between traditional realist theories and moral error theories on the one hand and expressivist theories on the other ought to be drawn by appealing to the kind of semantic explanations they invoke. As Baker sees it, however, what we are drawing in this way is not (or at least not simply) the distinction between referentialist and non-referentialist theories, but the distinction between realist and error theories on the one hand and quasi-realist theories on the other. Quasi-realists about ethics hold that there are no moral properties such as rightness and wrongness except in the minimalist's sense. To this extent, they are anti-realist about ethics. But this, they think, is compatible with speaking, with the realist, of there being moral properties, using the minimalist approach described above.

Given that he wants to distinguish between realism and quasi-realism, Baker faces the problem of what to say about Ralph Wedgwood's (2007) position. Wedgwood appears to think that one can explain the meaning of normative vocabulary without reference to normative properties but by appealing to its inferential connections to attitudes and action. Nonetheless, he holds that when people talk about normative properties they succeed in latching onto the real, non-natural normative properties. This seems to make

Wedgwood a full-blown realist, rather than a quasi-realist: unlike Blackburn (1993), he appears to believe that there are normative properties not simply in the appropriately minimal sense of *property*, which is all that is allowed by the quasi-realist.

Dreier (2004) already worried about classifying Gibbard and Wedgwood together, and more recently has said that he 'find[s] this rather mysterious' (Dreier 2018: 545). Indeed, unlike the example of moral error theories, Wedgwood's case threatens to undermine Dreier's meta-semantic version of the explanation explanation even as an account of the distinction between realist and anti-realist views. As Baker (2021: 2328–2330) notes, the move from a meta-semantic version of the explanation explanation to a semantic version allows him to deal with this case, since Wedgwood (see 2007: 49–50) appears to hold that the semantic explanations based on inferential role are not complete—he appears to think that appeal to normative properties is required to explain the meaning of moral vocabulary. Hence, Baker's explanation explanation correctly classifies Wedgwood as a realist.

However, it seems that one could imagine the position of someone who holds that the meaning of moral vocabulary may be completely given by its inferential connections to attitudes but nonetheless holds that our talk about moral properties succeeds in latching onto moral properties, not simply in the minimal sense of *property*. This character would appear, *contra* Baker, to be a realist. No such issue arises on our proposal. The semantic explanation explanation, as we see it, serves to distinguish expressivists from traditional realist theories and error theories on the basis of their adherence to non-referentialism. But non-referentialism by itself is compatible with both a realist and non-realist stance. This outcome is in line with the understanding of expressivism that we have been defending: ethical expressivism is neutral with respect to ontological questions about the existence of moral properties beyond what follows from minimalism. It is instead a thesis concerning the meaning and function of moral vocabulary. Nonetheless, the question of realism remains: can a sensible distinction be drawn between a realist non-referentialist and a quasi-realist non-referentialist? Neither the semantic explanation explanation nor Dreier's meta-semantic explanation explanation will do. We concur with Dreier that the issue is a mysterious one, and we will return to it in Chapter 7.

We have defended an *explanationist* reconstruction of the semantic thesis of expressivism about a certain class of linguistic items: according to the semantic thesis of expressivism, the meaning of those items can be explained in terms of speech acts expressing attitudes, without appealing to their referents. This distinguishes expressivism from referentialist views about meaning. But it also raises the question of what notion of explanation expressivists are invoking. Dreier (2004) argued that the relevant notion of explanation must be a metaphysical one, but this would seem at odds with the pragmatist understanding of the expressivist project we have been advocating. We are going to argue that the expressivist should understand the relevant notion of explanation as an *inferential* one. Unsurprisingly, this can be made precise by combining expressivism with another theory of meaning, centred around the notion of inference. This theory of meaning is inferentialism, to which we now turn.

2
Inferentialism

In this chapter, we introduce inferentialism, the view that the meaning of an expression is given or determined by its role in inferences. We begin by highlighting the main attraction of inferentialism, namely that it grounds meaning in our inferential practices. When understood as a semantic thesis, inferentialism, like expressivism, provides non-referential explanations of the meaning of expressions. When understood as a meta-semantic thesis, inferentialism provides non-representational explanations of what makes it the case that expressions have the meaning that they do.

Like expressivism, inferentialism comes in local and global varieties. The main focus of the chapter is on *logical* inferentialism, the view that the meaning of the logical constants is given by their introduction and elimination rules. The reason for this is that the application of inferentialism to logical expression is one of its central historical origins and has come to be considered the paradigmatic example of an inferentialist approach to meaning. We then discuss three key problems inferentialism faces: the Problem of Defective Concepts, the Problem of Constitutive Rules, and the Problem of Limited Applicability. We argue that, while inferentialists already have in house the resources to address the first two problems, they must join forces with expressivists in order to address the third problem. Moreover, joining forces with expressivists enables inferentialists to solve the first problem whilst allowing them to hold on to a classical notion of inference.

2.1 Inferentialist semantics

Inferentialism is an approach to the meaning of linguistic expressions that centres around their *use*. Thus, inferentialism belongs to the broad church of use theories of meaning, which take inspiration from Ludwig Wittgenstein's (1953) *dictum* that meaning is use. Typically, use theorists take only some aspects of the use of expressions to be relevant to their meaning.

Reasoning with Attitude: Foundations and Applications of Inferential Expressivism. Luca Incurvati and Julian J. Schlöder, Oxford University Press. © Oxford University Press 2023.
DOI: 10.1093/oso/9780197620984.003.0002

Inferentialists focus on the use of expressions in *inferences* or, as inferentialists like to put it, on their *inferential role*. However, not all uses of expressions in inferences appear to be relevant to their meaning—people make inferential mistakes, after all. Nonetheless, inferentialists contend, we can distinguish between proper and improper uses of expressions in inference.

> Content is understood in terms of properties of inference, and those are understood in terms of the norm-instituting attitudes of taking or treating moves [in linguistic interaction] as appropriate or inappropriate in practice. A theoretical route is accordingly made available from what people do to what they mean, from their practice to the contents of their states and expressions. (Brandom 1994: 134)

For example, it is proper to infer *p* from *p and q*, but improper to infer it from *p or q* (unless it is proper to rule out *q*). The next question is which proper uses of an expression are to be considered part of its inferential role. Robert Brandom (1994, 2000), one of the main proponents of inferentialism, identifies the inferential role of an expression with the totality of its uses in inferences, which results in a *holistic* version of inferentialism. By contrast, other inferentialists have adopted a *molecularist* approach, which takes the inferential role of an expression to consist in a proper subset of the totality of its uses.

One especially popular idea among molecularist inferentialists is that the proper uses of an expression that are relevant to its meaning can be given by means of *rules of inference*. For instance, the proper uses of conjunction are given by the following rules.

From *A* and *B* infer *A and B*.

From *A and B* infer *A*.

From *A and B* infer *B*.

Of course, sentences containing conjunctions can appear in other proper inferences, such as the inference from *A and B* and *If A and B, then C* to *C*. But the three rules above appear to be particularly central to how conjunction is used in that these are inferences in which a sentence containing a conjunction can occur specifically in virtue of containing a conjunction.

The idea that an expression's inferential role can be captured by means of inference rules has its origins in the application of inferentialism to logical vocabulary. We saw in the previous chapter that ethical expressivism has often provided the model for an expressivist approach to meaning. In a similar fashion, logical inferentialism has often provided the model for an inferentialist approach to meaning. The traditional version of logical inferentialism originates in the work of Gerhard Gentzen (1935). Gentzen articulated systems of natural deduction, in which the logical constants obey *introduction* and *elimination* rules.[1] The introduction rules for a constant specify conditions under which one may infer a sentence containing the constant as its main operator; the elimination rules specify what may be inferred from such a sentence. For instance, conjunction obeys the following rules, formalizing the rules for *and* given above.

$$(\wedge\text{I.})\ \frac{A \qquad B}{A \wedge B} \qquad (\wedge\text{E.}_1)\ \frac{A \wedge B}{A} \qquad (\wedge\text{E.}_2)\ \frac{A \wedge B}{B}$$

Gentzen accords a special status to the introduction rules.

> The introductions are, so to speak, the 'definitions' of the relevant sign, and the eliminations are, in the end, merely their consequences . . . N.B.: it is not necessary to appeal in this to a 'content sense' of the sign.
>
> (Gentzen 1935: 189, our translation)

Not only do logical constants obey introduction and elimination rules, Gentzen claims; their meaning may be fully specified by laying down the introduction rules, of which the elimination rules are mere consequences. In laying down introduction rules and determining corresponding elimination rules for a logical constant, one assigns a semantic value to it.

Traditional logical inferentialism. The semantic value of a logical constant is given by its inferential role, as captured by a set of inference rules representing how declarative sentences containing the logical constant as their main operator may be introduced or eliminated.

Gentzen contrasts this *proof-theoretic* mode of definition with the *model-theoretic* one, which takes the meaning of a logical constant to be given, for instance, by a suitable truth function.

[1] Natural deduction was independently discovered by Gentzen and Jaśkowski (1934). Jaśkowski's presentation was developed, in modern version, by Fitch (1952).

Gentzen's suggestion is, to our knowledge, the first explicit formulation of an inferentialist approach to meaning, according to which the meaning of an expression can be given ('defined') by merely laying down inference rules and without falling back on any supposed prior knowledge about its meaning ('content sense'). Thus, Gentzen's work marks the birth of the semantic doctrine of inferentialism.

Inferentialism (Semantic). The semantic value of an expression is given by its role in inferences.

Gentzen gives precedence to the introduction rules (see also Prawitz 1965; Read 2010): it is those rules that confer meaning upon a logical constant. Other inferentialists have given precedence to the elimination rules (Schroeder-Heister 1985). Other inferentialists still have given precedence to neither (Dummett 1991b): introduction and elimination rules jointly confer meaning upon a logical constant. Our approach will be to take both introduction and elimination rules to be constitutive of the meaning of an expression, but the discussion could be recast to conform with the other options.

Like expressivism, inferentialism comes in local and global versions. Local versions such as traditional logical inferentialism restrict the inferentialist approach to a specific class of linguistic expressions; global versions apply it to the whole of language. Global inferentialists who take a molecularist approach will insist that what the traditional logical inferentialist says about conjunction can be said, *mutatis mutandis*, about any meaningful linguistic item. How this could be the case is not always easy to see. For example, referring expressions are, for obvious reasons, often considered the paradigmatic example of expressions to be subject to referentialist treatment. Nonetheless, global inferentialists contend, we can distinguish between proper and improper inferences involving referring expressions as well. For instance, it is proper to infer from *Cicero is P* that *Tully is P* (for any predicate *P*) but improper to infer from *Cicero is P* that *Julius is P*. The meaning of the referring expression *Cicero* is given by the proper inferences that are suitably central to the use of *Cicero*. Something analogous can be said about the meaning of observational predicates, another class of expressions that might at first sight appear to resist inferentialist treatment. It is proper to infer from *a is red* that *a is coloured*, but improper to infer that *a is square* (for any referring expression *a*). The meaning of the predicate *red* is given by the proper inferences that are central to the use of *red*.

The inferences just listed take us from sentences to sentences: they are *language internal*. However, it might be objected, a theory of meaning should make it possible to explain how language succeeds in latching onto the world—to cross the gap between language and world. But inferentialist semantics, when understood as a global thesis, cannot do this if rules can only take us from some pieces of language to more pieces of language. Inferentialists may reject the presupposition of the objection that there is a gap between language and the world to be crossed in the first place (McDowell 1996), and hence insist that even the meanings of referring expressions and observational predicates can be given by language internal rules (Tanter Forthcoming). Alternatively, inferentialists can claim that the objection rests on too restrictive a notion of inference. Inference may take us from language to language, but it may also take us from language to world and from world to language. Non-linguistic information can render certain assertions correct, and the correctness of certain assertions may entail non-linguistic action (Sellars 1956; Dummett 1973a: 406). This may be captured by means of *language-entry* and *language-exit* rules. Language-entry rules tell us in which worldly circumstances it is proper to use a term. For instance, the meaning of *red* would partly be given by a rule stating that it is proper to assert *That's red* when presented with a red object. Language-exit rules establish a link between meaning and motivated action. For instance, the meaning of *red* would partly be given by a rule entailing that it is proper to get the red object upon being told *Get the red object.*

Like expressivism, inferentialism is often contrasted with referentialism and the referential semantics which underpins it. So far we have focused on inferentialism as a semantic thesis. So understood, inferentialism clearly stands in opposition to referentialism. It holds that the meaning of an expression can be completely explained without appealing to what the expression is traditionally taken to stand for. Hence, according to the characterization we gave in Section 1.10, it is a non-referentialist approach to meaning. However, inferentialism can also be understood as a doctrine at the level of *meta-semantics*. In the next section, we describe meta-semantic inferentialism and the extent to which it can be considered compatible with referentialism.

2.2 From semantics to meta-semantics and back

As noted in the previous chapter, referential semantic theories take referring expressions to denote individuals, predicates to denote functions, and sentences to denote truth values (possibly at contexts and indices). The denotation of a sentence is determined by the denotation of its components so that the truth value of, for instance, *Andrea rests* is 1 (at a context and index) if and only if the function denoted by *rests* applied to the denotation of *Andrea* returns 1. Details may differ. For instance, in some referential semantics sentences do not denote truth values but simply have them. Nonetheless, the basic idea remains the same: the truth value of a sentence such as *Andrea rests* is determined by the denotation of its components. Referential semantic theories extend this idea to more complex sentences by taking the logical connectives to denote *truth functions* so that the denotation of a complex sentence is determined by the denotation of its components.

$$\begin{aligned} [\![\text{and}]\!] &= \lambda A \lambda B.\min([\![A]\!], [\![B]\!]) \\ [\![\text{or}]\!] &= \lambda A \lambda B.\max([\![A]\!], [\![B]\!]) \\ [\![\text{not}]\!] &= \lambda A.1 - [\![A]\!] \end{aligned}$$

Under thc usual interpretation of the truth values in the referential semantics, these clauses state that a conjunction is true if and only if both conjuncts are true, a disjunction is true if and only if at least one disjunct is true, and a negation is true if and only if the negated sentence is false.

Referentialists go on to define validity in terms of truth preservation. As mentioned in the previous chapter, there are different options in this case too—for instance, one can take valid arguments to preserve truth *simpliciter* or truth at an index. Here the basic idea will suffice: A is a logical consequence of some set of sentences Γ just in case whenever all sentences in Γ receive the value 1, so does A. In symbols:

$$\Gamma \vDash A \text{ just in case, for all } B \in \Gamma, \text{ if } [\![B]\!] = 1 \text{ then } [\![A]\!] = 1$$

Thus, referentialists specify how the denotation of a sentence is determined by that of its components (the *semantics proper*) and define validity on the basis of this (the *postsemantics*). As this terminology introduced by MacFarlane (2003) suggests, validity comes after the assignment of meaning to sentences. But according to inferentialists, there is something wrong with this way of proceeding, when taken to reflect the proper order of semantic explanation: there is an important sense in which what follows from what is prior to the assignment of meaning to individual expressions of the language.

> The standard way [of referential semantics] is to assume that one has a prior grip on the notion of truth, and use it to explain what good inference consists in. . . . [I]nferentialist pragmatism reverses this order of explanation . . . . It starts with a practical distinction between good and bad inferences, understood as a distinction between appropriate and inappropriate doings, and goes on to understand talk about truth as talk about what is preserved by the good moves. (Brandom 2000: 12)

Consider again the case of conjunction. According to inferentialists, the meaning of conjunction is determined by its inferential role. Following Gentzen, we can take this inferential role to be captured by the facts that it is proper to infer *A and B* from *A* and *B* and that it is proper to infer *A* or *B* from *A and B*. But this means that what makes these inferences proper cannot be the meaning of conjunction, as the referentialist way of proceeding would seem to suggest. Rather, it is the fact that these inferences are proper that makes it the case that conjunction has the meaning that it does. Thus, by inferentialist lights, what referentialists call the semantics proper comes in the order of semantic explanation after what referentialists call the postsemantics.

To be sure, inferentialists do not *fully* reverse the order of semantic explanation suggested by referential semantic theories. Inferentialists typically define *A* to be a consequence of some set of sentences Γ just in case *A* can be obtained from Γ by successive applications of inference rules. This definition leaves open the structural properties of *successive application*, so there are choices to be made when it comes to defining validity, even if one takes the semantic values to be determined by inference rules. Nevertheless, inferentialists do disagree with the order of semantic explanation suggested by referential semantic theories. According to this order of explanation, *all* validity facts are settled after the semantic values are fixed. But according to inferentialists, at least some validity facts must be settled before fixing the semantic values, since these semantic values are in turn determined by validity facts.

Nonetheless, referentialists need not disagree with inferentialists about this. For claims about the order of explanation between inference and meaning are best understood not as semantic claims but as *meta*-semantic ones, concerning whether it is proper inference that determines the meaning of expressions or the other way around. Within the inferentialist camp, we must distinguish between the semantic thesis that the meaning of an expression is *given* by its inferential role and the meta-semantic thesis that the meaning of an expression is *determined* by its inferential role.

Inferentialism (Determination Thesis). The semantic value of an expression is determined by its inferential role.

While the semantic doctrine of inferentialism is an answer to the question of what the meanings of expressions *are*, the Determination Thesis of inferentialism is an answer to the Meaning Determination Question. Thus, referentialism and the semantic thesis of inferentialism make competing claims about what the meanings of natural language expressions are. The Determination Thesis of inferentialism, by contrast, is in principle compatible with taking the meanings of expressions to be given by referential semantic values. Inferentialists who endorse the Determination Thesis of inferentialism but not the semantic thesis merely insist that these referential semantic values are determined by proper use in inferences.

It is worth emphasizing what combining referentialism with the inferentialist Determination Thesis means for how to understand the relation between the semantics proper and postsemantics. If the inferentialist Determination Thesis holds, inferential practices determine the referential semantic values given in the semantics proper. But how are these semantic values determined? What is needed is an assumption that allows us to connect the domain of proper inference with the domain of reference. The typical assumption is that proper inference preserves truth. Consider, for example, the case of conjunction. Since the inference rules codify proper inference, the introduction rule for conjunction tells us that if *A* and *B* are true then so is *A and B*, and the elimination rules tell us that if *A and B* is true then so are *A* and *B*. Assuming the possible truth values are 0 and 1 under their usual interpretation, this means that the referential semantic value assigned to conjunction will be the function taking *A* and *B* to 1 just in case they both receive value 1. However, that proper inference preserves truth is what is supposed to follow from the definition of validity in the postsemantics of the referential semantic theory and the fact that proper inferences are valid. This means that the definition of validity in the postsemantics can be materially adequate, but it cannot reflect the order of semantic explanation embodied by the inferentialist Determination Thesis combined with referentialism: this proper order of explanation requires one to *assume* that proper inference preserves truth.

Matters are different for the position combining both the semantic thesis of inferentialism and its Determination Thesis. According to this position, meanings are determined by inferential roles in that they are *identical* to those roles. The combination of the inferentialist Determination Thesis and referentialism is less straightforward: a story needs to be told about how inferential practices determine non-inferential meanings, which will at least involve some additional assumptions connecting the domain of proper inference with the domain of reference. The referentialist, according to the semantic inferentialist, would need to provide some reason for why this detour is required. In the absence of such a reason, the combination of semantic inferentialism and the inferentialist Determination Thesis is to be preferred.

Moreover, the combination of the inferentialist Determination Thesis with referentialism might saddle inferentialists with theoretical commitments they might not be able to honour. In particular, it is not obvious that the assumption that proper inference preserves truth suffices to determine referential semantic values. Rudolf Carnap (1943) showed that the standard rules of the classical propositional calculus are compatible with assigning to *not* a function making both a sentence and its negation true, and with assigning to *or* a function making a disjunction true despite its disjuncts both being false. As it happens, the bilateral systems that will be our starting point in the next chapter succeed in solving Carnap's problem for classical propositional logic: any interpretation compatible with the inference rules of those systems will assign the standard truth functions to the connectives (see Smiley 1996; Incurvati and Smith 2009). The issue however becomes more complex when we move beyond the level of propositional logic (see Bonnay and Westerståhl 2016).

What reasons can be offered for resisting the move from the inferentialist Determination Thesis to the semantic thesis of inferentialism? As far as we are aware, two reasons have been offered. The first is due to Michael Dummett.

> [L]ogical constants figure equally in non-assertoric sentences. The meanings of the logical constants cannot, therefore, *consist* in their role in deductive inferences: they must have meanings of a more general kind, whereby they contribute to the meanings of sentences containing them just as other words do. (Dummett 1991b: 205)

Dummett's point is well taken, on the assumption that deductive inference can only involve assertoric sentences. But, as we will argue in the rest of the book, this assumption ought to be rejected: the theory of meaning we will develop takes the meaning of expressions to be given by their role in inferences involving not only assertions but also rejections and other speech acts.

The second reason for resisting the move from meta-semantics to semantics within inferentialism is conservatism (see Baker 2021: 2328). Referential semantics is at the centre of the success of current formal semantics, and the desire to reconcile formal semantics with inferentialism might lead one to conceive the latter simply as a thesis at the level of meta-semantics. We believe, however, that if inferentialism succeeds in providing sophisticated analyses of how the inference rules determine meaning, then it should reclaim this success as one that has impact on semantic research. There is, however, a more substantial problem in the vicinity, which affects semantic and meta-semantic versions of inferentialism alike, namely the problem of accounting for linguistic expressions beyond the core logical constants. We return to this problem in Section 2.6.

2.3 Why inferentialism?

So far, we have noted several choice points for inferentialist approaches to meaning, paying special attention to the distinction between the semantic thesis of inferentialism and its Determination Thesis at the meta-semantic level. Moreover, we have argued that, once the Determination Thesis is accepted, there seems to be little reason to resist accepting an inferentialist semantics. But why endorse the Determination Thesis in the first place or, more directly, an inferentialist semantics?

We noted above that, like expressivism, inferentialism is a non-referentialist approach to meaning in that it holds that the meaning of expressions subject to inferentialist treatment can be completely explained without appealing to the reference of those expressions. This means that, just like expressivism, inferentialism is neutral with respect to a host of ontological questions concerning what the expressions subjects to inferentialist treatment are typically taken to refer to (compare Read 2010: 557). To be sure, like sophisticated expressivists, contemporary inferentialists are wont to endorse some form of minimalism about truth, property, reference, and related notions (Chrisman 2008; Tiefensee 2019). Accordingly, arithmetical inferentialists can agree that it is true that the number of planets is eight, that there is an object which is identical to the number of planets, and that *eight* refers to this object. Similarly, and like sophisticated expressivists, ethical inferentialists can agree that it is true that lying is wrong, that there is a property of wrongness, and that *wrong* refers to this property.

Nonetheless, all that is forced upon us is a minimalist understanding of these claims about truth, reference, and existence. Inferentialism remains ontologically neutral with respect of the existence of mathematical objects or moral properties beyond what follows from minimalism. By contrast, referentialism, in providing semantic explanations in terms of reference, is committed to the existence of objects and properties beyond what is enforced by minimalism. This, we submit, puts referentialism on the back foot compared to inferentialism. In line with our discussion of the Pragmatist Razor in the previous chapter, semantics should not be in the business of settling metaphysical disputes which would appear to have no effect on the function and purpose of our discourse in the relevant area.

Let us take a closer look at the case of numerical expressions. Arithmetical inferentialists hold that the meaning of numerical terms can be completely explained in terms of inference rules, without recourse to their reference. One route for doing so would be to accept a version in terms of inference rules of the principle first considered by Frege in the *Grundlagen* (1884) and nowadays known as *Hume's Principle* (Boolos 1990). That is, an arithmetical inferentialist could take expressions of the form *the number of* to be governed by an elimination rule allowing us to move from *The number of Fs is the same as the number of Gs* to *There is a one-to-one correspondence between the Fs and the Gs*, and an introduction rule allowing us to move in the opposite direction. Since the non-self-identical objects are in one-to-one correspondence with non-self-identical objects, the introduction rule allows us to establish that the number of non-self-identical objects is identical to the number of non-self-identical objects, that is $0 = 0$. Given some form of minimalism about truth and reference, inferentialists can say that it is true that $0 = 0$ and that there is an object that *zero* refers to. This is compatible with insisting that the kind of reference that is thereby achieved is simply a *thin* one, as Dummett (1973a: chs 7 and 14, 1991a: chs 15–18) has argued, but also with holding that the reference secured by number words has to be construed realistically, as Wright (1983, 1995, 1998) has claimed. Inferentialism is neutral on the question whether number words refer to objects beyond a thin, minimal sense of *object*. Whether number words latch on an independently existing realm of objects is, from an inferentialist perspective, a question beyond the purview of semantics. This, our inferentialist insists, is how it should be: the answer to this question is immaterial to our practice involving the use of number words. This does not imply that there could not be additional *meta*-semantic grounds for thinking that the reference of number words ought to be construed thinly or thickly. We will return to the issue in Chapter 7.

Another advantage of the inferentialist approach to meaning lies in its methodology. Referentialist semantics is typically pursued in a bottom-up fashion by surveying a wide range of data and attempting to define truth conditions as appropriate generalizations that account for these data. Without any further constraints, it is wildly underdetermined what these truth conditions should be. The situation here should be familiar from the debate over the underdetermination of theory by data in the philosophy of science. Inferentialist semantics, by contrast, proceeds in a top-down manner by developing a theory which satisfies a number of theoretical constraints (in particular the proof-theoretic constraints we will discuss in the next section) and testing whether the theory matches the data.

In many cases, for instance when dealing with modal vocabulary, it is not straightforward to satisfy these constraints, witness the search for suitable natural deduction rules for the modal logics **S4** and **S5** (see, e.g., Poggiolesi and Restall 2012; Read 2015). Rather than being a hindrance, however, proof-theoretic constraints serve to narrow down the range of options available when developing a formal semantics. This represents a further advantage of the inferentialist approach over the referentialist one, which is instead presented with several competing candidates, all of which appear plausible given the linguistic data.

A final argument for an inferentialist approach to meaning comes from the relation between the theory of meaning and understanding. According to Dummett (1975b, 1976, 1991b: 107–113), a theory of meaning should be *full-blooded*: it should satisfy the requirement of being or yielding a theory of linguistic understanding. Using the terminology introduced in Chapter 1, a full-blooded theory of meaning is one that embodies or delivers an answer to the Meaning Knowledge Question in meta-semantics: it provides the means of explaining in virtue of what speakers have the semantic knowledge that they do. However, Dummett argues, referential semantic theories, in specifying the meaning of sentences in terms of truth conditions, fail to be full-blooded. They are *modest* in that they simply show or state what the meanings of expressions are. Thus, they fail to satisfy the requirement of full-bloodedness.[2]

Referential semantic theories specify, for each sentence of the language, its truth conditions on the basis of the meaning of its components. It might then seem that referential semantic theories do embody or deliver an answer to the Meaning Knowledge Question: to know the meaning of a sentence is to know its truth conditions. Or, as Wittgenstein famously wrote in the *Tractatus* (1922: §4.024), '[t]o understand a sentence means to know what is the case, if it is true'.[3]

However, according to Dummett, it is illusory to think that we have thereby made genuine progress on the Meaning Knowledge Question. Consider the case of the logical constants again. If to know the meaning of a sentence is to know its truth conditions, to know the meaning of *A and B* is to know the conditions under which *A and B* is true. Referential semantic theories specify these conditions to be exactly those in which *A* is true and *B* is true. Thus, speakers know the meaning of *A and B* in virtue of knowing that it is true just in case both *A* and *B* are. But, Dummett argues, this explanation is circular: we are specifying what it is to know the meaning of *A and B* in terms of what it is to know that *A and B* is true.[4] As Dummett puts it:

> A blanket account of understanding a statement as knowing what it is for it to be true is useless, because circular: it attempts to explain what it is to grasp a thought in terms of having a thought about that thought.
>
> (Dummett 2006: 78)

The point, as Dummett stresses, is not that to understand the explanation we must already possess the concept of conjunction. To require that a theory of meaning should be able to provide explanations of what meaning knowledge

[2] The question whether a theory of meaning ought to be full-blooded was at the centre of an extended exchange between Dummett and John McDowell. See McDowell 1987; Dummett 1987; McDowell 1997, 2007; Dummett 2007.

[3] We are deviating from the Ogden translation in rendering *Satz* as *sentence* instead of *proposition*.

[4] Dummett (1975a, 1991b) provided two other well-known objections to truth-conditional theories of meaning as theories of understanding, based on the possibility of acquiring and manifesting knowledge of truth conditions. In later work, however, Dummett (2006: 55) clarifies that '[n]either the objection arising from the manifestation nor that arising from the acquisition of the knowledge is central. The central objection is the circularity of a truth-conditional account'. Moreover, the objections based on acquisition and manifestation appear to target only the possibility of evidence-transcendent truth conditions—truth conditions whose obtaining is even in principle impossible to ascertain—whereas the objection from circularity is potentially more general.

consists in to someone who does not grasp any of the concepts expressed by the object language would be, as Dummett puts it, exorbitant. The point is rather that the proposed answer to the Meaning Knowledge Question tells us nothing about what it is to grasp the concept of conjunction. As Dummett puts it:

> A modest meaning-theory assumes not merely that those to whom it is addressed have the concepts expressible in the object language but that they require no explanation of what it is to grasp those concepts. A more robust conception of what it is to be expected of a meaning-theory is that it should, in all cases, make explicit in what a grasp of those concepts consists—the grasp which a speaker of the language must have of the concepts expressed by the words belonging to it. (Dummett 1991b: 108)

While the referentialist approach fails to yield a theory of understanding, Dummett continues, things are different for the inferentialist approach. Similarly to the case of referential semantic theories, inferentialist semantic theories appear to straightforwardly yield an answer to the Meaning Knowledge Question.

Inferentialism (Meaning Knowledge Thesis). A speaker knows the semantic value of an expression in virtue of grasping its inferential role.

However, the answer to the Meaning Knowledge Question provided by the inferentialist Meaning Knowledge Thesis does not suffer from the same shortcomings as the answer yielded by referentialist approaches. The answer does tell us what our grasp of the concepts expressible in the object language consists in. In the case of conjunction, for instance, the Meaning Knowledge Thesis tells us that in order to grasp the meaning of conjunction the speaker must be prepared to infer *A and B* from *A* and *B* and vice versa.

Indeed, some have suggested that it is 'hard to see what else could constitute meaning conjunction by "and" except being prepared to use it according to some rules and not others' (Boghossian 2011: 493). For this reason, if Dummett's requirement that a theory of meaning should be a theory of understanding, the view that an inferentialist account of the logical constants is the 'only game in town', as Murzi and Steinberger (2017: §2.1) put it, is widely accepted. But note that if Dummett's requirement is accepted as applying to semantic theories in general, the idea would appear to generalize. Dummett's argument that referential semantic theories cannot yield a theory of understanding is not, on the face of it, confined to the case of the logical constants. Thus, if successful, it applies to referentialist accounts in general.

We suspect that the majority of semanticists and philosophers of language would regard Dummett's requirement that a theory of meaning should yield a theory of understanding as still exorbitant. They would insist that a formal semantics can be supplemented with a theory of understanding, which explains what it is to grasp the truth conditions of a sentence. But inferentialists need not follow Dummett in regarding the fact that a theory of meaning provides an answer to the Meaning Knowledge Question as a necessary condition on the acceptability of a semantics. Instead, they can insist that, other things being equal, an inferentialist semantics is to be preferred to a referential semantics because it provides such an answer in a more direct fashion. Thus, the combination of referential semantics with a non-circular answer to the Meaning Knowledge Question is not to be ruled out as a non-starter. But if this combination is to be preferred to the combination of inferentialist semantics with the Meaning Knowledge Thesis of inferentialism, this has to be because other things are not equal—because inferentialism as a theory of meaning suffers from other problems or cannot explain all the phenomena that referential semantic theories explain.

2.4 The Problem of Defective Concepts

Perhaps the best-known problem for inferentialism and certainly one of the most influential ones is the *Problem of Defective Concepts*: that not every pair of introduction and elimination rules appears to confer a coherent meaning on the expression whose meaning is being defined. The problem was first pointed out by Arthur Prior (1960), who exhibited the connective tonk, governed by the following introduction and elimination rules.

$$(\mathsf{tonk\ I.})\ \frac{A}{A\ \mathsf{tonk}\ B} \qquad (\mathsf{tonk\ E.})\ \frac{A\ \mathsf{tonk}\ B}{B}$$

Adding tonk to a deductive system whose derivability relation is transitive—in which derivations can be freely chained together—allows one to infer any sentence from any sentence whatsoever.[5] As Prior put it, tonk is a runabout

[5] Formally, a derivability relation $\vdash$ is transitive just in case $A_1, \ldots, A_n \vdash B$ and $C_1, \ldots, C_n, B \vdash D$ jointly imply $A_1, \ldots, A_n, C_1, \ldots, C_n \vdash D$.

inference ticket. Prior concluded that the Problem of Defective Concepts sinks logical inferentialism: to single out which pairs of introduction and elimination rules confer a coherent meaning upon the logical constants involved, inferentalists must resort to good old fashioned referential semantics. In particular, it is easy to see that tonk does not correspond to any of the sixteen truth tables for binary connectives: tonk is bad because it makes no coherent contribution to the truth conditions of the sentences in which it occurs. A few inferentialists have insisted that tonk is, after all, a meaningful connective, since a system's derivability relation ought not to be transitive (Cook 2005; Ripley 2015).[6] Most inferentialists however reacted by arguing that while the Problem of Defective Concepts is a genuine one, it can be tackled by formulating *proof-theoretic* criteria on the admissibility of pairs of introduction and elimination rules: all and only sets of rules satisfying those criteria confer a coherent meaning on the relevant expressions. For inferentialists that give precedence to the introduction or to the elimination rules, the criteria must place constraints on how to obtain the remaining rules so that no defective concepts can arise; for inferentialists that give precedence to neither the introduction nor the elimination rules, the criteria place constraints on the relation that there must exist between introduction and elimination rules so that no defective concepts can arise.

Nuel Belnap (1962), in the first published response to Prior, argued that the problem with tonk is that it perturbs the antecedent context of derivability. Its addition to a deductive system whose derivability relation has certain basic structural rules—in particular, as we have seen, transitivity—results in a *non-conservative* extension: it allows us to prove sentences in the tonk-free language that were not previously provable. By contrast, the addition of conjunction to a standard natural deduction system without the conjunction rules gives rise to a conservative extension. Belnap thus proposed conservativeness as a proof-theoretic criterion for separating the good rules from the bad ones, and some inferentialists have followed suit (Dummett 1991b; Dicher 2016).[7] As Belnap already noticed, whether the addition of a pair

[6] Wansing (2006) exhibits a super-tonk connective, which leads to disaster even in non-transitive systems. Ripley (2015) argues that the rules of super-tonk do not have the right introduction and elimination form and so fail to be meaning-conferring. For a different defence of the meaningfulness of tonk, based on the possibility of a non-homophonic translation from languages containing tonk to tonk-free languages, see Warren 2015.

[7] Dummett calls conservativeness *total harmony*, to be distinguished from *intrinsic harmony*, to be discussed below. In addition to conservativeness, in the same paper Belnap also suggested the

of introduction and elimination rules results in a conservative extension depends on the system to which it is added. Conservativeness is thus a property of inference rules relative to a system (Steinberger 2011: 625). A proof-theoretic criterion for the admissibility of inference rules which is not sensitive to the system to which it is added is that of *harmony*, which is instead an intrinsic property of pairs of introduction and elimination rules.

The general idea of harmony is that there should be a certain balance between the introduction and the elimination rules for a constant: that the elimination rules should allow us to derive *no more* and *no less* than what can be derived from the premisses of the introduction rules. The general idea of harmony can be traced back to Gentzen, who clarifies his claim that the elimination rules are simply 'consequences' of the introduction rules to mean that

> when eliminating a sign, the relevant formula that contains it as its outermost sign may 'only be used according to the meaning it possesses in virtue of the introduction rule for this sign'.
>
> (Gentzen 1935: 185, our translation)

If, like Gentzen, one gives precedence to the introduction rules, one will see these rules as self-justifying and the elimination rules as justified by the fact that they are in harmony with the introduction rules. But one need not embrace the introduction-rules-first view to make sense of the requirement of harmony. In particular, if one gives precedence to the elimination rules, one will see these rules as self-justifying and the introduction rules as justified by the fact that they are in harmony with the elimination rules. If one gives precedence to neither, one will see introduction and elimination rules as jointly justified by the fact that they are in harmony with each other.

There are several conceptions of harmony in the literature. Perhaps the most familiar one goes back to Dag Prawitz (1965, 1974), who offered an interpretation of Gentzen's remarks in terms of the so-called *inversion principle*. Informally, the inversion principle states that the elimination rules must be inverses of the introduction rules in the following sense:

adoption of a uniqueness criterion, stating that two logical constants governed by the same rules must determine the same logical operator.

> the conclusion obtained by an elimination does not state anything more than what must have already been obtained if the major premiss of the elimination was inferred by an introduction. (Prawitz 1974: 246)

Prawitz formalizes the inversion principle in terms of the possibility of carrying out certain reduction procedures. When, in the course of a derivation, a logical constant is introduced and subsequently eliminated, we have a *detour*, and a *reduction procedure* is a procedure for removing such a detour. A special case of a detour is when the logical constant is eliminated at the very next step after its introduction. In this case, we have a *local maximum*, and a *local reduction procedure* is a reduction procedure for eliminating a local maximum.[8] Prawitz's formalization of the inversion principle then states that local maxima can always be eliminated via a local reduction procedure.[9] This property is known as *local soundness*.[10]

The conjunction rules are clearly locally sound. Suppose that, having obtained A and B via derivations Δ_1 and Δ_2, one goes on to derive $A \wedge B$ via an application of conjunction introduction and then concludes A via an application of conjunction elimination. This *detour* can be avoided by simply using the initial derivation Δ_1 of A.

$$\dfrac{\dfrac{\begin{matrix}\Delta_1 \\ A\end{matrix} \qquad \begin{matrix}\Delta_2 \\ B\end{matrix}}{A \wedge B}\,(\wedge\,\text{I.})}{A}\,(\wedge\text{E.}_1) \qquad \rightsquigarrow \qquad \begin{matrix}\Delta_1 \\ A\end{matrix}$$

By contrast, the rules for tonk are not locally sound. The local maximum created by the derivation of B from A via successive applications of the introduction and elimination rules for tonk cannot be eliminated.

[8] Gentzen calls local maxima *hillocks* (see von Plato and Gentzen 2008), while Dummett (1991b: 248–250) calls them *local peaks* and accordingly talks of *levelling* such peaks.

[9] Local reduction procedures are central to the possibility of obtaining normalization results, which state that derivations in certain natural deduction systems can always be converted into *detour*-free derivations. However, the availability of local reduction procedures is neither necessary nor sufficient for normalization. On the one hand, *detours* may occur other than those giving rise to local maxima. On the other, *detours* via local maxima may be avoidable despite the absence of local reduction procedures for the relevant logical constant. The latter point is especially important since some versions of the classical propositional calculus are normalizable (Prawitz 1965: 40) even though, as we will see, the classical rules for negation do not admit of a local reduction procedure.

[10] Dummett (1991b: 250) uses the label *intrisic harmony* for harmony as local soundness, while Steinberger (2013) uses the label *harmony-as-levelling procedures*.

$$\dfrac{\dfrac{\begin{array}{c}\vdots\\ A\end{array}}{A\ \text{tonk}\ B}\ (\text{tonk I.})}{B}\ (\text{tonk E.})$$

However, a constant being locally sound does not rule out that its elimination rules might allow us to derive less than what one could derive from the premisses of its introduction rules. Consider, for instance, the logical operator *smonjunction* characterized by the same rules as conjunction, except that one is only allowed to infer *A* from the smonjunction of *A* and *B*. The rules for smonjunction are locally sound, but the elimination rule does not allow us to derive everything that one could derive from *A* and *B*, since *B* itself is not derivable. For this reason, it has become customary to supplement local soundness with *local completeness* (Pfenning and Davies 2001) to develop a full-blown conception of harmony.[11] While local soundness demands that one should be able to locally eliminate detours, local completeness demands that one should be able to locally introduce them. Formally, let a *local expansion procedure* be a procedure for transforming a derivation of a sentence containing a constant as its main operator into a derivation of the same sentence via application of an elimination rule for the constant and applications of all introduction rules for the constant. Local completeness then states that derivations are always locally expandable. As witnessed by the following expansion procedure, the conjunction rules are locally complete.

$$\begin{array}{c}\vdots\\ A \wedge B\end{array} \quad \rightsquigarrow \quad \dfrac{\dfrac{\begin{array}{c}\vdots\\ A \wedge B\end{array}}{A}\ (\wedge\ \text{E.}) \qquad \dfrac{\begin{array}{c}\vdots\\ A \wedge B\end{array}}{B}\ (\wedge\ \text{E.})}{A \wedge B}\ (\wedge\ \text{I.})$$

It is easy to see that, by contrast, the rules for smonjunction are not locally complete.

The *local intrinsic conception of harmony* holds that the rules for a constant are harmonious just in case they are locally sound and complete.[12] This conception of harmony has been influential, but, as noted, one finds other conceptions in the literature. Similarly to local intrinsic harmony, Stephen

[11] Dummett (1991b: 287) already suggested that as well as being locally sound, the rules ought not to be unduly prohibitive. Dummett calls this requirement *stability*, but does not fully develop it.

[12] The term *local intrinsic harmony* is Francez and Dyckhoff's (2012).

Read's (2010) *general-elimination harmony* take its cue from Prawitz's inversion principle, but formalizes it so that for the elimination rules to be inverses of the introduction rules they must have a specific *form*. The resulting conception of harmony is a strengthening of local intrinsic harmony, in that if the rules for a constant are general-elimination harmonious, they are also locally intrinsically harmonious, but not vice versa. Neil Tennant (1997, 2020), for his part, develops a conception of harmony as deductive equilibrium. For convenience, in the remainder of the book, we will work with local intrinsic harmony, but the discussion straightforwardly carries over to general-elimination harmony and harmony as deductive equilibrium.

The proof-theoretic constraint of harmony seemingly rules out tonk. Moreover, it appears to rule intuitionistic logic in: in addition to the rules for conjunction, the rules for the other intuitionistic logical constants can also be shown to be locally intrinsically harmonious. The negation case, however, deserves closer inspection. The following are the standard intuitionistic rules for negation.

$$(\neg\text{I.})\ \dfrac{\begin{matrix}[A]\\ \vdots\\ \bot\end{matrix}}{\neg A} \qquad (\neg\text{E.})\ \dfrac{A \qquad \neg A}{\bot}$$

A natural deduction system for classical logic is then standardly obtained by adding to these rules an additional way of eliminating negation, such as the rule of double-negation elimination.

$$(\text{DNE})\ \dfrac{\neg\neg A}{A}$$

However, while the intuitionistic negation rules are locally sound, classical double-negation elimination gives rise to local maxima for which there is no local reduction procedure. An example is provided by the derivation of classical *reductio*, which allows one to conclude A from the fact that the supposition that $\neg A$ leads to absurdity.

$$\dfrac{\dfrac{\begin{matrix}[\neg A]^1\\ \vdots\\ \bot\end{matrix}}{\neg\neg A}\,(\neg\text{I.})^1}{A}\,(\text{DNE})$$

Similar considerations apply to the other standard ways of obtaining a natural deduction system for classical logic. The harmony constraint appears to sanction intuitionistic logic and rule out classical logic. This is the core of Dummett's (1991b) *proof-theoretic* argument for intuitionistic logic: since meaning has to be grounded in our practices, Dummett contends, we should be inferentialists; but since inferentialism must comply with the harmony constraint, we should also be intuitionists.

In the next chapter, we argue that inferentialists can have their cake and eat it too: by helping themselves to ingredients from the expressivist programme, inferentialists can reconcile the harmony constraint with classical logic.

2.5 The Problem of Constitutive Rules

We have seen that inferentialism is naturally associated with a particular answer to the Meaning Knowledge Question in meta-semantics. This is the answer provided by the inferentialist Meaning Knowledge Thesis, according to which a speaker knows the semantic value of an expression in virtue of grasping its inferential role. Moreover, in the inferentialist tradition, it is commonplace to take the inferential role of an expression to be captured by means of inference rules, which can therefore be taken to be *constitutive* of the meaning of the expression. Thus, inferentialism's answer to the Meaning Knowledge Question becomes that a speaker knows the semantic value of an expression in virtue of grasping the inference rules that are constitutive of its meaning. However, Williamson (2008) has presented a forceful challenge to the idea that semantic knowledge could ever consist in grasp of an expression's constitutive rules, no matter what those rules are taken to be. This is the *Problem of Constitutive Rules*.

Williamson's challenge takes the form of a general challenge to any theory of semantic knowledge that posits a link between understanding and assent, an *understanding-assent link* for short. A widespread view in philosophy, initially advanced by Paul Boghossian (1996), has it that Willard Van Orman Quine's (1951) attacks on the notion of analyticity refute the idea that there are *metaphysically* analytic truths: that there are sentences that are true in virtue of meaning alone. However, according to the widespread view, Quine's attacks fail to undermine the idea that there are *epistemically* analytic sentences: that there are sentences that can be known in virtue of grasping their meaning alone. These are the sentences that one must assent to in order to count as understanding the meaning of a certain expression, provided that one also understands the remaining expressions in the sentence. However, Williamson argues, for any expression, there can be no particular sentence to which assent is required in order to count as understanding that expression. Suppose, for instance, one takes *vixen* to be synonymous with *female fox* and takes knowledge of the meaning of *vixen* to consist in knowing this fact. Then, one's theory of semantic knowledge sanctions the following understanding-assent link.

UA-vixen. Necessarily, whoever understands the sentence *Every vixen is a female fox* assents to it.

But, Williamson contends, there are counterexamples to UA-vixen. For instance, someone might understand *vixen* but refuse to assent to *Every vixen is a female fox* on the grounds that they hold the view that quantified statements involving borderline cases do not have a definite truth value and that there are borderline cases of being a vixen. As implausible as this view may be, it can be held by someone who grasps the meaning of *vixen*.

By subscribing to the view that a speaker knows the semantic value of an expression in virtue of grasping the inference rules that are constitutive of its meaning, inferentialists appear to subscribe to understanding-assent links. Suppose, for instance, that inferentialists take the meaning of *if* to be partly constituted by the rule of *modus ponens*. Then, Williamson argues, the inferentialist theory of semantic knowledge sanctions the following understanding-assent link:

UA-if. Necessarily, whoever understands *modus ponens* assents to it.

But, Williamson claims, there are counterexamples to UA-if. Indeed, there is a real-life one in this case. The philosopher Vann McGee (1985) presented a purported counterexample to *modus ponens*, which he therefore considers an invalid rule. Far from not understanding *if*, McGee is a leading expert on its meaning. And anyone who knows him will confirm that he is as skilled as anyone in using the word *if* in conversation. So there is someone who understands *modus ponens* but does not assent to it. Moreover, Williamson continues, blaming the particular choice of *modus ponens* as a rule constitutive of the meaning of *if* will not do: for any putative meaning-constitutive rule, one can find or construct cases in which someone understands the rule but does not assent to it. Hence, Williamson concludes, the inferentialist account of semantic knowledge fails.

Care is needed about how to understand assent to *modus ponens* in UA-if from an inferentialist point of view. Clearly, not everyone who understands the meaning of *if* assents to *modus ponens* in the sense that they endorse its schematic formulation by means of meta-linguistic variables. People may understand the meaning of *if* without being familiar with meta-linguistic variables. However, in saying that a speaker knows the semantic value of an expression in virtue of grasping the inference rules constitutive of its meaning, inferentialists typically mean that knowing the semantic value of an expression requires that the speaker implicitly accept those rules. This implicit acceptance is manifested by the fact that, in their linguistic practice, they accept applications of concrete instances of the rules. Thus, in particular, someone who understands the meaning of *if* may nonetheless be taken to assent to *modus ponens* in the sense that they accept concrete instances of *modus ponens*.

With this clarification on board, we can now see what it takes for McGee to be a genuine counterexample to UA-if. McGee's purported counterexample to *modus ponens* takes the form of a scenario in which one has strong evidence for the premisses of the argument from *If a Republican wins the election, then if it's not Reagan who wins it will be Anderson* and *A Republican will win the election* to *If it's not Reagan who wins, it will be Anderson*, but no evidence for its conclusion (McGee 1985: 462). For McGee to be a genuine counterexample to UA-if, McGee must be right about the theoretical claim that the argument is a genuine instance of *modus ponens*, which has been disputed (Stojnić 2017). For if the argument considered by McGee is not a genuine instance of *modus ponens*, then all that follows is that McGee *thinks* that he is refusing to assent to *modus ponens*. It may be that McGee's inferential behaviour shows that he accepts all *actual* concrete applications of *modus ponens*. Similarly, McGee must be right about the theoretical claim that the fact that one may have strong evidence for the premisses of an argument while having no evidence for its conclusion shows that the argument is invalid, which has also been disputed (Bledin 2015) and which will not be the case in the multilateral framework to be developed in this book. For if an argument can be valid without preserving evidence, then again all that follows is that McGee *thinks* that he is refusing to assent to *modus ponens*. It may be that McGee's inferential behaviour shows that he accepts all concrete applications of *modus ponens* as legitimate, despite not taking them to preserve evidence.

It is likely that Williamson wants us to think of McGee's case as one in which his inferential behaviour shows that he refuses to assent to *modus ponens*. He writes that '[p]resumably, [McGee] refuses to make some inferences by *modus ponens*' (Williamson 2003: 251–252). As mentioned, doubts can be raised as to whether this is actually the case. But perhaps this does not matter too much: like UA-vixen, UA-if is a necessary claim and so we do not need a real-life counterexample to it. And it seems at least plausible that there might be an expert on the meaning of *if* whose inferential behaviour shows that they refuse to assent to *modus ponens*. However, the existence of such an expert, possible or actual, does not undermine the inferentialist theory of semantic knowledge.

Let us first consider the case in which the expert's inferential behaviour shows that they do not accept common-or-garden applications of *modus ponens* such as the inference from *If it is raining, then the streets are wet* and *It is raining* to *The streets are wet*. In this case, we would be entitled to question their grasp of the meaning of *if*, their expertise notwithstanding (compare Warren 2021: 9227–9229). The more interesting case is one in which the expert's inferential behaviour shows that they do not accept *recherché* applications of *modus ponens*. The putative instance of *modus ponens* at stake in McGee's counterexample appears to be one such application, since it involves nested conditionals, which are notoriously difficult to interpret (see Chapter 8). If the differences between ordinary speakers and the expert crop up only in *recherché* cases, it might seem implausible to claim that the expert has a deviant grasp of the meaning of *if*.

How can inferentialists deal with differences in use in *recherché* cases? One option is for inferentialists to retreat and take the meaning-constitutive rules for *if* to only involve common-or-garden applications of *modus ponens*. Williamson (2008: 96) claims that this is hopeless, since we can consider cases in which an expert's inferential behaviour shows that they do not accept such applications. However, as noted, in such cases it seems reasonable to question the expert's grasp of the meaning of *if*.

But another option is for inferentialists to put their foot down and insist that differences in *recherché* applications of *modus ponens* do make for a difference in the understanding of *if*. Williamson contends that differences in use in *recherché* cases cannot make for a difference in meaning on the grounds that the expert and ordinary speakers fully participate in a social practice involving the relevant expression.

> The social determination of meaning requires nothing like an exact match in use between different individuals; it requires only enough connection in use between them to form a social practice. Full participation in that practice constitutes full understanding. That is why there is no litmus test for understanding. (Williamson 2008: 98)

To see how inferentialists should respond, it will be helpful to first consider how holist inferentialists address a related problem. Holist inferentialists identify an expression's inferential role with the totality of its uses. Since there are differences between the totality of uses of individual speakers, holist inferentialists claim that, strictly speaking, different speakers mean different things by the same expression. This might seem implausible, but holist inferentialists typically insist that differences in use do not hamper participation in a shared social practice, as long as there is sufficient overlap among speakers in the totality of uses of an expression, which then corresponds to a sufficient overlap in the meaning of that expression (Brandom 2007: 663–670). Inferentialists of all brands can give the same kind of response to Williamson's contention that it is implausible that differences in use in *recherché* cases give rise to differences in meaning because such differences would prevent participation in a shared social practice. Inferentialists can insist that the relevant differences in meaning do not hamper participation in a shared social practice because the overlap in the other uses of the expression that are relevant to its meaning is sufficient to ensure participation in the practice. Indeed, this seems very plausible when we are only dealing with differences in meaning that arise in *recherché* cases: differences with regards to the application of *modus ponens* in such cases do not hamper participation in a shared social practice involving *if* precisely because such differences concern *recherché* cases and are therefore unlikely to come up in ordinary linguistic practice.

Where does this leave the Problem of Constitutive Rules? For McGee to be a counterexample to UA-if, it is not enough that he thinks that he has found an invalid instance of *modus ponens*. Rather, his inferential behaviour must show that he refuses to assent to *modus ponens*. But even if his inferential behaviour is of the appropriate kind, or if we think that there might be an expert on the conditional who displays the appropriate inferential behaviour, this does not refute the inferentialist theory of semantic knowledge. If the expert's inferential behaviour shows that they do not accept *modus ponens* in common-or-garden cases, it is plausible for inferentialists to take the expert's grasp of *if* to be deviant. If the expert's inferential behaviour shows that they do not accept *modus ponens* in *recherché* cases only, inferentialists can either take the meaning of *if* to be constituted only by common-or-garden applications of *modus ponens* or insist that the expert's grasp is nonetheless deviant, without the expert's participation in ordinary linguistic practice being precluded.

Of course, and as Williamson emphasizes, the inferentialist theory of semantic knowledge must not foreclose the possibility of disagreements about the behaviour of natural language expressions. But again, there is no reason to think that this will be the case. The inferentialist theory of semantic knowledge is in fact compatible with several types of disagreements about the behaviour of natural language expressions. First, it is compatible with theoretical disagreements about what conclusions to draw from the linguistic data about the behaviour of an expression, such as whether a certain piece of data which appears to be a counterexample to *modus ponens* is in fact an instance of *modus ponens*. Second, if the meaning of an expression is constituted by the rules governing its common-or-garden variety cases, the inferentialist theory of semantic knowledge is compatible with disagreements about how to use the expression in *recherché* cases to which the meaning-constitutive rules do not apply. Third, the inferentialist theory of semantic knowledge is compatible with disagreements about what the meaning-constitutive rules are, even if there is no disagreement about the data. An analogy with disagreements about the rules of grammar might be helpful. Presumably, native speakers agree on the grammaticality of common-or-garden variety sentences, but may disagree on whether a certain sentence instantiates a particular purported rule of grammar, on whether a *recherché* sentence is grammatical, and on how to explicitly formulate the rules of grammar. Additionally, the inferentialist theory of semantic knowledge is compatible with normative disagreements about what the meaning-constitutive rules ought to be—about how to revise our usage and hence the meaning of an expression because of theoretical considerations. Although we cannot provide a full-blown defence of the possibility of disagreements about the behaviour of natural language expressions from an inferentialist standpoint, we hope that our discussion has made it clear how such a defence would proceed (see Warren 2021 for further discussion on the relationship between understanding-assent links and disagreement).

2.6 The Problem of Limited Applicability

Earlier we argued that, like expressivism, inferentialism is best understood as appealing to the pragmatist constraint that semantic explanations should not go beyond the function and purpose of the relevant expressions in our practices. The inferentialist explains the meaning of an expression in terms of what we do with it, and so complies with the pragmatist constraint. The referentialist, by contrast, has to offer a story linking reference to an expression's functional role. All things being equal, the inferentialist's semantic explanations are to be preferred. In a similar spirit, we argued that inferentialist semantics immediately yields an answer to the Meaning Knowledge Question: a speaker knows the semantic value of an expression in virtue of grasping its inferential role. Referentialists, by contrast, must supplement their modest answer to the Meaning Knowledge Question—that knowing the meaning of a sentence consists in knowing its truth conditions—with an explanation of what it is to know the truth conditions of a sentence. Again, all things being equal, the inferentialist's meta-semantic explanations are to be preferred.

However, not all things would appear to be equal. Referentialism is the philosophical view underwriting most linguistic research in semantics and a wide range of linguistic phenomena have received referentialist treatment. Inferentialism, for its part, has been successfully applied to the case of the core logical constants, an area of discourse where the view, as noted, has much initial appeal. However, inferentialism has rarely progressed beyond the case of the core logical constants, and it is not clear whether the inferentialist strategy to explain their meaning may be generalized to other areas of discourse. While referentialism has undoubtedly proven its efficacy in linguistic research, inferentialists have failed to apply their approach beyond a rather restricted domain. This is the *Problem of Limited Applicability*.

The Problem of Limited Applicability threatens to undermine inferentialism's credentials as a serious contender in the theory of meaning. We take Williamson to express a common sentiment when he writes that

> [i]f you want an explicit theory of how some particular linguistic construction contributes to the meanings of sentences in which it occurs, the inferentialist is unlikely to have one. (Williamson 2010: 23)

Williamson is a staunch defender of referentialism. But the Problem of Limited Applicability is also acknowledged by inferentialists such as Greg Restall—rightly, to our mind. As Restall notes, we need from inferentialists

> more work on the range of applications in the theory of meaning for speech acts beyond assertion and concepts beyond the core logical constants. (Restall 2016)

Inferentialists must do better. In this book, we take up the challenge. We provide inferentialist accounts of a variety of linguistic phenomena, including epistemic modals, moral vocabulary, the truth predicate, conditionals, and probability operators. The problem of applic*ability* turns out to be a problem of applic*ation*. The overwhelming focus on referentialist semantics over inferentialist semantics is a product of circumstance, not of inferentialism being intrinsically lacking.

Our inferentialist accounts retain the basic model of linguistic explanation provided by logical inferentialism: the meaning of expressions will be given by harmoniously related introduction and elimination rules. To this extent, we agree with Brandom that inferentialists should

> look to the contents of logical concepts as providing the key to understanding conceptual content generally. (Brandom 2007: 161)

However, we contend, inferentialists should also embrace elements of the expressivist programme in order to go beyond the core logical constants. By helping themselves to the richness of speech acts and the attitudes they express, inferentialists can give their due to linguistic phenomena that have been up to now within the purview of referentialist semantics alone.

Thus, that between expressivism and inferentialism is a marriage made in heaven. By becoming inferentialists, expressivists can solve the Frege–Geach Problem, as well as clarify the kind of semantic explanation they offer. By becoming expressivists, inferentialists can solve the Problem of Limited Applicability. Our next task is to show how to combine the two views.

3
Inferential Expressivism

Inferential expressivism combines aspects of traditional expressivism and inferentialism. Traditional expressivism holds that the meaning of certain linguistic items is given by the attitudes they are used to express. Inferentialism holds that the meaning of certain linguistic items is given by the inferences they license. Inferential expressivism holds that the meaning of certain linguistic items is given by the inferences to attitude expressions they license.

In this chapter, we introduce inferential expressivism. Our starting point is the meaning of negation. We argue that the bilateralist programme in logic—which holds that the meaning of the logical constants is given by conditions on assertion and rejection—is best regarded as a form of inferential expressivism about negation. And while traditional expressivism about negation falls prey to the Frege–Geach Problem, its inferential counterpart has the resources to solve it. This will be a recurring theme: inferential expressivism about a certain domain retains the appeal of expressivism whilst avoiding the pitfalls of its traditional versions. We go on to develop a version of bilateral logic in which *strong* and *weak* rejections are distinguished. This opens the door to *multilateralism*, the view that the meaning of linguistic expressions is given by conditions on more than two distinct speech acts.

3.1 Traditional expressivism about negation

The pragmatist Huw Price (1990) offers a functionalist story about negation. Price argues that there must be a primitive operation of rejection, since without such an operation we could not inform someone that our claims are incompatible with theirs. If someone asserts some A and we try to refute them by asserting a contrary B, we may fail because our interlocutor may not realize that A and B are incompatible. Even if our interlocutor understands the truth table for negation and we assert *not* A, Price continues,

Reasoning with Attitude: Foundations and Applications of Inferential Expressivism. Luca Incurvati and Julian J. Schlöder, Oxford University Press. © Oxford University Press 2023.
DOI: 10.1093/oso/9780197620984.003.0003

we may still fail if our interlocutor does not realize that truth and falsity are incompatible. Thus, to inform someone of a perceived incompatibility, we need a primitive operation to *register* this incompatibility. This operation, Price concludes, is rejection and the meaning of negation is given by its function of indicating rejection. Similar accounts of negation have been offered by other pragmatists, including Charles Sanders Peirce (1905), Frank Ramsey (1927), Wilfrid Sellars (1969), Blackburn (1988), and Brandom (1994).[1]

As Price notes, an analogous argument can be given for the need for a primitive attitude of *disbelief*, expressed by negation, alongside belief: we need such a primitive attitude to register the incompatibility of a sentence with what we believe. Thus, Price's account would appear to be a form of traditional expressivism about negation, and indeed one motivated by the Pragmatist Razor. Meta-semantically, the function of negative statements is to register a perceived incompatibility. By the Pragmatist Razor, it follows that, semantically, negation indicates rejection, the speech act expressing disbelief and 'a perfectly general means of registering and pointing out the incompatibility' (Price 1990: 224).

Once one observes that Price's functionalist account would seem to be a form of traditional expressivism about negation, it should come as no surprise that it has to contend with the Frege–Geach Problem. Indeed, when Frege (1919) originally presented the problem, his target was precisely the view that negation indicates rejection. Towards a *reductio*, Frege supposes that *not* is a force indicator for rejection. He then considers the following seemingly valid inference.

(1) a. If the accused was not in Berlin, he did not commit the murder.
b. The accused was not in Berlin.
―――――――――――――――――――――――――
c. He did not commit the murder.

Frege points out that the antecedent of the conditional *If the accused was not in Berlin, he did not commit the murder* cannot be a rejection of *The accused was in Berlin*, since somebody uttering the conditional might believe that

[1] For an enlightening history of rejection in the American pragmatist tradition, see Beisecker 2019. On Cambridge pragmatism, including Ramsey, Blackburn, and Price, see Misak 2016.

the accused was in Berlin. Thus, the *not* in the antecedent of the conditional must *modify* the claim that the accused was in Berlin (instead of indicating its force). Suppose we insist that in unembedded contexts such as the second premiss, *The accused was not in Berlin*, is a rejection of *The accused was in Berlin*. Then the second premiss does not coincide with the antecedent of the first premiss, and so the inference cannot be validated by *modus ponens*. For the inference to be an instance of *modus ponens*, the *not* in the second premiss must modify the claim that the accused was in Berlin. But then *not* does not indicate rejection.

The Frege–Geach Problem sinks traditional expressivism about negation. Frege concludes that negation is not to be explained in terms of rejection. Instead, he says, we should explain rejection in terms of negation: to reject a sentence is simply to assert its negation. Frege appeals here to parsimony. By explaining rejection in terms of negation and assertion we only need two primitives instead of the three we would need if we took rejection also to be a primitive.

For his part, Price acknowledges that he is

> putting to one side a venerated objection to an account of negation in terms of denial, namely Frege's argument that such a view cannot make sense of embedded negation. (Price 1990: 225)

But he insists that the argument is 'powerless' if we allow a negative statement to express both a rejection and the assertion of a negation. This would seem to make his view a form of hybrid expressivism about negation. As we noted in Chapter 1, however, a hybrid approach fails to properly deal with the Frege–Geach Problem. In the next section, we explain how expressivists about negation can do better by helping themselves to the resources of inferentialism.[2]

[2] Ripley (2020: 54) also finds it plausible to solve the Frege–Geach Problem by endorsing a form of hybrid expressivism about negation. Indeed, in addition to taking rejections to also be negative assertions, he takes assertions to also be negative rejections: any time one performs an assertion or a rejection, one performs both. Our objection to the hybrid solution to the Frege–Geach Problem carries over to Ripley's approach. Nonetheless, the spirit of Ripley's approach is correct: Price's insight is that assertion and rejection should be on equal footing, but his hybrid solution to the Frege–Geach Problem does not do justice to this insight. Below, we will show that we can do justice to Price's insight and solve the Frege–Geach Problem without multiplying the speech acts performed with a linguistic utterance.

3.2 Bilateralism and Frege–Geach

Frege's conclusion that the meaning of negation is not to be explained in terms of rejection is too hasty. As we are now going to show, *bilateralism* offers a way to solve the Frege–Geach Problem whilst giving its due to the idea that the meaning of negation is explained in terms of rejection.

According to bilateralism, the meaning of the logical constants is given by conditions on the speech acts of assertion and rejection. The meaning-conferring conditions are formulated by means of inference rules in a natural deduction system. Bilateralism is therefore a version of inferentialist semantics: the meaning of an expression is given by its role in inferences. Formally, *sentences* are obtained from a countable set of propositional atoms $p_1, \ldots, p_n$, conjunction $\wedge$ and negation $\neg$ in the usual way, and are denoted by upper-case Latin letters. We abbreviate $\neg(\neg A \wedge \neg B)$ as $A \vee B$ and $\neg(A \wedge \neg B)$ as $A \supset B$. *Formulae* are sentences prefixed (or *signed*) with *force markers* for assertion and rejection, and are denoted by lower-case Greek letters.[3] The formula $+A$ represents the assertion of the sentence A; the formula $-A$ represents its rejection.

The meaning of negation can still be explained in terms of rejection, in keeping with Price's observations about the function of the negation operator. But rather than *indicating* rejection, negation is *inferentially explained* in terms of rejection. In particular, the rules for negation in standard bilateral systems (Smiley 1996; Rumfitt 2000) allow one to move from rejection to negative assertion and vice versa, and from assertion to negative rejection and vice versa.

$$(+\neg\text{I.})\ \frac{-A}{+\neg A} \qquad (+\neg\text{E.})\ \frac{+\neg A}{-A} \qquad (-\neg\text{I.})\ \frac{+A}{-\neg A} \qquad (-\neg\text{E.})\ \frac{-\neg A}{+A}$$

As noted by Ian Rumfitt (2000), these rules satisfy the usual proof-theoretic criteria on the admissibility of inference rules, in particular the harmony constraint. As in traditional logical inferentialism, the meaning-conferring rules for negation divide into rules for its introduction and elimination. But we now have multiple contexts in which an operator can be introduced or from which it can be eliminated: rejected contexts as well as customary asserted ones. The rules fix the meaning of negation in that they state when

[3] We will use the term *force indicator* to talk about natural language expressions that are used to indicate force, and the term *force marker* for their formal counterparts.

one can infer the assertion or rejection of a negative and what one can infer from such an assertion or rejection.

But not only is bilateralism a form of logical inferentialism; it is also, we contend, a form of expressivism about negation. For the meaning of negation is explained in terms of rejection, the speech act expressing the attitude of disbelief. Bilateralism is not, to be sure, a form of *traditional* expressivism about negation, since *not* does not indicate the speech act of rejection; it is, nevertheless, a form of *inferential* expressivism, since the meaning of negation is inferentially explained *in terms* of rejection. And rejection is the speech act expressing the attitude of disbelief, where this attitude, to stress, is not to be reduced and indeed is not reducible to negative belief. As Price pointed out, the same arguments for the existence of a primitive speech act of rejection alongside assertion are easily converted into arguments for the existence of a primitive attitude of disbelief alongside belief, the attitude expressed by assertions.

Negation has been explained in terms of rejection, but for Price's functionalist account to be vindicated, rejection must be able to serve as a primitive operation to register incompatibility. This can be achieved by laying down rules ensuring that assertion and rejection are incompatible. The appropriate rules are Rejection, which permits to infer absurdity from having asserted and rejected the same sentence, and the Smileian *reductio* rules, named after Timothy Smiley (1996), which state how to discharge an inferred absurdity.[4]

$$\text{(Rejection)}\ \frac{+A \qquad -A}{\bot} \qquad (\text{SR}_1)\ \frac{\begin{matrix}[+A]\\ \vdots \\ \bot\end{matrix}}{-A} \qquad (\text{SR}_2)\ \frac{\begin{matrix}[-A]\\ \vdots \\ \bot\end{matrix}}{+A}$$

These rules are known as *coordination principles*, in that they do not characterize the meaning of an operator, but govern the interaction of assertion and rejection. The coordination principles characterize assertion and rejection as contrary expressions of attitudes: Rejection states that it is inconsistent to assert and reject the same sentence;[5] the Smileian *reductio* principles (SR)

[4] In bilateral logics, it is customary to treat $\bot$ as a punctuation sign indicating that a logical dead end has been reached (Tennant 1999). This sign is therefore not prefixed with a force marker (Rumfitt 2000).

[5] This does not mean that one *cannot* simultaneously believe and disbelieve the same sentence (for example, one can compartmentalize). Rather, two attitude expressions are inconsistent when

state that whenever it is inconsistent to express one attitude towards some sentence, one is committed to expressing the other one.

All that is left is to lay down the rules giving the meaning of conjunction. Here, the bilateralist can simply follow the traditional logical inferentialist: the bilateral rules for asserted conjunction are obtained by simply prefixing each sentence in the standard conjunction rules with an assertion sign. As we show in the Appendix, this suffices, in the presence of the other rules, to yield the appropriate rules for rejected conjunction.

$$(+\wedge\text{I.})\ \frac{+A \qquad +B}{+A \wedge B} \qquad (+\wedge\text{E.}_1)\ \frac{+A \wedge B}{+A} \qquad (+\wedge\text{E.}_2)\ \frac{+A \wedge B}{+B}$$

It is easy to see that the logic of assertion of the resulting system is classical. In particular, if negation is defined by the quartet of rules above and the interaction of assertion and rejection is governed by the coordination principles, then classical *reductio* and double negation elimination are valid in the logic of assertion. The bilateralist has inferentially defined classical logic, contrary to the received wisdom that associates inferentialism with intuitionistic logic (see Chapter 2).

We can now see how bilateralism deals with the Frege–Geach Problem. This will provide a template for how inferential expressivism, more generally, can address the problem. Bilateralists agree with Frege (1919) that *not* cannot be taken to indicate rejection, since when one utters *If the accused was not in Berlin, he did not commit the murder*, the first occurrence of *not* does not express disbelief. But the *not* occurring unembedded in the premiss *The accused was not in Berlin* does not indicate rejection either. So the conclusion *The accused did not commit the murder* follows by *modus ponens*, just as Frege claimed it should. Nevertheless, bilateralists explain negation in terms of rejection, vindicating the pragmatist insight that *not* is used to register incompatibilities. Note that bilateralists, like traditional expressivists about negation, only need two primitives, namely assertion and rejection. Thus, considerations of parsimony do not favour Frege's reductionist account, which reduces rejection to negative assertion, over the bilateralist account, which gives the meaning of negation in terms of conditions on assertion and rejection.

our linguistic norms make it improper to express both, similar to how they make it improper to infer *A* from *A or B*.

Bilateralism takes at its heart the idea that rejection is not to be reduced to negative assertion. Rejection is a primitive speech act, on a par with assertion, and disbelief is a primitive attitude, on a par with belief. In this book, we will focus on two dimensions of speech acts: the attitudes they express and their essential effect on the conversation. We have seen how the first aspect can be understood bilaterally: assertions express beliefs, rejections express disbeliefs. But the second aspect can be understood bilaterally too, by appropriately extending Robert Stalnaker's model of conversation (Stalnaker 1978, 2002).

According to this model, conversation takes place against a backdrop of shared beliefs, the *common ground*.[6] In Stalnaker's view, the essential effect of an assertion is a proposal to update the common ground. Such a proposal may be accepted by all conversation participants, in which case the common ground is updated accordingly. Assertions may be used for a variety of purposes—to inform, to convince, to testify, to boast, and many more—but, in the Stalnakerian picture, the function of managing the common ground is essential and underwrites all uses of assertion.

But what about rejections? On a reductionist picture *à la* Frege, the essential effect of a rejection would simply be a proposal to perform a negative update to the common ground—a proposal to add *not p* to it. We contend that the bilateralist should also adopt a bilateral conception of the essential effects of assertion and rejection. The common ground contains the sentences supposed to be believed by all participants in a conversation. Assertions express beliefs and are therefore used to propose updates to the shared beliefs, the common ground. But rejections do not express beliefs—they express disbeliefs. We therefore take them to be used to propose updates to the *negative common ground*, which consists of all sentences supposed to be disbelieved by all participants in a conversation. The essential effect of a rejection is a proposal to update the negative common ground.

The (positive) common ground and the negative common ground are mutually constraining. It cannot be simultaneously the case that all

[6] On the standard account, the common ground is a set of propositions, closed under logical inference. Since we talk about assertions and rejections of *sentences*, we also treat the common ground as a set of sentences, similarly closed under logical inference. The discussion could be recast in terms of propositions, understood in the minimal sense of *proposition*. It should also be noted that, strictly speaking, Stalnaker treats the common ground in terms of shared acceptances, where an acceptance is an attitude broader than belief, held towards a proposition for the purposes of the conversation (2002: 715–720). Here, the differences between acceptance and belief can be set aside, and our discussion could be easily recast in terms of acceptances. Similar considerations apply to the other shared attitudes that we will be concerned with.

participants are assumed to believe some sentence and also assumed to disbelieve it. Linguistic items add further constraining structure. When a sentence is in the negative common ground, one can infer, via the bilateral negation rules, that its negation is in its positive counterpart. This is just the familiar principle that the common ground is closed under inference, but extended to the bilateral structure of positive and negative common ground.[7] The general idea is that linguistic coordination goes beyond just coordinating on beliefs. For every attitude we can express, we can coordinate on jointly supposing to have this attitude. We will return to this point in Chapter 5.

3.3 Inferential expressivism

We are now in a position to describe how our preferred understanding of bilateralism yields a general characterization of inferential expressivism. Traditional expressivism takes terms receiving expressivist treatment to indicate the expression of an attitude. Thus, traditional expressivism about negation takes *not* to indicate rejection, so that an utterance of *Grass is not green* expresses the disbelief that grass is green. The link between *not* and the attitude expression of disbelief is *direct.*

By contrast, bilateralism, understood as a form of inferential expressivism about negation takes *not* to be an operator which does not indicate the expression of disbelief: an utterance of *Grass is not green* expresses the belief that grass is not green. In this sense, inferential expressivism about negation is a form of sophisticated expressivism and indeed endorses the minimalism about belief that we discussed in Section 1.10: a sincere utterance of an embeddable declarative sentence serves to perform an assertion and thus expresses a belief. Nonetheless, inferential expressivism about negation takes the meaning of negation to be inferentially explained in terms of rejection, which expresses the attitude of disbelief. The link between *not* and rejection is *indirect* and inferential. To wit, from an utterance of *Grass is not green,* which expresses the belief that grass is not green, one can infer the rejection of *Grass is green,* which expresses the disbelief that grass is green. Inferential expressivism therefore retains the negative thesis of traditional expressivist and inferentialist approaches to meaning that the meaning of an expression

[7] As we explain in the next section, the relevant notion of inference is to be understood in terms of preservation of commitment.

is not given by what it stands for. However, it adds the following positive thesis about what the meaning of vocabulary receiving inferential expressivist treatment consists in.

Inferential expressivism (Semantic). The semantic value of a term is given by the inferential relations that an attitude expression towards a sentence containing the term bears to other attitude expressions in virtue of the fact that the sentence contains the term.

For example, according to the inferential expressivist about negation, the semantic value of *not* is given by the inferential relations an assertion or rejection of a sentence containing *not* bear to other assertions or rejections in virtue of the fact that that sentence contains *not*. These inferential relations consist in the fact that from the rejection of *A*, one can infer the assertion of *not A* and vice versa, and that from the assertion of *A*, one can infer the rejection of *not A* and vice versa.

It may appear as if an intractable notion of inference is operative in how we explain the meanings of terms receiving inferential expressivist treatment. It is implausible to think that someone who expresses a particular attitude also expresses all attitudes that follow from it. Someone uttering *not A* does not express all attitudes that follow from disbelief towards *A*. It is equally implausible to think that they must *have* these attitudes. Arbitrarily many attitude expressions follow from just a single one. By uttering *Grass is not green*, one expresses belief that grass is not green, from which the expression of disbelief that grass is green follows. But from this utterance also follows the expression of belief that grass is not green and blue, that grass is not green and pink, and so on. As Restall (2005) notes, there is a mind-cluttering problem here (see Harman 1986). Since arbitrarily many attitude expressions follow from a single attitude expression, inferential expressivism seems to entail that anyone who asserts a single sentence has unboundedly many attitudes.

We submit that the proper understanding of *inferred attitude expressions* in inferential expressivism is *social*. The meaning-conferring inferential relations between attitude expressions are those that unpack the attitudes one is *committed* to having. Someone expressing an attitude need not immediately draw all the associated inferences; in fact, they cannot do so, due to cognitive limitations. Nonetheless, they are *committed* to expressing all the attitudes that can be inferred from that attitude expression in the following sense: if it is pointed out to them that from an attitude they have expressed some further attitude expression follows, they must express this further attitude or admit to a mistake. But if they are not challenged in this way, they may not draw the inference and are not required to express or even have the corresponding attitude.

In Chapter 1, we outlined an account of attitude expression whereby in expressing a certain attitude one explicitly commits to having that attitude. Thus, in uttering *Grass is not green*, one expresses belief towards grass not being green and thereby explicitly commits to believing that grass is not green. Alongside this explicit commitment, however, one also binds oneself by further commitments. Thus, for instance, one is also committed to expressing disbelief that grass is green. The expression of belief that grass is not green stands in particular inferential relations to other attitude expressions. Commitments are closed under these relations.

This avoids the mind-cluttering problem. For example, one of the two bilateral rules of conjunction elimination states that $+A \wedge B$ entails $+A$. This rule preserves commitment, since whenever a speaker has explicitly committed to believing $A \wedge B$, they are also committed to believing A. This means that if A is up for discussion, they must express belief towards A or admit to a mistake and retract their commitment to believing $A \wedge B$. Similarly, if they explicitly committed to believing $\neg A$, they are committed to disbelieving A and must overtly express this attitude when challenged (or admit a mistake). Our understanding of inference bears certain similarities to a principle described by Catarina Dutilh Novaes (2015) to establish a link between logic and belief. Someone asserting a particular sentence need not believe all sentences that logically follow, but they are nonetheless committed to conceding all the consequences of their overt commitment.

Given this conception of inference, we can recognize the semantic inferential expressivist thesis as the outcome of applying the Pragmatist Razor to the following meta-semantic thesis, an answer to the Functional Role Question.

Inferential expressivism (Meta-semantic). The functional role of terms receiving inferential expressivist treatment is to register commitments to certain attitudes.

In Chapter 1 we argued that the expressivist is best seen as subscribing to the Pragmatist Razor: if an expression has a certain function, its meaning should be explained by appealing exclusively to features needed to account for this function. When applied to the meta-semantic claim that sentences serve to undertake commitments towards attitude expressions, the Razor delivers the semantic claim of inferential expressivism: that the meaning of a linguistic term receiving inferential expressivist treatment is given by the characteristic inferences towards attitude expressions which the occurrence of that term licenses. These inferences codify the commitments undertaken by the use of the term. This preserves the core of the semantic insights of traditional expressivism and inferentialism. The traditional expressivist insight was that the meaning of certain terms is given by their contribution to the attitudes a sentence containing them is used to express; and the inferentialist insight was that the meaning of certain terms is given by their contribution to the characteristic inferences that a sentence containing them features in. Inferential expressivism combines these insights into the idea that the meaning of a term is given by its contribution to what commitments towards attitude expressions one can infer from the use of a sentence containing it.

Like the inferentialist, the inferential expressivist holds that the referentialist terminology of 'postsemantics' suggests the wrong order of explanation. In the correct order of semantic explanation, we take certain inferences as proper—those that unpack the commitments a speaker incurs by expressing attitudes—and these inferences determine the semantic values of expressions. Likewise, inferential expressivism offers answers to the Meaning Determination and Meaning Knowledge Questions in meta-semantics that are akin to the inferentialist answers. To wit, the meaning of expressions is determined by inferential relations between attitude expressions and to know these meanings is to grasp these relations. We may now see that in addition to preserving the core of the semantic insights of traditional expressivism and inferentialism, inferential expressivism preserves their core *meta*-semantic insights too. In keeping with traditional expressivism, it explains the function of statements in terms of commitments to expressing a variety of attitudes, including non-doxastic ones, thereby making room for the richness of our inner lives and linguistic interactions. In keeping with inferentialism, it explains why expressions have the meanings that they do and what the semantic knowledge of speakers consists in in terms of the inferential role of expressions, now understood in terms of the commitments incurred by speakers in virtue of their use of language.

In addition to bilateralism, there are precedents to the idea of combining expressivism and inferentialism in the meta-ethics literature (Chrisman 2008; Tiefensee 2016, 2021). Moreover, Baker and Woods (2015: 414) claim that to avoid the pitfalls of traditional expressivism, 'the more complex, regimented features of meaning should be grounded in the communicative role of the sentences', where an inferentialist (meta-)semantics is one possible way to spell out such communicative roles. In Chapter 5, we will apply inferential expressivism to the case of moral vocabulary to develop what is, to our knowledge, the first formal semantic framework which includes an inferential expressivist account of moral language.

3.4 Mixed inferences and Frege–Geach

The inferential expressivist holds that *not* does not indicate rejection. Rather, its meaning is inferentially explained in terms of rejection. But if *not* does not indicate rejection, it may seem a simple act of faith to believe in the existence of a primitive speech act of rejection. Smiley (1996) and Rumfitt (2000) argue that, in fact, to find examples of rejection in natural language we need to look no further than negative answers to self-posed polar questions: although *not* does not indicate rejection, *no* does. Similarly, *yes* indicates assertion.[8] The resulting picture is as follows.

(2) a. Is it the case that A? Yes! *asserts* A, $+A$
  b. Is it the case that A? No! *rejects* A, $-A$
  c. Is it the case that not A? Yes! *asserts not* A, $+\neg A$

Thus, bilateralists have at their disposal a linguistic realization of rejection as a speech act distinct from negative assertion: *No!* in response to a yes-or-no question is an unembeddable force indicator and *not* is an embeddable operator suitably related to rejection.

However, taking *no* to be a force indicator gives rise to a revenge version of the Frege–Geach argument. As Smiley notes, answers to self-posed polar questions can figure in inferences. Now consider the following inference, similar to the original Frege–Geach inference, except that the second premiss and the conclusion are rejections instead of negative assertions.

[8] The idea to express speech acts by answering self-posed questions was anticipated by Frege (1919, 1923). Wittgenstein (1953: §22) also remarks that one can realize an assertion as a question with an added affirmative.

(3) a. Is it the case that, if the accused was not in Berlin, then he did not commit the murder? Yes!
b. Is it the case that the accused was in Berlin? No!
c. Is it the case that he committed the murder? No!

Prima facie, this inference is valid. As in the case of the original Frege–Geach inference, the *not* in the first premiss must *modify* the claim that the accused was in Berlin, but the second premiss is taken to express disbelief towards *the accused was in Berlin* via the use of the force indicator *no*. Thus, (3) cannot be validated by *modus ponens* unless the *no* in the second premiss also modifies the claim that the accused was in Berlin to its negation, in which case it is not a force indicator. The problem in this case is that the inferential expressivist wants to insist that, unlike *not*, *no* is indeed a force indicator. So how can the revenge Frege–Geach inference be valid?

The answer is that to validate this inference one need not consider it a direct application of *modus ponens* (Smiley 1996). It suffices that negation and rejection be suitably related. In particular, the rules for negation allow one to pass from the second premiss—the rejection of *The accused was in Berlin*—to the assertion of *The accused was not in Berlin*. By *modus ponens* for asserted conditionals, one can then infer the assertion of *The accused did not commit the murder*. This, again by the rules of negation, implies the rejection of *The accused committed the murder*, as desired. Again, Frege is right that when one utters *If the accused was not in Berlin, he did not commit the murder*, the first occurrence of *not* does not indicate rejection. But this utterance *does* commit the speaker to rejecting *The accused committed the murder* should they reject *The accused was in Berlin*. The inference rules giving the meaning of negation in terms of rejection ensure this. Inferential expressivism maintains that negative assertions license rejections in unembedded contexts but recognizes Frege's point that this is not so in embedded contexts. By explaining the meaning of negation in terms of its inferential relation to rejection, as spelled out by the bilateral negation rules, inferential expressivism can account for the behaviour of negation in all contexts and keep its meaning constant across these contexts.

According to inferential expressivism, the inference (3) is *mixed* in that it involves both assertion and rejection. Examples of other prima facie valid mixed inferences are not hard to come by. The following inference pattern is discussed by Smiley (1996).

(4) a. Is it the case that, if *A*, then *B*? Yes!
 b. Is it the case that *B*? No!
 ---
 c. Is it the case that *A*? No!

Someone who identifies rejection with negative assertion would consider this inference a case of *modus tollens*. But one need not theorize about negation at all to recognize (4) as valid. Inferential expressivists, for their part, can account for mixed inferences such as (4) by appealing to the incompatibility between assertion and rejection as codified by the coordination principles and without any mention of negation. For suppose that someone asserts *If A, then B* and rejects *B*. The fact that they have asserted *If A, then B* means that should they also assert *A* they would be committed, via *modus ponens*, to asserting *B*. But the Rejection rule tells us that this would be absurd, given that they have rejected *B*. By Smileian *reductio*, we conclude that they are committed to rejecting *A*.

Thus, taking *no* to indicate rejection gives rise to a revenge Frege–Geach Problem, but inferential expressivism has the resources to solve this version of the problem as well. However, taking *no* to indicate rejection also leads to the problem of weak rejections, first raised by Imogen Dickie (2010). It is to this problem that we now turn.

3.5 Evidence and the problem of weak rejections

Several inferentialists take legitimate inference to preserve evidence. Prawitz holds that a deductive inference is legitimate if

> a subject who makes the inference and has evidence for its premisses thereby gets evidence for the conclusion (Prawitz 2015: 73)

and since

> assertions are evaluated among other things with respect to the *grounds* or *evidence* the speakers have for making them, we may also say that the aim of...inferences is to make assertions *justified*. (Prawitz 2015: 71)

On this view, evidence, assertion and inference are tightly connected. Evidence justifies assertions and is preserved in legitimate inference, which serves to justify further assertions. Dummett (1991b: 176) similarly claims that 'deductive argument... preserves some property of statements that renders an assertion of them correct'. This property is evidence, and inference proceeds by 'rearranging' the evidence justifying the premisses to obtain the evidence that justifies the conclusion.[9]

On this account of inference, if the premisses of a valid argument are licensed by the evidence, so is the conclusion. Dickie argues that inspection of negative answers to polar questions reveals that, unlike assertions, many rejections are *unspecific* with respect to what kind of evidence they carry. The rejection of a sentence may be licensed by evidence for its falsity, but need not be, as witnessed by the following examples. The first two are due to Dickie, the third is based on an example of Grice (1991: 82).

(5) Is it the case that Homer wrote the *Iliad*? No! Homer did not exist.

(6) Is it the case that Homer was a unicorn? No! There is no such thing as the property of being a unicorn.

(7) Is it the case that X will win the election? No! X or Y will win.

These rejections are *weak*: they are not licensed by evidence for the falsity of the rejected sentence. For instance, in (7) the speaker is not rejecting *X will win the election* on the basis of evidence that X will not win, but on the basis of evidence that X or Y will win. Similar considerations apply to the other examples. Dickie concludes that the lack of specificity of rejections makes them unsuitable to serve as premisses and conclusions in an evidence-preserving proof theory.

Bilateralists might insist that they only intended to talk about *strong* rejections, that is rejections that carry evidence for the negation of the rejected sentence. This, Dickie continues, would not get bilateralists out of their predicament. For if the sign − stands for strong rejections only, then the Smileian *reductio* principle (SR_1) does not preserve evidence and is therefore invalid. For instance, it would be absurd for someone having evidence that Homer does not exist to assert that Homer wrote the *Iliad*. By Smileian *reductio*, they might then reject *Homer wrote the Iliad*. If this rejection

[9] Dummett conceives of pieces of evidence as particular arrangements of observations and mathematical constructions, corresponding to canonical processes of verification (1973b: 308; 1991b: 176, 211, 317–318; see also Dickie 2010: 164). However, such a verificationist conception of evidence is not a necessary component of the view that inference preserves evidence.

is strong and inference preserves evidence, they would have evidence for *Homer did not write the Iliad.* But one may have evidence for *Homer did not exist* without having evidence for *Homer did not write the Iliad.* The problem is that, like rejections, inferences towards absurdity are evidentially unspecific. It may be absurd for a speaker to assert A on the basis of evidence for its falsity. Such evidence licenses the strong rejection of A. But it may also be absurd for a speaker to assert A on the basis of entirely different kinds of evidence, for instance on the basis of higher-order evidence that they lack evidence for A. Such evidence licenses the weak rejection of A, but not its *strong* rejection. Hence, (SR_1) is not valid when $-$ is a sign for strong rejection.

Bilateralists are confronted with a dilemma. If Smileian *reductio* is valid for their rejection sign, then the notion of rejection encompasses weak rejections, which cannot serve as premisses and conclusions in an evidence-preserving proof theory. If Smileian *reductio* is not valid for their rejection sign, then the standard *reductio* rule is not valid either.[10] Either way, bilateralists cannot have a classical evidence-preserving proof theory.

In the next section, we argue that bilateralists should tackle the dilemma by becoming *multilateralists* and having *two* rejection signs. For one of these signs, Smileian *reductio* is valid. This means that the sign stands for rejections that can be weak. Nonetheless, weak rejections can be accommodated within proof theory by taking inference to preserve commitment instead of evidence. For the other rejection sign, Smileian *reductio* is not valid. However, this is the case only if we allow inferences to absurdity that fail to preserve evidence. By formulating versions of Smileian *reductio* restricted to evidence-preserving inferences to absurdity, multilateralists can have a classical evidence-preserving proof theory as a fragment of their extended proof theory that includes weak rejections and preserves commitment. This extended proof theory is itself almost classical: although *reductio* fails because of the presence of weak rejections, the proof theory validates all classically valid arguments.

[10] This can be seen as follows. Suppose that the assertion of A leads to absurdity. By *reductio*, we can infer the assertion of $\neg A$. If negative assertion implies rejection—as it should—then we can obtain the rejection of A. That is, (SR_1) is valid. Thus, if Smileian *reductio* is invalid, then so is *reductio*.

3.6 From bilateralism to multilateralism

Dickie (2010) observed that rejections are evidentially unspecific: it is indeterminate on the basis of what kind of evidence they are made. For this reason, rejections cannot serve as premisses and conclusions in an evidence-preserving proof theory. But it does not follow from this that rejections cannot serve as premisses and conclusions in legitimate inference. Indeed, it appears that they can. The following examples of mixed inferences are a case in point.

(8) a. Is it the case that if Homer wrote the *Odyssey*, then he wrote the *Iliad*? Yes!
b. Did Homer write the *Iliad*? No.
―――――――――――――――
c. Did Homer write the *Odyssey*? No.

(9) a. Is it the case that if Y will lose the election, then X will win the election? Yes!
b. Is it the case that X will win the election? No.
―――――――――――――――
c. Is it the case that Y will lose the election? No.

The first inference seems valid regardless of the evidence on the basis of which the rejection in the second premiss is made—whether it is made on the basis of evidence that Homer did not write the *Iliad* or evidence that, say, Homer did not exist. Similarly, the second inference seems valid regardless of whether the rejection in the second premiss is made on the basis of evidence that X did not win the election or evidence that X or Y will win it. How can we account for the validity of these inferences?

Dickie situates her discussion in the context of an antecedent literature (notably, Dummett 1991b) that focuses on the evidence on the basis of which assertions are made, but we have been stressing the alternative option of focusing on the commitments with which assertions are associated (as done by, e.g., Brandom 1994). When we introduced the main idea of inferential expressivism in Section 3.3, we emphasized that the inference rules should be understood as preserving commitments to attitude expressions, which is not the same as preserving evidence. Given that a speaker has expressed certain attitudes, the meaning-conferring inference rules allow us to determine to which further attitude expressions they are committed.

As noted, this understanding of inference allows us to make sense of strong rejections: in uttering *not A*, someone implicitly commits themselves to having disbelief towards *A* and to overtly expressing this attitude when challenged (or admit a mistake). But understanding inference in terms of commitment preservation also allows us to make room for weak rejections. When one weakly rejects *A*, one is expressing one's *refraining* from believing that *A*. For instance, in uttering *Is it the case that X will win the election? No! X or Y will win* the speaker is not expressing disbelief or negative belief that X will win the election; rather, they are expressing their refraining from believing that X will win. As Dickie noted, Smileian *reductio* is valid for rejections that can be weak. If the supposition that someone has expressed belief that *A* leads to absurdity, one can infer that they are committed to refraining from believing that *A*. Conversely, if the supposition that someone has expressed refraining from believing that *A* leads to absurdity, one can infer that they are committed to believing that *A*. In addition, it is absurd to express, at the same time, belief that *A* and refraining from believing that *A*. Thus, the coordination principles preserve commitment when taken to be about weak rejections.

Taking ⊖ as a sign for weak rejections, we can therefore take the following rules to be part of a commitment-preserving proof theory.

$$\text{(Weak Rejection)}\ \frac{+A \qquad \ominus A}{\bot} \qquad (\text{SR}_1)\ \frac{\begin{array}{c}[+A]\\ \vdots\\ \bot\end{array}}{\ominus A} \qquad (\text{SR}_2)\ \frac{\begin{array}{c}[\ominus A]\\ \vdots\\ \bot\end{array}}{+A}$$

Using these coordination principles, we can account for the validity of the mixed inferences (8) and (9), in the same way in which we earlier could account for the validity of the inference pattern (4) using the coordination principles for strong rejection. Even if, as Dickie maintains, weak rejections cannot feature as premisses and conclusions in an evidence-preserving proof theory, they can serve as premisses and conclusions in a *commitment-preserving* proof theory.

We have argued that weak rejection expresses the speaker's refraining from believing the sentence being rejected. So understood, weak rejections can serve as premisses and conclusions in a commitment-preserving proof theory. But what is their essential effect on the conversation? Recall that, according to Stalnaker, the essential effect of an assertion is a proposal to update the common ground. For the common ground to be updated, the proposal must be accepted by all participants in the conversation. However, not every update proposal is acceptable to all conversation participants. So to fully manage the common ground speakers need a mechanism to *prevent* such an update. Weak rejections provide such a mechanism.

A weak rejection expresses the speaker's refraining from believing. Thus, if one of the conversation participants weakly rejects a sentence, the sentence cannot be supposed to be believed by all conversation participants, which is required for the common ground to be updated. So, in weakly rejecting a sentence, a speaker prevents a common ground update with that sentence. That such an update is prevented is the essential effect of weak rejection. Of course, one might also attempt to prevent an update by asserting a sentence incompatible with the one considered for update, for instance its negation. But then something more than preventing an update has happened. It would be mistaken to dispense with weak rejection because one has negative assertion. Similarly, it would be mistaken to dispense with weak rejection because one has strong rejection. Stalnaker appears to agree:

> It should be made clear that to reject an assertion is not to assert or assent to the contradictory of the assertion, but only to refuse to accept the assertion. If an assertion is rejected, the context [common ground] remains the same as it was. (Stalnaker 1978: 87, fn. 9)

However, although weak rejections can serve as premisses and conclusions in an evidence-preserving proof theory, they do not validate all of the bilateral negation rules. In particular, it does not follow from the fact that one is committed to refraining from believing A that one is committed to believing *not* A. And it does not follow from the fact that one is committed to refraining from believing *not* A that one is committed to believing A. Thus, (+¬I.) and (−¬E.) do not preserve commitment for weak rejections. Nonetheless, like the other bilateral negation rules, they do preserve commitment for strong rejections. Reserving the − sign for strong rejections, we can therefore take the following rules to give the inferential expressivist meaning of negation.

$$(+\neg\text{I.})\ \frac{-A}{+\neg A} \qquad (+\neg\text{E.})\ \frac{+\neg A}{-A} \qquad (-\neg\text{I.})\ \frac{+A}{-\neg A} \qquad (-\neg\text{E.})\ \frac{-\neg A}{+A}$$

The logic is therefore, *multilateral*: it includes an assertion sign +, a weak rejection sign ⊖, and a strong rejection sign −. The logic comprises the bilateral coordination principles for weak rejections and rules giving the meaning of negation in terms of strong rejections.

These rules, however, do not suffice to deliver the classical laws of negation since we have only laid down coordination principles for weak rejections. This means that we have not yet solved the problem of weak rejections. Dickie's first observation was that rejections are evidentially unspecific. Understanding inference in terms of commitment preservation allows one to include rejections in one's proof theory despite their unspecificity. Dickie's second observation was that inferences towards absurdity are likewise unspecific, which means that Smileian *reductio* for strong rejections does not preserve evidence. Given that we are no longer understanding inference in terms of evidence preservation, it does not follow that Smileian *reductio* for strong rejections is invalid. Nonetheless, it would *seem* to follow that the multilateralist cannot have a classical *evidence-preserving* proof theory, in line with Dickie's observations. Moreover, the same example we used above to show that Smileian *reductio* for strong rejections does not preserve evidence shows that Smileian *reductio* for strong rejections does not preserve commitment either: it is absurd for someone committed to believing *Homer did not exist* to commit to believing *Homer wrote the Iliad*, but it does not follow that they are committed to disbelieving *Homer wrote the Iliad*. Since it is the combination of the bilateral negation rules with the coordination principles for strong rejections that delivers the classical laws of negation, it appears that the multilateralist cannot have a classical *commitment-preserving* proof theory either.

Not all is lost, however. Even if Smileian *reductio* does not preserve evidence for strong rejections when unspecific inferences to absurdity are countenanced, it may still be valid when such unspecific inferences are excluded. It turns out that adding appropriately restricted versions of Smileian *reductio* for strong rejections to the commitment-preserving proof theory suffices to validate all classically valid arguments. Moreover, the evidence-preserving fragment of the resulting proof theory obeys classical logic. Let us elaborate.

When dealing with the unspecificity of rejection, we isolated the evidentially specific instances, namely the strong rejections, and noted that the bilateral negation rules preserve commitment, despite the fact that they fail to do so for rejections *tout court*. Similarly, we can isolate the evidentially specific inferences towards absurdity and note that, for those inferences,

Smileian *reductio* for strong rejection is valid in that it preserves commitment and indeed evidence, despite the fact that it fails to do so for inferences towards absurdity *tout court*. The inferences towards absurdity that are evidentially specific include at least the inferences that preserve evidence, since an inference that proceeds from evidentially specific premisses and preserves evidence cannot reach an evidentially unspecific conclusion. Thus, writing $\vdots*$ for an evidence-preserving derivation, the following *Smileian reductio** rules, which together with the Strong Rejection rule form the *coordination principles**, preserve both commitment and evidence.

$$\text{(Strong Rejection)}\ \frac{+A \qquad -A}{\bot} \qquad (\mathrm{SR}_1{*})\ \frac{\begin{matrix}[+A]\\ \vdots *\\ \bot\end{matrix}}{-A} \qquad (\mathrm{SR}_2{*})\ \frac{\begin{matrix}[-A]\\ \vdots *\\ \bot\end{matrix}}{+A}$$

These rules are not subject to Dickie's counterexamples to Smileian *reductio* for strong rejections. For example, the inference towards absurdity from *Homer did not exist* and the assumption *Homer wrote the Iliad* does not preserve evidence. Homer's existence is a precondition for intelligible talk of evidence for him having written the *Iliad*. Thus if *Homer did not exist* is a premiss, *Homer wrote the Iliad* cannot occur in an evidence-preserving argument at all. This is not to say that it would be incorrect to infer absurdity from *Homer did not exist* and the assumption *Homer wrote the Iliad.* This inference is valid in that it preserves commitment. For someone who commits to believing that Homer wrote the *Iliad* is committed to believing that Homer exists, which is incompatible with commitment to believing that Homer did not exist. But it does not preserve evidence, so it excluded from Smileian *reductio** and one cannot infer the strong rejection of *Homer wrote the Iliad* on its basis.

The coordination principles govern the interaction of assertion and weak rejection while the coordination principles* govern the interaction of assertion and strong rejection. These principles jointly yield that the strong rejection of a sentence implies its weak rejection, as shown by the following derivation.

$$\cfrac{\cfrac{-A \qquad [+A]^1}{\bot}\ \text{(Strong Rejection)}}{\ominus A}\ (\mathrm{SR}_1)^1$$

By strongly rejecting *A*, one expresses disbelief towards *A*. In so doing, one also commits oneself to refraining from believing *A*. This is how it should be, since the weak rejection of a sentence is already incompatible with its assertion. For Price's functionalist insight about negation to be preserved, this incompatibility must be inherited by negative assertion and the fact that strong rejection implies weak rejection guarantees this: when one negatively asserts a sentence, by the negation rules one commits to disbelieving it, and hence, by the derivation above, to refraining from believing it.

The Smileian *reductio** rules are formulated by restricting their inferences to absurdity to those that preserve evidence. The next question is which inferences preserve evidence and how they can be characterized in a way that can be formally captured in an inference rule. Within the confines of the language of propositional logic, the answer is simple. Failures of evidence preservation may only arise because of the presence of weak rejections. Thus, if an inference involves no weakly rejected premisses, it is guaranteed to preserve evidence. We can therefore phrase the Smileian *reductio** rules as follows.

$$(\mathrm{SR}_1{}^*)\ \frac{\begin{array}{c}[+A]\\ \vdots\\ \bot\end{array}}{-A}\ \text{if no premisses signed with } \ominus \text{ were used to derive } \bot$$

$$(\mathrm{SR}_2{}^*)\ \frac{\begin{array}{c}[-A]\\ \vdots\\ \bot\end{array}}{+A}\ \text{if no premisses signed with } \ominus \text{ were used to derive } \bot$$

As a result, we obtain the signed version of the standard *reductio* rule.

$$(\text{Signed } \textit{Reductio})\ \frac{\begin{array}{c}[+A]\\ \vdots\\ \bot\end{array}}{+\neg A}\ \text{if no premisses signed with } \ominus \text{ were used to derive } \bot$$

Let *Basic Multilateral Logic* (BML) be the natural deduction calculus consisting of the bilateral negation rules for strong rejection, the bilateral conjunction rules, the coordination principles for weak rejection and the coordination principles* for strong rejection. In the appendix to this chapter, we show how to derive rules governing the Boolean logical constants under all signs of this calculus. We write $\vdash^{\mathsf{BML}}$ for the derivability relation of BML and $\models^{\mathsf{CPL}}$ for the consequence relation of classical propositional logic. In the Appendix, we also prove the following result.

Theorem 3.6.1. $\Gamma \models^{\mathsf{CPL}} A$ *if and only if* $\{+B \mid B \in \Gamma\} \vdash^{\mathsf{BML}} +A$.

Thus, the valid arguments in the logic of assertion are exactly the classically valid arguments. Formulating versions of Smileian *reductio* that are valid for strong rejection enables the multilateralist to vindicate all the classically valid arguments. However, since the expressive power of BML as a whole goes beyond that of classical logic due to the presence of weak rejections, it does not follow that all classical *meta-rules* are valid. For example, it is not in general the case that if $\Gamma, +A \vdash^{\mathsf{BML}} \bot$, then $\Gamma \vdash^{\mathsf{BML}} +\neg A$ if Γ contains weakly rejected premisses. Thus the unrestricted meta-rule of *reductio* fails in BML. As shown by the derivability of Signed *Reductio* in BML, however, this failure can only arise in the presence of unspecific inferences to absurdity. Similar considerations apply to the rules of proof by cases and conditional proof for the material conditional: their unrestricted versions fail in BML, but this failure can only arise in the presence of unspecific inferences in their subderivations. The evidence-preserving fragment of the proof theory is fully classical.

Once we extend the object language beyond propositional logic, it becomes more difficult to determine which inferences preserve evidence. In the next chapters, we will consider examples of inferences that do not preserve evidence involving expressions such as epistemic modals and the truth predicate. We will see that by validating these inferences in our base system (since they preserve commitment) but excluding them from Smileian *reductio** (since they do not preserve evidence), we can obtain systems which account for phenomena which are usually thought to be incompatible with classical logic while also preserving the intuitive appeal of the classical laws of negation. Moreover, similarly to the case of weak rejection, this will allow us to isolate fully classical evidence-preserving fragments of these systems.

For now, let us take stock. Bilateralists distinguish between rejection, expressing disbelief, and negative assertion, expressing negative belief. This raises the challenge of producing a means of realizing rejections which is clearly distinct from negative assertion. Bilateralists met the challenge by arguing that rejections may be realized by *no* answers to self-posed polar questions. This way of meeting the challenge, we noted, gave rise to a revenge version of the Frege–Geach Problem. Bilateralists, we argued, have the resources to solve this problem. However, taking *no* to indicate rejection gives rise to a further problem, the problem of weak rejections. To solve *this* problem, we argued, bilateralists must become multilateralists and countenance a sign for weak rejection *alongside* their sign for rejections expressing disbelief—the strong rejections. But then, the initial challenge of producing a means of realizing strong rejections rears its head again: how can strong rejections be linguistically realized, if not by using *not* or by using *no*?

3.7 Linguistic realization of strong rejection

The system we arrived at at the end of the previous section is *trilateral*: it includes three distinct speech act signs: +, standing for assertion, which expresses belief; −, standing for strong rejection, which expresses disbelief; and ⊖, standing for weak rejection, which expresses refraining from believing. Assertions may be realized by answering *yes* to self-posed polar questions, and weak rejections may be realized by answering *no*. We must explain how strong rejections may be realized. We are going to argue that they too may be realized by *no* answers to self-posed polar questions. In fact, by default, *no* answers to self-posed polar questions realize a strong rejection. But, per se, the conventional meaning of *no* only enforces a weak reading of the rejection being performed.

To explain in detail our proposal, it will be helpful to look at the relation between assertion and strong rejection, and between assertion and weak rejection. The coordination principles imply that assertion and weak rejection stand in a relation akin to that of *contradictoriness* in the classical square of opposition: it is consistent to assert a sentence just in case it is absurd to weakly reject it. By contrast, the coordination principles* imply that assertion and strong rejection stand in a relation akin to that of *contrariness* in the classical square of opposition: it is always absurd to both assert and strongly reject the same sentence; but it may be consistent to neither assert nor strongly reject the same sentence, for instance if one is agnostic about it.

Now, linguistic evidence supports the claim that natural language has a general preference for contrary negation over contradictory negation (Horn 1989: chs 4–5). The most familiar case is the *neg-raising* phenomenon: in certain attitude reports, a syntactically wide-scoped negation can be *raised* to take narrow scope.[11] For instance, an utterance of *Alice doesn't believe that it is raining* is by default taken to convey that Alice has the belief that it is not raining. Letting **B** stand for the belief operator, this reading of what the utterance conveys can be formalized as $\mathbf{B}\neg A$. However, the surface form of the sentence would only seem to warrant a reading according to which Alice lacks the belief that it is raining, formalized as $\neg\mathbf{B}A$. Thus, assuming that the question at hand is whether Alice believes that it is raining, $\mathbf{B}A$, we have as a default interpretation its contrary $\mathbf{B}\neg A$ instead of its contradictory $\neg\mathbf{B}A$, despite the latter being the interpretation that the surface form of the sentence would seem to license.[12]

A preference for contrariness similarly accounts for how strong rejections are realized. Linguistic conventions only ensure that negative answers to self-posed polar question are read as weak rejections, in line with what we said in the previous section. However, a preference for contrariness in natural language means that the default interpretation of such negative answers is as realizing a strong rejection.

A prediction of this proposal is that plain *no* answers to self-posed polar questions are interpreted as strong rejections. This prediction is borne out. Consider the following three rejections.

(10) Will X win the election? No, she won't.

(11) Will X win the election? No, X or Y will win.

(12) Will X win the election? No.

The first rejection is strong: the speaker is naturally interpreted as expressing disbelief towards X winning the election, thereby committing to believing, by the bilateral negation rules, that X will not win. The second rejection, by contrast, is weak: the speaker undertakes no such commitment, and is therefore best interpreted as expressing refraining from believing that X will

[11] Horn 1978 is an early overview on the relevant linguistics literature. Horn 2020 is a more recent survey article.

[12] There are competing models of the exact mechanism behind neg-raising, roughly dividing into syntactic (see, e.g., Collins and Postal 2014) and semantic/pragmatic approaches (see, e.g., Bartsch 1973; Gajewski 2007). Recently, however, even defenders of the syntactic approach have acknowledged that some semantic/pragmatic component is needed to explain the full range of phenomena surrounding neg-raising, thus recognizing, in effect, that a general preference for contrariness is present in natural language (see Collins and Postal 2018). Whether actual syntactic movement of negation is needed to account for some of the phenomena surrounding neg-raising, such as the distribution of negative polarity items, is an issue we do not need to take a stand on.

win the election. One might have thought that the plain *no* in the third rejection would be ambiguous between a weak and strong rejection, but this is not so: it is naturally read as having the same conversational effect as the strong rejection in the first example. The *no* in the second example is read as a weak rejection only because of the added clause *X or Y will win*. Thus, analogously to the case of neg-raising, strong rejections are the preferred readings of *no*. By default, a *no* answer is read as a strong rejection, as in the third example, but this reading is usually cancellable. The phrase *X or Y will win* in the second example cancels the preferred reading, so that the *no* is read on the basis of just what the linguistic conventions enforce, namely as a weak rejection.

A general linguistic preference for contrariness over contradictoriness suffices to explain how strong rejections are realized. But if we consider the commitments associated with assertion, weak rejection, and strong rejection, we see that the similarities between the case of strong and weak rejection and neg-raising go deeper still. Since assertion expresses belief and expression of an attitude implies commitment to having that attitude, asserting A explicitly commits one to believing that A. Working in a modal logic of commitment in which *is committed to* is formalized as $\Box$ (Asher and Lascarides 2008; Schlöder et al. 2018), we can therefore formalize the commitment profile of asserting A as $\Box\mathbf{B}A$. Similarly, strongly rejecting A explicitly commits one to disbelieving A. Letting $\mathbf{D}$ stand for the disbelief operator, we can therefore formalize the commitment profile of strong rejection as $\Box\,\mathbf{D}A$. Finally, weakly rejecting A explicitly commits one to refraining from believing A. On the plausible assumption that to be committed to refraining from doing something one must be committed to not doing it, it follows that weakly rejecting A explicitly commits one to not believing A. We can therefore formalize (at least part of) the commitment profile of weak rejection as $\Box\neg\mathbf{B}A$.

Now, the bilateral negation rules tell us that the commitments undertaken by strongly rejecting a sentence and asserting its negation are equivalent. In particular, whenever one asserts $\neg A$ and therefore explicitly commits to $\mathbf{B}\neg A$, one also implicitly commits to $\mathbf{D}A$. Thus, the difference between the commitments undertaken by performing a weak rejection and those undertaken by performing a strong one can be formally captured in terms of what scope a negation takes over the belief operator. This is in line with the terminology adopted by Rumfitt (1997), who distinguishes between external (wide-scoped) rejection and internal (narrow-scoped) rejection. Thus, analogously to the case of neg-raising, the commitment undertaken by a weak rejection—to not believing A—is by default raised to the commitment undertaken by a strong rejection—to believing the negation of A.

The situation is therefore as follows. Assertions are realized by *yes* answers to self-posed polar questions; rejections are realized by *no* answers. By default, *no* answers realize a strong rejection. But this preferred reading is sometimes cancelled by some suitable continuation of the negative answer, in which case the *no* realizes a weak rejection. The three force markers of Basic Multilateral Logic denote real linguistic phenomena.

We started the chapter by arguing that bilateralism allows one to deal with the Frege–Geach Problem while preserving the insights of traditional expressivism about negation. Indeed, bilateralism can be understood as a form of inferential expressivism, the view that meanings are given in terms of inferential relations between attitude expressions. Bilateralism, however, faces the problem of weak rejections. We showed that this problem can be addressed by moving from bilateralism to multilateralism. It turns out that the benefits of the move from bilateralism to multilateralism go well beyond this: by working with a multilateral framework, the inferential expressivist can solve the Problem of Limited Applicability for inferentialism. We begin in the next chapter by presenting an inferential expressivist account of epistemic modals.

3.8 Appendix

All constants under all signs

In the first part of this appendix, we present the trilateral Boolean calculus, including both the primitive rules of BML and the derived rules that govern the remaining traditional Boolean logical constants and their behaviour under all signs. To illustrate how the calculus works, we exhibit a selection of the derivations of the non-primitive rules. Figures 3.1, 3.2, 3.3, and 3.4 display the calculus, including derived rules for the Boolean connectives, where $A \vee B$ abbreviates $\neg(\neg A \wedge \neg B)$ and $A \supset B$ abbreviates $\neg(A \wedge \neg B)$. Rules with bold labels are primitive, all others are derivable from the primitive rules.

$$(\mathbf{SR_1}^*)\ \frac{\begin{array}{c}[+A]\\ \vdots{*}\\ \bot\end{array}}{-A} \qquad (\mathbf{SR_2}^*)\ \frac{\begin{array}{c}[-A]\\ \vdots{*}\\ \bot\end{array}}{+A} \qquad (\mathbf{SR_1})\ \frac{\begin{array}{c}[+A]\\ \vdots\\ \bot\end{array}}{\ominus A} \qquad (\mathbf{SR_2})\ \frac{\begin{array}{c}[\ominus A]\\ \vdots\\ \bot\end{array}}{+A}$$

$$(\textbf{Strong Rejection})\ \frac{+A \quad -A}{\bot} \qquad (\textbf{Weak Rejection})\ \frac{+A \quad \ominus A}{\bot}$$

Fig. 3.1. Coordination principles in Basic Multilateral Logic. The notation $\vdots{*}$ denotes a derivation using no premisses signed with $\ominus$.

$$(+\neg\mathbf{I.})\ \frac{-A}{+\neg A} \qquad (+\neg\mathbf{E.})\ \frac{+\neg A}{-A}$$

$$(+\wedge\mathbf{I.})\ \frac{+A \quad +B}{+A \wedge B} \qquad (+\wedge\mathbf{E.}_1)\ \frac{+A \wedge B}{+A} \qquad (+\wedge\mathbf{E.}_2)\ \frac{+A \wedge B}{+B}$$

$$(+\vee\text{I.}_1)\ \frac{+A}{+A \vee B} \qquad (+\vee\text{I.}_2)\ \frac{+B}{+A \vee B} \qquad (+\vee\text{E.})\ \frac{+A \vee B \quad \begin{array}{c}[+A]\\ \vdots{*}\\ +C\end{array} \quad \begin{array}{c}[+B]\\ \vdots{*}\\ +C\end{array}}{+C}$$

$$(+\supset\text{I.})\ \frac{\begin{array}{c}[+A]\\ \vdots{*}\\ +B\end{array}}{+A \supset B} \qquad (+\supset\text{E.})\ \frac{+A \supset B \quad +A}{+B}$$

Fig. 3.2. Operations under + in Basic Multilateral Logic. The notation $\vdots{*}$ denotes a derivation using no premisses signed with $\ominus$.

$$(-\neg\mathbf{I.})\ \frac{+A}{-\neg A} \qquad (-\neg\mathbf{E.})\ \frac{-\neg A}{+A}$$

$$(-\wedge\text{I.}_1)\ \frac{-A}{-A \wedge B} \qquad (-\wedge\text{I.}_2)\ \frac{-B}{-A \wedge B} \qquad (-\wedge\text{E.})\ \frac{-A \wedge B \quad \begin{array}{c}[-A]\\ \vdots{*}\\ +C\end{array} \quad \begin{array}{c}[-B]\\ \vdots{*}\\ +C\end{array}}{+C}$$

$$(-\vee\text{I.})\ \frac{-A \quad -B}{-A \vee B} \qquad (-\vee\text{E.}_1)\ \frac{-A \vee B}{-A} \qquad (-\vee\text{E.}_2)\ \frac{-A \vee B}{-B}$$

$$(-\supset\text{I.})\ \frac{+A \quad -B}{-A \supset B} \qquad (-\supset\text{E.}_1)\ \frac{-A \supset B}{+A} \qquad (-\supset\text{E.}_2)\ \frac{-A \supset B}{-B}$$

Fig. 3.3. Operations under − in Basic Multilateral Logic. The notation $\vdots{*}$ denotes a derivation using no premisses signed with $\ominus$.

$$(\ominus\neg\text{I.})\ \frac{+A}{\ominus\neg A} \qquad\qquad (\ominus\neg\text{E.})\ \frac{\begin{matrix}\vdots *\\ \ominus\neg A\end{matrix}}{+A}$$

$$(\ominus\wedge\text{I.}_1)\ \frac{\ominus A}{\ominus A\wedge B} \quad (\ominus\wedge\text{I.}_2)\ \frac{\ominus B}{\ominus A\wedge B} \quad (\ominus\wedge\text{E.})\ \frac{\ominus A\wedge B \qquad \begin{matrix}[\ominus A]\\ \vdots\\ +C\end{matrix} \qquad \begin{matrix}[\ominus B]\\ \vdots\\ +C\end{matrix}}{+C}$$

$$(\ominus\vee\text{I.})\ \frac{\begin{matrix}\vdots *\\ \ominus A\end{matrix} \qquad \begin{matrix}\vdots *\\ \ominus B\end{matrix}}{\ominus A\vee B} \qquad (\ominus\vee\text{E.}_1)\ \frac{\ominus A\vee B}{\ominus A} \quad (\ominus\vee\text{E.}_2)\ \frac{\ominus A\vee B}{\ominus B}$$

$$(\ominus\supset\text{I.})\ \frac{+A \qquad \ominus B}{\ominus A\supset B} \qquad (\ominus\supset\text{E.}_1)\ \frac{\begin{matrix}\vdots *\\ \ominus A\supset B\end{matrix}}{+A} \quad (\ominus\supset\text{E.}_2)\ \frac{\ominus A\supset B}{\ominus B}$$

Fig. 3.4. Operations under $\ominus$ in Basic Multilateral Logic. The notation $\vdots *$ denotes a derivation using no premisses signed with $\ominus$.

We begin with the operational rules under $+$. The following derivation witnesses the derivability of the rule $(+\vee\text{I.}_1)$. The case of $(+\vee\text{I.}_2)$ is analogous.

$$\dfrac{\dfrac{+A \qquad \dfrac{\dfrac{\dfrac{[-\neg(\neg A\wedge\neg B)]^1}{+\neg A\wedge\neg B}\,(-\neg\text{E.})}{+\neg A}\,(+\wedge\text{E.}_1)}{-A}\,(+\neg\text{E.})}{\bot}\,(\text{Strong Rejection})}{+\neg(\neg A\wedge\neg B)}\,(\text{SR}_2)^1$$

The following derivation witnesses the derivability of $(+\vee\text{E.})$. Note that to use (SR_2) in the last step, the derivations of $+C$ from the assumption $+A$ and of $+C$ from the assumption $+B$ must not use premisses signed with $\ominus$, as is stated in the derived rule in Figure 3.2.

$$\dfrac{\dfrac{\dfrac{+\neg(\neg A\wedge\neg B)}{-(\neg A\wedge\neg B)}\,(+\neg\text{E.}) \qquad \dfrac{\dfrac{\dfrac{\dfrac{[-C]^1 \qquad \begin{matrix}[+A]^2\\ \vdots *\\ +C\end{matrix}}{\bot}\,(\text{Strong Rej.})}{-A}\,(\text{SR}_1)^2}{+\neg A}\,(+\neg\text{I.}) \qquad \dfrac{\dfrac{\dfrac{[-C]^1 \qquad \begin{matrix}[+B]^3\\ \vdots *\\ +C\end{matrix}}{\bot}\,(\text{Strong Rej.})}{-B}\,(\text{SR}_1)^3}{+\neg B}\,(+\neg\text{I.})}{+\neg A\wedge\neg B}\,(+\wedge\text{I.})}{\bot}\,(\text{Strong Rejection})}{+C}\,(\text{SR}_2{}^*)^1$$

We can now derive the rules for the material conditional ⊃ under +.

$$
\dfrac{\dfrac{\dfrac{\dfrac{[-\neg(A\wedge\neg B)]^1}{+A\wedge\neg B}\,(-\neg\mathrm{E.})}{+A}\,(+\wedge\mathrm{E.}_1)}{\begin{array}{c}\vdots *\\ +B\end{array}} \qquad \dfrac{\dfrac{\dfrac{[-\neg(A\wedge\neg B)]^1}{+A\wedge\neg B}}{+\neg B}\,(+\wedge\mathrm{E.}_2)}{-B}\,(+\neg\mathrm{E.})}{\dfrac{\bot}{+\neg(A\wedge\neg B)}\,(\mathrm{SR}_2{}^*)^1}\,(\text{Strong Rejection})
$$

$$
\dfrac{\dfrac{+\neg(A\wedge\neg B)}{-A\wedge\neg B}\,(+\neg\mathrm{E.}) \qquad \dfrac{+A \qquad \dfrac{[-B]^1}{+\neg B}\,(+\neg\mathrm{I.})}{+A\wedge\neg B}\,(+\wedge\mathrm{I.})}{\dfrac{\bot}{+B}\,(\mathrm{SR}_2{}^*)^1}\,(\text{Strong Rejection})
$$

We now turn to the operational rules under − and ⊖. The operational rules for ∧ under − are derived analogously to the way the rules for ∨ under + are derived above. It is then straightforward to derive the rules for ∨ under −. The rules for ⊃ under − are derived as follows, with ($-\supset\mathrm{E.}_2$) being derived analogously to ($-\supset\mathrm{E.}_1$).

$$
\dfrac{\dfrac{[+\neg(A\wedge\neg B)]^1 \qquad +A}{+B}\,(+\supset\mathrm{E.}) \qquad -B}{\dfrac{\bot}{-\neg(A\wedge\neg B)}\,(\mathrm{SR}_2{}^*)^1}\,(\text{Strong Rejection})
$$

$$
\dfrac{\dfrac{\dfrac{-\neg(A\wedge\neg B)}{+A\wedge\neg B}\,(-\neg\mathrm{E.})}{+A}\,(+\wedge\mathrm{E.}_1) \qquad [-A]^1}{\dfrac{\bot}{+A}\,(\mathrm{SR}_2{}^*)^1}\,(\text{Strong Rejection})
$$

With regards to the behaviour of connectives under ⊖, we begin with the rules for ¬.

$$
\dfrac{\dfrac{+A}{-\neg A}\,(-\neg\mathrm{I.}) \qquad [+\neg A]^1}{\dfrac{\bot}{\ominus\neg A}\,(\mathrm{SR}_1)^1}\,(\text{Strong Rejection}) \qquad\qquad \dfrac{\begin{array}{c}\vdots *\\ \ominus\neg A\end{array} \qquad \dfrac{[-A]^1}{+\neg A}\,(+\neg\mathrm{I.})}{\dfrac{\bot}{+A}\,(\mathrm{SR}_1{}^*)^1}\,(\text{Weak Rejection})
$$

The derivations of the rules for ∧ under ⊖ are analogous to the case of ∨ under +. We present the derivation of (⊖∧E.) in full to make clear why no restrictions are needed in its subderivations.

$$\dfrac{\dfrac{\ominus A \wedge B \qquad \dfrac{\dfrac{\dfrac{[\ominus C]^1 \qquad \begin{matrix}[\ominus A]^2\\ \vdots\\ +C\end{matrix}}{\bot}\,\text{(Weak Rejection)}}{+A}\,(\mathrm{SR}_2)^2 \qquad \dfrac{\dfrac{[\ominus C]^1 \qquad \begin{matrix}[\ominus B]^3\\ \vdots\\ +C\end{matrix}}{\bot}\,\text{(Weak Rejection)}}{+B}\,(\mathrm{SR}_2)^3}{+A \wedge B}\,(+\wedge\mathrm{I.})}{\bot}\,\text{(Weak Rejection)}}{+C}\,(\mathrm{SR}_2)^1$$

The derivations of the rules for $\vee$ under $\ominus$ are analogous to the case of $\vee$ under $-$, except that the subderivations can be unrestricted because we can use the coordination principles instead of the coordination principles*. Finally, the rules for $\supset$ under $\ominus$ are derived analogously to the ones for $\supset$ under $-$.

Classicality

In the second part of this appendix, we prove Theorem 3.6.1, which shows that the logic of assertion of Basic Multilateral Logic validates all and only the classically valid arguments.

Theorem 3.6.1. $\Gamma \vDash^{\mathsf{CPL}} A$ *iff* $\{+B \mid B \in \Gamma\} \vdash^{\mathsf{BML}} +A$.

Proof. The left-to-right direction follows from the fact that the derived rules in Figure 3.2 together with the following derived rules axiomatize classical logic under $+$.

$$(\mathrm{DNI})\ \dfrac{+A}{+\neg\neg A} \qquad (\mathrm{DNE})\ \dfrac{+\neg\neg A}{+A} \qquad (\mathit{Reductio})\ \dfrac{+A \supset \neg A}{+\neg A}$$

Their derivations are as follows.

$$\dfrac{\dfrac{+A}{-\neg A}\,(-\neg\mathrm{I.})}{+\neg\neg A}\,(+\neg\mathrm{I.}) \qquad \dfrac{\dfrac{+\neg\neg A}{-\neg A}\,(+\neg\mathrm{E.})}{+A}\,(-\neg\mathrm{E.})$$

$$\dfrac{\dfrac{\dfrac{+A \supset \neg A \qquad \dfrac{[-\neg A]^1}{+A}\,(-\neg\mathrm{E.})}{+\neg A}\,(-\supset\mathrm{E.}) \qquad [-\neg A]^1}{\bot}\,\text{(Strong Rejection)}}{+\neg A}\,(\mathrm{SR}_2{}^*)^1$$

For the right-to-left direction, replace $+A$ with A and $-A$ or $\ominus A$ with $\neg A$ in each rule of BML. It is easy to see that the resulting rules are valid in classical propositional logic. Now, if $\{+B \mid B \in \Gamma\} \vdash^{\mathsf{BML}} +A$, then A is derivable from Γ in the rewritten calculus, which means that $\Gamma \models^{\mathsf{CPL}} A$. □

Soundness and completeness

In the third part of this appendix, we prove that BML is sound and complete with respect to a suitable model theory. Let an *ω-pointed model* be a mapping V from ω to models of classical propositional logic satisfying the following conditions.

- For any $x \in \omega$, $V \models_x A$ if and only if $V(x)$ is a model of classical propositional logic in which A holds.
- $V \not\models \bot$.
- $V \models +A$ if and only if $\forall x \in \omega : V \models_x A$.
- $V \models -A$ if and only if $\forall x \in \omega : V \not\models_x A$.
- $V \models \ominus A$ if and only if $\exists x \in \omega : V \not\models_x A$.

We write $\Gamma \models^{\mathsf{BML}} \varphi$ when Γ is a set of signed sentences, φ is a signed sentence and for all ω-pointed models V, if $V \models \psi$ for all $\psi \in \Gamma$, then $V \models \varphi$. We begin by proving the soundness of BML with respect to this model theory.

Theorem 3.8.2 (Soundness). *Let Γ be a set of signed sentences and φ be a signed sentence. If $\Gamma \models^{\mathsf{BML}} \varphi$, then $\Gamma \vdash^{\mathsf{BML}} \varphi$.*

Proof. This is a standard induction on the length of derivations. The only interesting cases are the coordination principles*. We prove the soundness of the rule (SR_1*). Assume $\Gamma \vdash^{\mathsf{BML}}_{D} -A$ by an application of (SR_1*), i.e., $\Gamma \cup \{+A\} \vdash^{\mathsf{BML}}_{D'} \bot$ where D' uses only asserted premisses from Γ. Let Γ' be the set of asserted formulae in Γ. Then $\Gamma' \cup \{+A\} \vdash^{\mathsf{BML}}_{D'} \bot$. Assume that $\Gamma' \not\models^{\mathsf{BML}} -A$. Then there is a model V of Γ' and a point $y \in \omega$ such that $V \models_y A$. Construct an ω-pointed model V' in which every point is y, i.e., for any x and atom p, $V' \models_x p$ just in case $V \models_y p$. Since Γ' contains only asserted formulae, for each member $+B$ of Γ', $V \models_y B$. Therefore, $V' \models \Gamma'$. Moreover, since $V \models_y A$, $V' \models +A$. But by the induction hypothesis, the shorter derivation of $\bot$ from $\Gamma' \cup \{+A\}$ is sound, so there cannot be such a V'. Contradiction. □

We now turn to the proof of completeness.

Theorem 3.8.3 (Completeness). *Let* Γ *be a set of signed sentences and* φ *be a signed sentence. If* $\Gamma \vdash^{\mathsf{BML}} \varphi$*, then* $\Gamma \vDash^{\mathsf{BML}} \varphi$.

We first need two lemmata.

Lemma 3.8.4. *If* Γ *is* BML*-consistent* (i.e., $\Gamma \nvdash^{\mathsf{BML}} \bot$)*, then* $\Gamma' = \{B \mid +B \in \Gamma\}$ *is classically satisfiable.*

Proof. Assume Γ' is not classically consistent. Then there is a proof of $\bot$ from the premisses in Γ'. By Theorem 3.6.1 this proof can be carried out in BML with the premisses from Γ, so $\Gamma \vdash^{\mathsf{BML}} \bot$. Therefore, by contraposition, if Γ is BML-consistent, Γ' is classically consistent. By the satisfiability theorem in classical logic, if Γ is BML-consistent, Γ' is classically satisfiable. □

From this we immediately obtain the following.

Lemma 3.8.5. *Let* Γ *contain no weakly rejected formulae and* A *be any sentence. If* $\Gamma \cup \{\ominus A\}$ *is* BML*-consistent, then* $\{\neg A\} \cup \{B \mid +B \in \Gamma\} \cup \{\neg B \mid -B \in \Gamma\}$ *is satisfiable in classical propositional logic.*

Proof. By the previous lemma, it suffices to show that $\Gamma \cup \{+\neg A\}$ is BML-consistent. Assume it is not. Then $\Gamma, +\neg A \vdash^{\mathsf{BML}} \bot$, i.e., $\Gamma \vdash^{\mathsf{BML}} \ominus\neg A$ by (SR_1). Since Γ only contains asserted formulae, $\Gamma \vdash^{\mathsf{BML}} +A$ by the rule ($\ominus\neg$E.), displayed in Figure 3.4. This contradicts the assumption that $\Gamma \cup \{\ominus A\}$ is consistent. □

Proof of Theorem 3.8.3. We first prove a model existence result: for every BML-consistent Γ, there is an ω-pointed model of Γ. Let Γ^s be the set of formulae in Γ that are signed with $+$ or $-$ and let $\{\ominus B_i \mid i \in \omega\}$ be an enumeration of the formulae in Γ that are signed with $\ominus$. For all $i \in \omega$, write Γ^i for $\Gamma^s \cup \{\ominus B_i\}$. Since $\Gamma^i \subseteq \Gamma$ for all i, the Γ^i's are BML-consistent. By Lemma 3.8.5, there are models I_i of classical propositional logic that satisfy $\{\neg B_i\} \cup \{B \mid +B \in \Gamma^s\} \cup \{\neg B \mid -B \in \Gamma^s\}$. Then define an ω-pointed model V by setting $V(x) = I_x(p)$ for all $x \in \omega$. With this construction, clearly, $V \vDash \Gamma$.

Completeness now follows by a case distinction on the sign of φ.

- Assume $\Gamma \nvdash^{\mathsf{BML}} +A$. Then $\Gamma \cup \{\ominus A\}$ is consistent and so there is a V with $V \vDash \Gamma \cup \{-A\}$. Hence, $\Gamma \nvDash^{\mathsf{BML}} +A$.
- Assume $\Gamma \nvdash^{\mathsf{BML}} -A$. Then $\Gamma \cup \{\ominus\neg A\}$ is consistent and so there is a V with $V \vDash \Gamma \cup \{\ominus\neg A\}$. Hence, $\Gamma \nvDash^{\mathsf{BML}} -A$.
- Assume $\Gamma \nvdash^{\mathsf{BML}} \ominus A$. Then $\Gamma \cup \{+A\}$ is consistent and so there is a V with $V \vDash \Gamma \cup \{+A\}$. Hence, $\Gamma \nvDash^{\mathsf{BML}} \ominus A$.

□

4
Epistemic Modals

In Chapter 2, we presented the Problem of Limited Applicability for inferentialism. According to this problem, inferentialism runs into difficulties when one attempts to extend it beyond the core logical constants. In this chapter, we take a first step towards addressing the Problem of Limited Applicability by developing an inferentialist expressivist account of epistemic modality. In the previous chapter, we explained the meaning of the operator *not* in terms of the speech act of strong rejection. We now explain the meaning of epistemic *might* in terms of *weak assertion*, a speech act whose existence we argue for on the basis of linguistic evidence. We show that our account of *might* provides a solution to certain well-known puzzles about the semantics of modal vocabulary whilst validating all classical arguments. This demonstrates that the inferential expressivist approach to meaning can be successfully extended beyond the core logical constants.

The extension of bilateralism to multilateralism is a key ingredient of our approach. By introducing signs for additional speech acts within our semantic framework we can explain the meaning of additional operators in terms of those speech acts. A general, *multilateral* methodology for inferential expressivist semantic theorizing emerges. To inferentially explain the meaning of a linguistic item in terms of an attitude expression, we must (i) identify the linguistic realization of this expression, (ii) expand the coordination principles so as to specify how this attitude expression interacts with other attitude expressions, (iii) lay down rules that codify the inferential relations between the linguistic item and the attitude expression, and (iv) determine which new inferences preserve evidence to appropriately restrict the coordination principles*. This chapter illustrates the multilateral methodology by applying it to the case of epistemic modals.

4.1 Traditional expressivism about *might*

Modal expressions such as *might* and *must* can be used epistemically, as when one says that Goldbach's Conjecture might be true and it might be

Reasoning with Attitude: Foundations and Applications of Inferential Expressivism. Luca Incurvati and Julian J. Schlöder, Oxford University Press. © Oxford University Press 2023.
DOI: 10.1093/oso/9780197620984.003.0004

false. In broad outline, the orthodox approach to epistemic modality takes *might A* to say that *A* is compatible with some contextually determined body of knowledge (Kratzer 1977; DeRose 1991). The orthodox approach thus equates the truth conditions of *It might be raining* with those of *For all I know, it is raining* and *I don't know that it is not raining*. The epistemic modal *must* is then simply treated as the dual of *might*: a sentence of the form *must A* is true at a context just in case it is not the case that *not A* is true at that context.

The orthodox approach is typically coupled with a descriptivist understanding of the constitutive function of epistemic modal talk. On the orthodox approach, epistemic modal talk serves to describe the world, albeit a special aspect of it: in uttering a sentence having an epistemic modal as its main operator, the speaker describes her own epistemic position. Thus, *It might be raining* serves to communicate the information that the speaker does not know that it is not raining.

The last two decades have witnessed the appearance of a number of challenges to the orthodox approach. Two types of cases have been especially prominent. The first, originally described by Yalcin (2007), concerns the difference in behaviour between *might A* and *for all I know, A* (or, equivalently, *I don't know that not A*) in certain embedded contexts. It is a familiar phenomenon that it is infelicitous to utter a *Moorean sentence* such as (1). As one would expect given the orthodox approach to the meaning of *might*, it is similarly infelicitous to utter a *Yalcinean* sentence such as (2).

(1) # It is raining and for all I know it is not.

(2) # It is raining and it might not be.

However, whilst the infelicity of (1) disappears in suppositional contexts, this is not so for (2).

(3) Suppose that it is raining and for all I know it is not.

(4) # Suppose that it is raining and it might not be.

The same difference in behaviour between *might* and *for all I know* can be observed when we embed Moorean sentences and Yalcinean sentences in conditional antecedents.

The following is a natural diagnosis of the difference between *might* and *for all I know*. It is pragmatically incoherent to describe the world as being a certain way while also describing oneself as not knowing that the world is that way. But the world being a certain way is of course compatible with one not knowing that it is that way, which is why this state of affairs can be supposed to obtain. Moorean sentences are not semantically contradictory. The fact that Yalcinean sentences remain odd in suppositional contexts is evidence against a pragmatic explanation of their infelicity and in support of the idea that they are semantically contradictory. But where does this contradictoriness come from? The expressivist has an answer ready at hand. The constitutive function of epistemic modal talk is not that of *describing* one's epistemic state, but to *express* it. For Yalcin, in particular, in uttering *It is raining and it might not be* expresses, at the same time, belief that it is raining and compatibility of one's state of mind with *It is not raining*. And this *is* contradictory.[1]

We now turn to the second challenge to the orthodox approach. The challenge is based on the phenomenon of *modal disagreement*. Consider the following dialogue.

(5) *Alice*: I can't find the keys.
Bob: They might be in the car.
Alice: No, they are not in the car. I just checked.

Alice and Bob appear to disagree. However, it is difficult to locate a proposition about a single body of knowledge that Bob is warranted in asserting and Alice is warranted in rejecting. In particular, if we take the contextually determined body of knowledge to be that of the speaker, as an analysis of *might* in terms of *for all I know* would seem to have it, there seems to be no conflict between Bob's assertion that the keys might be in the car and Alice's assertion that they are not. Bob has said something about his body of knowledge and Alice has said something about hers. But this is not a disagreement.

The situation is reminiscent of the difficulties of speaker subjectivism in accounting for moral disagreement, which we described in Chapter 1 and constituted one of the motivations for ethical expressivism. In an analogous fashion, the difficulties of the orthodox approach in accounting for modal disagreement lend support to a traditional expressivist analysis of *might*: in

[1] Later in this chapter, we will depart from this particular form of expressivist explanation of epistemic modal claims, over and above the fact that we favour an inferentialist understanding of expressivism.

asserting that the keys might be in the car, Bob is expressing an attitude which is incompatible with the attitude expressed by Alice when she asserts that the keys are not in the car.

The traditional expressivist analysis of *might* is however undermined by its embeddability behaviour (von Fintel and Gillies 2007). One way to appreciate this point is to observe that the embeddability of *might* under conditional antecedents can be used to run a version of the Frege–Geach argument (Swanson 2010; MacFarlane 2014).

(6) a. If it might be raining, we might go to the cinema.
b. It might be raining.
――――――――――――――――――――
c. We might go to the cinema.

Since this inference appears to be valid, the Frege–Geach argument can be performed as usual by observing that the *might* in the conditional antecedent cannot indicate an attitude expression and concluding that the *might* in the second premiss must modify the content of *It is raining* instead of expressing compatibility of one's state of mind with its negation.

In response to difficulties of this sort, some linguists and philosophers have advocated what is, in effect, a hybrid expressivist analysis of *might*. In addition to asserting a modal sentence, an utterance of *might A* serves to raise a possibility (Swanson 2006; Portner 2007) or proffering and giving advice (von Fintel and Gillies 2007). For familiar reasons, however, a hybrid approach fails to properly deal with the Frege–Geach Problem. As in the case of negation, expressivists can do better by marrying their approach with the resources of inferentialism: as we shall see, the inferential expressivist account of *might* retains the advantages of the traditional expressivist analysis while making sense of its embedding behaviour.

4.2 *Might* and *perhaps*

It is widely agreed that *might* must modify content because of its embeddability behaviour (von Fintel and Gillies 2007; Swanson 2010; MacFarlane 2014). We concur, but it does not follow that *every* piece of epistemic vocabulary modifies content. Indeed, we will now present linguistic data in support of the claim that the epistemic adverb *perhaps* is a *force modifier*: its occurrence in an utterance modifies the speech act that would otherwise be performed with that utterance, but not its content. Analogous data could be presented in defence of the view that *maybe* is a force modifier, but we restrict attention to *perhaps.*

It seems that *perhaps* and *might* have been conflated at times. For instance, Benjamin Schnieder (2010) uses examples involving *perhaps* to comment on Yalcin's puzzle about the embedding behaviour of sentences of the form *A and it might be that not A*, and Joshua Crabill (2013) makes use of Schnieder's insights, but replaces *perhaps* with *might* again. At first glance, *perhaps* and *might* are rather similar, besides some syntactic differences.

(7) a. Perhaps it is raining.
    b. It might be raining.

(7a) and (7b) can be justifiably uttered in exactly the same circumstances. We take this to show that they are inferentially equivalent: in a fixed context, the same inferences can be drawn from an utterance of (7a) and an utterance of (7b). Moreover, *perhaps* and *might* both lead to Yalcin's puzzle.

(8) a. #Suppose it is raining and perhaps it is not raining.
    b. #Suppose it is raining and it might not be raining.

Schnieder (2010) seeks to explain the oddness of (8a) as a special case of the fact that *perhaps* does not embed under *suppose* at all.

(9) # Suppose perhaps it is raining.

(9) sounds rather odd.[2] Hence, Schnieder argues, it is not surprising that (8a) sounds odd as well. Schnieder takes the fact that *perhaps* fails to embed under *suppose* to provide evidence that its function is to express an attitude. This is in agreement with what we argue below.

But what does this mean for *might*? To start with, the fact that *Perhaps it is raining* and *It might be raining* are inferentially equivalent does not mean that *might* and *perhaps* have identical embedding behaviour. For instance, while (9) is infelicitous, (10) appears to be fine.

[2] Using particular prosody, *perhaps* appears to sound quite alright under *suppose*: *suppose, perhaps, it is raining*, where the commas are read as short pauses, is fine. However, this is a mere syntactic variation of *perhaps suppose it is raining*, where *perhaps* takes scope over *suppose*.

(10) Suppose it might be raining.

Crabill (2013) claims that (10) is infelicitous, but we suspect this is due to his identification of *might* with *perhaps*. Here are two natural cases of embeddings of *might* under *suppose*.

(11) a. Biologists supposed it might be a gene like the one causing Burkitt's lymphoma that made cells lose control of their proliferation.
(*New York Times*, 'Geneticists' Target: Immortal Cells', 22 December 1992)

b. The standard model …is presumably closer to the truth about fundamental particles than [earlier theories]. At least, it makes sense to suppose that it might be.
(*Stanford Encyclopedia of Philosophy*, 'Truthlikeness', Oddie 2016)

These examples provide evidence that (10) is generally acceptable. It follows that Schnieder and Crabill cannot solve Yalcin's puzzle. For if their strategy to explain the infelicity of (8a) applied to (8b) as well, (10) should be as infelicitous as (9).

Moreover, *perhaps* does not embed in conditional antecedents, whereas *might* does.

(12) # If perhaps it is raining, I'd better take an umbrella.

(13) If it might be raining, I'd better take an umbrella.

Finally, Eric Swanson (2010) argues against expressivist treatments of *might* by observing that it embeds under quantifiers. In this respect too, *perhaps* differs from *might*.

(14) Every day it might be raining.

(15) # Every day perhaps it is raining.

Because of its embedding behaviour, *perhaps* is sometimes taken to belong to the category of *speaker-oriented adverbs* like *frankly*, *fortunately*, or *evidently*, which speakers use to comment on their utterances (Bellert 1977; Mittwoch 1977; Ernst 2001, 2009). However, it is not correct to say that *perhaps* is used to *comment* on the performance of a speech act. If one says *frankly A*, *fortunately A*, or *evidently A*, all the effects of asserting *A* still obtain, but this is not the case for *perhaps A*. For instance, on a commitment account of assertion (Brandom 1983), uttering *frankly A* commits one to *A*, and on a knowledge-norm account (Williamson 2000), uttering *frankly A* requires one to know that *A*. But if one says *perhaps A*, one is *not* thereby committed to *A* or required to know that *A*.

Moreover, speaker-oriented adverbs cannot co-occur. For instance, *frankly fortunately it is raining* and *fortunately evidently it is raining* sound bad.[3] However, speaker-oriented adverbs *can* co-occur with *perhaps*. Here are two examples.

(16) a. Frankly, perhaps Route 4 isn't what Ms. Milby needs to investigate.
(*Washington Post*, 'Commuter Advice from Several Directions', 6 December 2001)
b. Frankly it's perhaps now too late.
(*BBC*, 'George Low Stabbing: Cyprus Murder Suspect "Set Free"', 6 July 2017)

Thus, *perhaps* fits with neither speaker-oriented adverbs (such as *frankly*) nor embeddable operators (such as *might*). But there is a third option. Whilst *frankly* is used to *comment* on the performance of a speech act, *perhaps* is used to *modify* the speech act performed. Hence, in (16b), *frankly* serves to comment on the performance of the speech act obtained by modifying with *perhaps* an assertion of *It's now too late*.

One might object that *perhaps* cannot be a force modifier because it embeds in conditional *consequents*.

(17) If it is going to rain, perhaps we should stay in.

However, such an embedding is compatible with *perhaps* not being an embeddable operator. For instance, *frankly* also embeds in this way.

(18) If it is going to rain, frankly we should stay in.

But nobody would conclude that *frankly* is not a speaker-oriented adverb. Instead, conditionals such as (18) are best analysed as *conditional*

[3] Certain paraphrases, however, are acceptable, such as *Frankly, it is fortunate that it is raining*. This does not affect the argument, since *fortunate* is an adjective, not a speaker-oriented adverb.

performances of speech acts (see Chapter 8). This is compatible with the occurrence of a force-modifying expression in the consequent of (17): *perhaps* modifies the speech act that is being conditionally performed.

The evidence for *perhaps* being a force modifier we have presented so far is that *perhaps* exhibits the embedding behaviour that one would expect of a particle operating exclusively at the speech act level, and that the role of *perhaps* cannot be reduced to that of commenting on one's performance of a speech act. We are now going to add to this evidence the observation that in certain non-assertoric speech acts, *perhaps* also modifies force. First, consider a natural use of *perhaps* in a polar question.

(19) Is it perhaps resin?
(British National Corpus, file KCV, line 4908)

Examining the potential positive answers to the question in (19) reveals that this question puts *Is it resin?* under discussion. Thus, *perhaps* does not modify the content of the question.

(20) a. Is it perhaps resin?
b. Yes, it is.
?c. Yes, perhaps it is.
#d. Yes, but perhaps it is something else.

If *perhaps* were to modify the content of the question in (19), the appropriate positive answer to (20a) would be (20c). But the proper answer is (20b).[4] Moreover, the infelicity of (20d) indicates that *yes* here targets *it is resin*, since it cannot be felicitously followed by *perhaps it is something else* (whereas *Perhaps it is resin* could be). We conclude that *perhaps* in (20a) affects the question's force, but not its content: it seems to make it a biased (Bellert 1977) or tentative question instead of a neutral polar question.

What about uses of *might* in polar questions? If *might* is an embeddable operator, we should expect it to modify the content of a question. This prediction appears to be borne out.

[4] While (20c) does not strike us as downright infelicitous, it appears to be facetious or mocking. The appropriate answer using *perhaps* appears to be *I don't know, perhaps it is.*

(21) a. Might it be resin?
?b. Yes, it is.
c. Yes, it might be.
d. Yes, but it might be something else.

The preferred answer to the question in (21a) seems to be (21c).[5] This indicates that the question concerns a content modified by *might*.[6] Accordingly, (21d) is felicitous: if *yes* targets *it might be resin*, it should be compatible with *it might be something else*, as is indeed the case.

Now, consider an utterance in imperative mood containing *perhaps* and contrast it with the same utterance without *perhaps*.

(22) Perhaps check with the Seahawks.
(*New York Times*, 'Seahawks Beckon, and U.S.C. Cringes', 8 January 2010)

(23) Check with the Seahawks.

These two utterances seem to express the same *content*, but with different *forces*: (23) can have the force of a command, whereas (22) is more of a suggestion. Thus, in (22) *perhaps* again appears to modify force.

We conclude that *perhaps* modifies force rather than content. It modifies polar questions to biased or tentative questions, commands to suggestions, and has an analogous function when applied to assertions.

Now, recall from the previous chapter that the bilateralists suggested that an assertion can be realized by posing a question to oneself and answering *yes*, as in (24a): *perhaps* can be used to modify such an answer, as in (24b), so as to perform the speech act we call *weak assertion*. For clarity's sake, we will often call *strong assertion* what is usually referred to as *assertion*.

(24) a. Is it raining? Yes.
b. Is it raining? Perhaps yes.

[5] Again, (21b) is not downright infelicitous, but it seems to *overanswer* the question.

[6] One can use *might* in highly polite requests, such as *Might you have time tomorrow?*. Here, answering *yes* seems to mean that one *does* have time tomorrow. We suspect that this is a conventionalized form of politeness. In any case, this does not affect our argument. All we intend to show is that *perhaps* functions as a force modifier. Some overlap with certain uses of *might* is compatible with this claim.

(25) a. Perhaps it is raining.
b. Is it raining? Perhaps.

The forms in (25) are linguistic variants of (24b): in (25a), *it is raining* would otherwise be a strong assertion but is modified into a weak assertion by the presence of *perhaps*; (25b) is the result of eliding the *yes* in (24b). Thus, (24b), (25a), and (25b) all serve to perform the weak assertion of *It is raining*. This is a different speech act than the strong assertion of *It might be raining*. There is, however, a close connection between *might* and *perhaps*: as mentioned, they seem interchangeable in non-embedded contexts. This means that the weak assertion of *It is raining* and the strong assertion of *It might be raining* are inferentially equivalent. In what follows, we shall exploit this inferential equivalence to give an account of *might* in terms of weak assertion.

4.3 Weak assertion

We have gathered evidence for the existence of a distinctive speech act which we call *weak assertion* and which can be realized by using *perhaps* in what would otherwise be a strong assertion or by responding with *perhaps* to a self-posed polar question. However, we still need to explain what this speech act *is*.

As in the case of strong assertion, strong rejection, and weak rejection, we focus on two dimensions of weak assertion: the attitude it expresses and its essential effect on the conversation. Consider the following dialogical variant of an example from Chapter 3.

(26) a. *Alice*: X will win the election.
b. *Bob*: No, X or Y will win.

Bob is here conveying both refraining from believing that X will win and refraining from disbelieving that X will win. His refraining from believing that X will win is expressed with the particle *no*, which serves to perform a weak rejection of *X will win the election*. His refraining from disbelieving that X will win is merely conveyed via a pragmatic implicature: if Bob did not intend to indicate that he refrains from disbelieving that X will win, he should have explicitly asserted that Y will win. Indeed, this implicature can be cancelled.

(27) a. *Alice*: X will win the election.
b. *Bob*: No, X or Y will win. In fact, Y will.

Is there a speech act that serves to express (and not merely convey) one's refraining from disbelieving? We contend that there is, and that it is the speech act of weak assertion. For instance, Bob may weakly assert that X will win the election and thereby express his refraining from disbelieving that X will win the election. That this is not a pragmatic implicature can be seen by observing that this effect of his utterance cannot be cancelled.

(28) Perhaps X will win the election. # In fact, X will not win.

We are now in a position to also describe the essential effect of weak assertion on the conversation. Suppose someone performs a strong rejection. The essential effect of this act is a proposal to update the negative common ground, which contains the sentences that are supposed to be disbelieved by all conversation participants. But if one of the conversation participants weakly asserts a sentence, the sentence cannot be supposed to be disbelieved by all conversation participants: one of the participants has expressed their refraining from disbelieving the sentence. So, in weakly asserting a sentence, a speaker prevents a negative common ground update with that sentence. That such an update is prevented is the essential effect of weak assertion.

We therefore have four speech acts, expressing distinct attitudes and giving rise to distinct effects on the conversation. In particular, by performing one of these speech acts, the speaker expresses a particular attitude and thereby takes a public stance on the admissibility of a sentence into the common ground. By strongly asserting *A*, one expresses belief towards *A* and thereby proposes to add *A* to the common ground (or accepts a previous proposal to this effect). By strongly rejecting *A*, one expresses disbelief towards *A* and thereby proposes to add *A* to the negative common ground (or accepts a previous proposal to this effect). By weakly asserting *A*, one refrains from expressing disbelief towards *A* and thereby prevents *A* from being added to the negative common ground. By weakly rejecting *A*, one refrains from expressing belief towards *A* and thereby prevents *A* from being added to the common ground. Jointly, these speech acts allow precise management of what is accepted—and not accepted—into the common ground.

Our account of weak assertion is reminiscent of the account of epistemic modality suggested in passing by Dummett, who follows Frege in using $\vdash$ as an assertion sign.

> Expressions of epistemic modality do not ordinarily occur within the scope of sentential operators, and are best understood, not as contributing to the senses of the sentences they govern ...but as an expression of the force with which those sentences are uttered. When 'may' expresses epistemic modality, 'It may be the case that A' is best understood as meaning 'Not ($\vdash$ (not A))'. (Dummett 1973a: 330)

Negating a negative assertion could be reasonably seen as expressing refraining from disbelief. However, there are two key differences between our analysis and Dummett's suggestion. First, as we shall see, we take the analysis to be correct for *perhaps* rather than for *might* to account for the meaning of epistemic modals in embedded contexts. Second, our multilateral approach does not force us to embark on the difficult task of making sense of negating force markers.

4.4 Coordinating the speech acts

Bilateralism takes the meaning of linguistic items to be given by conditions on strong assertion and strong rejection. In the previous chapter, we extended bilateralism to multilateralism to account for the phenomenon of weak rejection. We now extend the multilateral framework further so as to also encompass weak assertion. This allows us to codify conditions that give the inferential expressivist meaning of *not* and *might*. In particular, the meaning of *might* is inferentially explained in terms of weak assertion. Weak assertion is therefore prior to *not* and *might* in the order of semantic explanation.

Our multilateral logic is cast in a language $\mathcal{L}_\Diamond$ consisting of the constants $\wedge$, $\neg$, and $\Diamond$, formalizing *and*, *not*, and *might*. As before, we use upper-case Latin letters to denote *sentences* in this language and lower-case Greek letters to denote *formulae*, which are obtained by prefixing sentences with force markers. The force markers are now $+$ for strong assertion, $-$ for strong rejection, $\ominus$ for weak rejection, and $\oplus$ for weak assertion.

In keeping with our multilateral methodology, our first task is to give principles to coordinate weak assertion with the other speech acts. In the previous chapter, we took the relation between strong assertion and weak rejection to be codified by the following coordination principles.

$$\text{(Weak Rejection)}\ \frac{+A \qquad \ominus A}{\bot} \qquad \text{(SR}_1\text{)}\ \frac{\begin{array}{c}[+A]\\ \vdots\\ \bot\end{array}}{\ominus A} \qquad \text{(SR}_2\text{)}\ \frac{\begin{array}{c}[\ominus A]\\ \vdots\\ \bot\end{array}}{+A}$$

Now, weak assertion stands to strong rejection as weak rejection stands to strong assertion. It is absurd to express, at the same time, disbelief towards *A* and one's refraining from disbelieving *A*. Moreover, if the supposition that someone has expressed disbelief towards *A* leads to absurdity, one can infer that they are committed to refraining from disbelieving *A*. Conversely, if the supposition that someone has expressed refraining from disbelieving *A* leads to absurdity, one can infer that they are committed to disbelieving *A*. Similar considerations apply when we consider speech acts from the point of view of their effect on the conversation. It is absurd to propose an update to the negative common ground and to prevent such a proposal. And if it is absurd for someone to propose an update to the negative common ground, this must be because they are already committed to preventing such an update. Finally, if it is absurd for someone to prevent an update to the negative common ground, this must be because they are already committed to proposing or accepting such an update. The fact that instead of belief and positive common ground, we are dealing with disbelief and negative common ground does not affect the cogency of these arguments. We can therefore codify the interaction of weak assertion and strong rejection by means of the following rules.

$$\text{(Weak Assertion)}\ \frac{\oplus A \qquad -A}{\bot} \qquad \text{(SR}_3\text{)}\ \frac{\begin{array}{c}[\oplus A]\\ \vdots\\ \bot\end{array}}{-A} \qquad \text{(SR}_4\text{)}\ \frac{\begin{array}{c}[-A]\\ \vdots\\ \bot\end{array}}{\oplus A}$$

From now on, we will call *coordination principles* the rules governing the interaction of strong assertion and weak rejection as well as the rules governing the interaction of strong rejection and weak assertion. In addition to the coordination principles, we also have the coordination principles*, governing the interaction of strong assertion and strong rejection (where, as before, $\vdots *$ denotes an evidence-preserving derivation).

$$\text{(Strong Rejection)}\ \frac{+A \qquad -A}{\bot} \qquad (\mathrm{SR}_1{}^*)\ \frac{\begin{array}{c}[+A]\\ \vdots *\\ \bot\end{array}}{-A} \qquad (\mathrm{SR}_2{}^*)\ \frac{\begin{array}{c}[-A]\\ \vdots *\\ \bot\end{array}}{+A}$$

In the previous chapter, we identified weak rejections as a source of potential evidential unspecificity. For this reason, when formulating Basic Multilateral Logic, we banned the use of weak rejections as premisses in evidence-preserving inferences. We are now extending the multilateral framework with *might* and the speech act of weak assertion. Thus, there are further potential sources of evidential unspecificity. We return to this issue in Section 4.6, after discussing the rules for *might*.

The coordination principles characterize the interaction of weak rejection with strong assertion and of weak assertion with strong rejection. The coordination principles* characterize the interaction of the strong speech acts. Together, these principles determine all interactions between the four speech acts. Figure 4.1 depicts the situation. The four speech acts stand in the classical square of opposition, with the strong speech acts standing in a

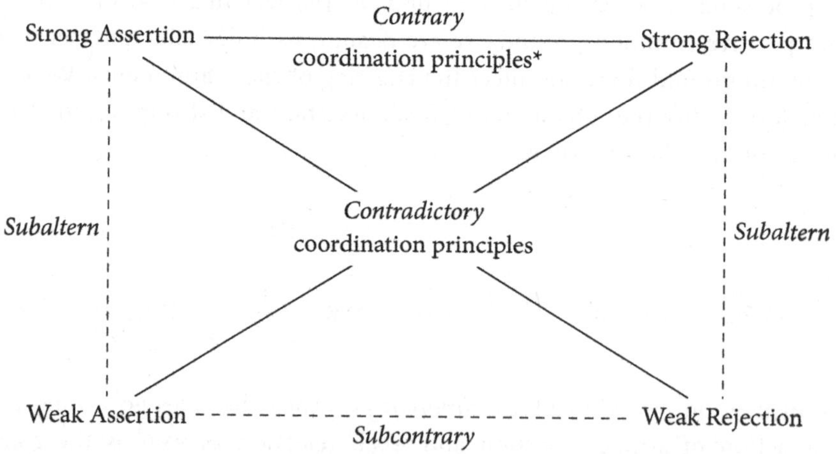

Fig. 4.1. Opposition in multilateral logic. Dashed lines are derivable.

relation akin to contrariness and opposing weak/strong pairs standing in a relation akin to contradictoriness.

Let us consider the case of strong assertion and weak rejection; the case of strong rejection and weak assertion is similar. The Weak Rejection inference rule states that it is absurd to strongly assert and weakly reject the same sentence. The Smileian *reductio* rules state that it is absurd to be committed to neither believing that *A* nor refraining from believing that *A*. That is akin to contradictoriness. It does not follow that one must either believe or refrain from believing what every sentence says. It only means that, once the issue of *A* is raised, one must explicitly express either the belief that *A* or one's refraining from believing that *A*. The point can be appreciated by observing that Smileian *reductio* is equivalent to the following rule of ***bilateral excluded middle***.

$$\frac{\begin{array}{c}[+A]\\ \vdots\\ \varphi\end{array}\qquad\begin{array}{c}[\ominus A]\\ \vdots\\ \varphi\end{array}}{\varphi}$$

The rule states that one is already committed to whatever attitude one would be committed to by both asserting and weakly rejecting the same content. This does not mean that, for every sentence, one must either assert it or weakly reject it. Bilateral excluded middle may tell us that someone is committed to φ even though they have neither asserted nor weakly rejected *A*.

The coordination principles* characterize strong assertion and strong rejection as contrary. The Strong Rejection rule states that it is absurd to strongly assert and strongly reject the same sentence. The Smileian *reductio** rules state that it is absurd to be committed to neither believing nor disbelieving the same content when one's evidence makes one of the attitudes absurd. The remaining relations—that weak assertion and rejection are subcontraries, and that each of them is subaltern to its strong counterpart—follow from the coordination principles and the coordination principles*. The interaction of weak assertion and weak rejection as subcontraries is codified by the following derivable rule.

$$(\mathrm{SC})\ \frac{\begin{array}{c}[\oplus A]\\ \vdots\\ \varphi\end{array}\qquad\begin{array}{c}[\ominus A]\\ \vdots\\ \varphi\end{array}}{\varphi}$$

And the following rules characterize weak assertion as subaltern to strong assertion. Analogous rules can be derived characterizing weak rejection as subaltern to its strong counterpart.

$$(\mathrm{SA}_1)\ \frac{+A}{\oplus A} \qquad (\mathrm{SA}_1{}^*)\ \frac{\oplus A \qquad \begin{array}{c}[+A] \\ \vdots * \\ +B\end{array}}{\oplus B}$$

The rule (SA_1) states that strongly asserting some sentence A commits one to refraining from disbelieving that A. The rule (SA_2) states that weak assertion is closed under strongly asserted implication. This ensures that inferences such as the one from *If A, then B* and *Perhaps A* to *Perhaps B* are valid. The restrictions on (SA_2) follow from the way this rule is derived from Smileian *reductio**, but can also be independently justified.

The derivation of $+B$ from $+A$ in (SA_2) preserves evidence. This means that evidence licensing the strong assertion of A is also evidence licensing the strong assertion of B or, equivalently, that evidence licensing the strong rejection of B is also evidence licensing the strong rejection of A by *modus tollens*. Now, the weak assertion of A, formalized by the premiss $\oplus A$, expresses one's refraining from disbelieving that A. Thus, the premiss rules out having evidence licensing the strong rejection of A (which expresses disbelief that A) and hence, by the subderivation, rules out having evidence licensing the strong rejection of B. But ruling out evidence licensing the strong rejection of B (which expresses disbelief that B) is just the desired conclusion $\oplus B$. Note that if the derivation from $+B$ to $+A$ did not preserve evidence, one could only conclude that asserting A commits one to believing B or, equivalently, that weakly rejecting B commits one to refraining from believing A. And this does not suffice to obtain the conclusion of (SA_2). The premiss $\oplus A$ rules out disbelieving that A, but it does not rule out refraining from believing that A. One can coherently refrain from both believing and disbelieving what a sentence says. Indeed, this seems to be what the agnostic does (see Ferrari and Incurvati 2022).

4.5 The meaning of *might*

We now proceed to the next step in our multilateral methodology: giving rules codifying the relation between *might*, formalized by $\Diamond$, and the speech act of weak assertion, denoted by ⊕. As noted earlier, when *might* and *perhaps* take scope over the same non-embedded clause, they can be interchanged without affecting the inferential meaning of the sentence. Thus, ◇s can be introduced by moving from weak to strong assertion, and can be eliminated symmetrically.

$$(+\Diamond\text{I.})\ \frac{\oplus A}{+\Diamond A} \qquad (+\Diamond\text{E.})\ \frac{+\Diamond A}{\oplus A}$$

Two additional rules for ◇ account for the fact that *Perhaps it might be raining* is inferentially equivalent to *Perhaps it is raining.*

$$(\oplus\Diamond\text{I.})\ \frac{\oplus A}{\oplus\Diamond A} \qquad (\oplus\Diamond\text{E.})\ \frac{\oplus\Diamond A}{\oplus A}$$

These rules imply that iterating *might* does not affect the content of an utterance (see also Yalcin 2007; Willer 2013). However, this only applies when the context does not indicate that multiple occurrences of *might* are to be understood with reference to different common grounds. A case based on DeRose 1991: 584–585 will clarify the situation. Suppose a medical test has been run but the results are not yet known. A negative result rules out John having the disease; a positive result leaves that possibility open but does not prove it. Responding to a friend asking for information, John's partner Jane says:

> ANSWER-1. We haven't got the results yet. It might be the case that John might have the disease. We'll know whether he might have it when we get the results.

Jane asserts *It might be the case that John might have the disease* but seems unwilling to assert *John might have the disease.* This answer seems entirely appropriate. However, it seems equally appropriate for Jane to assert *John might have the disease* (DeRose 1991: 583).

> ANSWER-2. John might have the disease. He has some of the symptoms. We won't get the test results until tomorrow.

In ANSWER-2, it is clear that *might* is to be understood with reference to the current common ground cg_c. This suggests that in ANSWER-1 Jane is unwilling to assert *John might have the disease* because this occurrence of *might* is understood with reference to a *different* common ground. Since the fact that the common ground will be updated with the test results is very salient, this is naturally taken to be cg_a, the common ground after the results are known. Thus, in ANSWER-1, Jane asserts *It might*^{cg_c} *be that John might*^{cg_a} *have the disease.* In so doing, she prevents from making it common ground that, once the results are known, she will prevent from making it common ground that John does not have the disease. This seems to be a correct reading of her utterance and does not imply that John might^{cg_a} have the disease, even in the presence of the $\oplus\Diamond$ rules.

We now turn to negation. The bilateral rules for negation specified the meaning of negation under strong assertion and strong rejection by allowing one to move from one speech act to the other by means of negation. The rules ensured that the right commitment relations exist between disbelief and negative belief and between belief and negative disbelief. As a result, a proposal to update the negative common ground with a sentence also commits one to accepting a proposal to update the common ground with that sentence's negation; and a proposal to update the common ground with a sentence also commits one to accepting a proposal to update the negative common ground with that sentence's negation. The same is the case for the rules for negation under weak assertion and weak rejection. Accordingly, the following rules are now derivable.

$$(\ominus\neg\text{I.})\ \frac{\oplus A}{\ominus\neg A} \qquad (\ominus\neg\text{E.})\ \frac{\ominus\neg A}{\oplus A}$$

$$(\oplus\neg\text{I.})\ \frac{\ominus A}{\oplus\neg A} \qquad (\oplus\neg\text{E.})\ \frac{\oplus\neg A}{\ominus A}$$

The rules enforce the right commitment relations between refraining from disbelieving and refraining from negatively believing, and between refraining from believing and refraining from negatively disbelieving. As a result, preventing an update of the negative common ground with a sentence commits one to preventing an update of the common ground with that sentence's negation; and preventing an update of the common ground with a sentence commits one to preventing an update of the negative common ground with that sentence's negation.

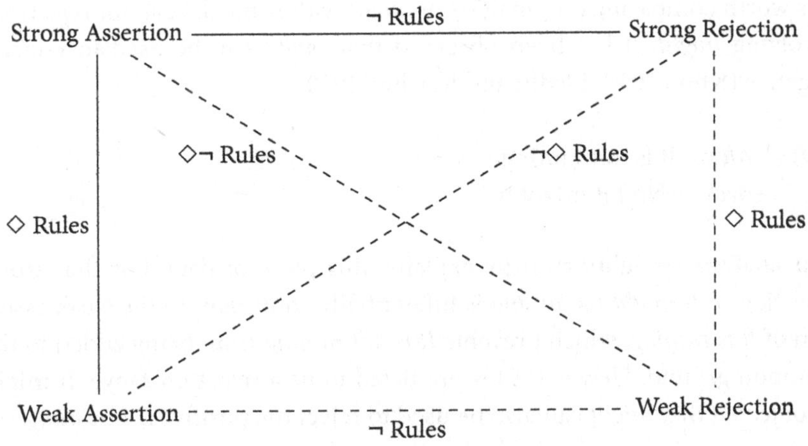

Fig. 4.2. Operations in multilateral logic. Dashed lines are derivable.

Like the bilateral rules for ¬, the rules for ◇ are obviously harmonious, since the elimination rules are the direct inverses of the introduction rules. Together, these rules determine how to introduce and eliminate both ◇ and ¬ under all four speech acts. Figure 4.2 provides an overview.

The figure makes it clear that to move from strong rejection to weak rejection one can go via strong assertion and weak assertion using the rules we have given. One thereby obtains the following derivable rules for rejections of ◇s.

$$(\ominus\Diamond\text{I.})\ \frac{-A}{\ominus\Diamond A} \qquad (\ominus\Diamond\text{E.})\ \frac{\ominus\Diamond A}{-A}$$

Analogously, one can derive the following rules for moving between strong assertion and weak rejection, and between strong rejection and weak assertion.

$$(+\Diamond\neg\text{I.})\ \frac{\ominus A}{+\Diamond\neg A} \qquad (+\Diamond\neg\text{E.})\ \frac{+\Diamond\neg A}{\ominus A}$$

$$(-\neg\Diamond\text{I.})\ \frac{\oplus A}{-\neg\Diamond A} \qquad (-\neg\Diamond\text{E.})\ \frac{-\neg\Diamond A}{\oplus A}$$

It is worth comparing our analysis of *might* with some data about rejections involving *might*. It has been observed that *might* can be used to reject a negative (Khoo 2015; Bledin and Rawlins 2020).

(29) *Alex*: It is not raining.
Becky: (No,) it might be.

Our analysis straightforwardly explains this piece of data. For the strong assertion of *It might be raining* is inferentially equivalent to the weak assertion of *It is raining*, which prevents *It is not raining* from being added to the common ground. Hence, (29) is predicted to be a rejection move. It might be objected that *might* can also be used to reject the positive *It is raining*.

(30) *Alex*: It is raining.
Becky: (No,) it *might* be.

However, this rejection is *pragmatic*: *might* conversationally implicates *not surely*, just as *some* in the following example implicates *not all* (Khoo 2015; Schlöder and Fernández 2015, 2019).

(31) *Alex*: Alicia ate all the cookies.
Becky: (No,) she ate *some* cookies.

That these are conversational implicatures is evinced by the fact that they can be cancelled.

(32) a. She ate some cookies—in fact, she ate all of them!
b. It might be raining—in fact, it is raining!

Thus, the fact that *might* can be used to reject a positive should not be mistaken for evidence regarding its semantic contribution.

4.6 Weak assertions, epistemic modal assertions, and evidence

We have extended the language of Basic Multilateral Logic with a force marker $\oplus$ for weak assertion and with rules for epistemic *might*, formalized as $\Diamond$. Being an extension of BML, the system we are developing here includes the coordination principles*, which are restricted to evidence-preserving inferences. Thus, we must now tackle the last step in our multilateral methodology towards an inferential expressivist account of epistemic modality: determine which of the new inferences involving weak assertion or *might* preserve evidence.

The existence of weak rejections showed that rejections in the broad sense are evidentially unspecific: they can be made on the basis of evidence for the negation of the rejected sentence, but they need not be. The situation in the case of assertions is similar: assertions in the broad sense can be made on the basis of evidence for the asserted sentence—in which case they are strong—but they need not be. To show this, it suffices to adapt some of the examples we used earlier to display the unspecificity of rejections.

(33) Did Homer write the *Iliad*? Perhaps! If he existed.

(34) Will X win the election? Perhaps! X or Y will win.

The weak assertions in these examples do not carry evidence for the weakly asserted sentence: one cannot justify a strong assertion of that sentence on their basis. The weak assertion in the first example is made on the basis of there being evidence for the sentence if a certain condition is satisfied; the weak assertion in the second example is made on the basis of evidence for some other sentence. The fact that weak rejections are not made on the basis of evidence for the rejected sentence meant that they had to be excluded from the Smileian *reductio** rules. Since weak rejections behave similarly with respect to evidence, they too have to be excluded from the Smileian *reductio** rules.

$$(\mathrm{SR}_1{}^*)\ \frac{\begin{matrix}[+A]\\ \vdots\\ \bot\end{matrix}}{-A}\ \text{if no premisses signed with } \ominus \text{ or } \oplus \text{ were used to derive } \bot$$

$$(\mathrm{SR}_2{}^*)\ \frac{\begin{matrix}[-A]\\ \vdots\\ \bot\end{matrix}}{+A}\ \text{if no premisses signed with } \ominus \text{ or } \oplus \text{ were used to derive } \bot$$

We are not done yet. For a strong assertion of a sentence containing *might* as its main operator—call it *an epistemic modal assertion*—may also be licensed by unspecific evidence, as shown by the following examples.

(35) Might Homer have written the *Iliad*? Yes! If he existed.

(36) Might X win the election? Yes! It will be X or Y.

There is a tension here. On the one hand, strong assertions are always licensed by specific evidence, in keeping with the connection between assertion and evidence. In the case of an epistemic modal assertion, this is specific evidence for an epistemic modal sentence. Indeed, epistemic modal assertions can occur in evidence-preserving inferences. Consider the following situation. We know that John diligently watches the forecast and loathes getting wet. This evidence licenses the assertion of *If it might rain, John has an umbrella*. Moreover, the forecast has given a moderate chance of rain. This evidence licenses the assertion of *It might rain*. In such as situation, we may justifiably assert that John has an umbrella on the basis of the evidence at our disposal. This is because the inference from *If it might rain, John has an umbrella* and *It might rain* to *John has an umbrella* preserves evidence.

On the other hand, with respect to evidence for the sentence *under* the epistemic modal, epistemic modal assertions display the same evidential unspecificity of weak assertions. In example (35), the strong assertion of *Homer might have written the Iliad* is made on the basis of there being evidence for the sentence under *might* if a certain condition is satisfied. In example (36), the strong assertion of *X might win the election* is made on the basis of evidence not for the sentence under *might*, but for some other sentence. For this reason, a rule of inference accessing a sentence under *might* is not evidence-preserving. In the previous chapter, we argued that Smileian *reductio* for strong rejections should be restricted to evidence-preserving inferences towards absurdity. It may be absurd to assert *Homer wrote the Iliad* not only because one has evidence for *Homer did not write the Iliad*, but also because one has evidence for *Homer did not exist*. However, in the latter case, it would be mistaken to conclude that Homer did not write the *Iliad*. This is because, in that case, the inference towards absurdity does not preserve evidence. We now see that we must restrict Smileian *reductio* for strong rejections further. For although it is absurd to assert *Homer wrote the Iliad* when there is evidence for *Homer might not have written the Iliad*, it would be mistaken to conclude that Homer did not write the *Iliad*. In this case too, the inference towards absurdity does not preserve evidence.

We must therefore formulate Smileian *reductio** so as to exclude inferences that access a sentence below an epistemic modal. Formally, this can be achieved by excluding inferences that use rules for the elimination of $\Diamond$. These rules are used to access a sentence under *might* and therefore do not preserve evidence. Thus, we obtain the following version of Smileian *reductio** for a multilateral logic involving force markers for weak assertion and rejection as well as an epistemic modal operator.

$$\begin{array}{c}[+A]\\ \vdots\\ (SR_1{}^*)\ \dfrac{\bot}{-A}\end{array}\ \text{if the derivation of } \bot \text{ does not use } (+\Diamond\text{E.}), (\oplus\Diamond\text{E.}) \text{ or premisses signed with } \ominus \text{ or } \oplus$$

$$\begin{array}{c}[-A]\\ \vdots\\ (SR_2{}^*)\ \dfrac{\bot}{+A}\end{array}\ \text{if the derivation of } \bot \text{ does not use } (+\Diamond\text{E.}), (\oplus\Diamond\text{E.}) \text{ or premisses signed with } \ominus \text{ or } \oplus$$

Epistemic modal assertions are permitted as premisses under Smileian *reductio**, but the modal cannot be eliminated. Importantly, this restriction does not prevent epistemic modal premisses from featuring in *reductio* arguments. For example, from the premisses *John does not have an umbrella* and *It might be raining* one can derive, by Smileian *reductio**, that *It is not the case that if it might be raining, John has an umbrella.*

There is a precedent to the claim that certain inferences involving epistemic modal operators fail to preserve evidence. Moritz Schulz (2010) gives the example of the rule of *epistemic strengthening*: the rule allowing us to infer *It must be that A* from *A*. This rule is prima facie plausible, but when formalized as A entailing $\Box A$ in a classical modal logic, it immediately leads to a modal collapse. For assume $\Diamond A$ and for *reductio* that $\neg A$. By the formalized version of epistemic strengthening, it follows that $\Box\neg A$, which is inconsistent with $\Diamond A$. By *reductio*, we can conclude that A. Thus $\Diamond A$ entails A, trivializing the modal. Schulz argues that one should have never considered epistemic strengthening to be valid in the first place, since it does not preserve evidence. Suppose, for instance, that one sees that the lights are on. Then one's evidence seems to license *They are home* but not *They must be home* (see Bledin and Lando 2018 for analogous examples). Schulz concludes that epistemic strengthening must be rejected.

Schulz is correct that epistemic strengthening does not preserve evidence, but his conclusion that this inference rule is invalid is hasty. For its validity only requires that it preserve commitment. If we treat *must* as the dual of *might* and hence abbreviate $\neg\Diamond\neg$ with $\Box$, we can derive formal versions of epistemic strengthening as well as its inverse *epistemic weakening* (the rule allowing us to infer A from *It must be that A*) using the coordination principles.

$$(+\Box\text{I.})\ \frac{+A}{+\Box A} \qquad (+\Box\text{E.})\ \frac{+\Box A}{+A}$$

Since these are derived rules, they are valid and hence preserve commitment. The rule (+□E.) also preserves evidence, since it may be derived using evidence-preserving rules. By contrast, the following derivation of (+□I.) uses a rule that does not preserve evidence, namely (⊕◇E.).

$$\dfrac{\dfrac{+A \qquad \dfrac{\dfrac{\dfrac{[\ominus\neg\Diamond\neg A]^1}{\oplus\Diamond\neg A}\,(\ominus\neg\text{E.})}{\oplus\neg A}\,(\oplus\Diamond\text{E.})}{\ominus A}\,(\oplus\neg\text{E.})}{\bot}\,(\text{Weak Rejection})}{+\neg\Diamond\neg p}\,(\text{SR})^1$$

Thus, epistemic strengthening does not preserve evidence. In the Appendix, we prove a soundness theorem that, as we will explain shortly, entails that the use of a ◇-Elimination rule to derive epistemic strengthening is necessary. It follows that the derived rule (+□E.) cannot be used in Smileian *reductio** proofs. This blocks the derivation of $+A$ from $+\Diamond A$, averting the modal collapse.

Everything is now in place for the inferential expressivist account of epistemic modality. Let *Epistemic Multilateral Logic* (EML) be the natural deduction system consisting of the bilateral rules for conjunction and negation from the previous chapter, the rules for epistemic modality from the previous section, the coordination principles governing the interaction of strong assertion and weak rejection, the coordination principles governing the interaction of strong rejection and weak assertion, and the coordination principles*. Figure 4.3 displays the rules of EML. In the Appendix, we show how to derive rules governing the Boolean logical constants and the modal under all signs of EML. The system can be naturally extended to the first-order level by appropriately extending the language and adding rules for the universal quantifier (see Aloni et al. Forthcoming for details).

Epistemic Multilateral Logic extends the modal logic **S5** (Incurvati and Schlöder 2022a), that is the normal modal logic containing the reflexivity

Coordination

$$(\text{Strong Rejection})\ \frac{+A \quad -A}{\bot} \qquad (\mathrm{SR}_1{}^*)\ \frac{\begin{array}{c}[+A]\\ \vdots{*}\\ \bot\end{array}}{-A} \qquad (\mathrm{SR}_2{}^*)\ \frac{\begin{array}{c}[-A]\\ \vdots{*}\\ \bot\end{array}}{+A}$$

$$(\text{Weak Rejection})\ \frac{+A \quad \ominus A}{\bot} \qquad (\mathrm{SR}_1)\ \frac{\begin{array}{c}[+A]\\ \vdots\\ \bot\end{array}}{\ominus A} \qquad (\mathrm{SR}_2)\ \frac{\begin{array}{c}[\ominus A]\\ \vdots\\ \bot\end{array}}{+A}$$

$$(\text{Weak Assertion})\ \frac{\oplus A \quad -A}{\bot} \qquad (\mathrm{SR}_3)\ \frac{\begin{array}{c}[\oplus A]\\ \vdots\\ \bot\end{array}}{-A} \qquad (\mathrm{SR}_4)\ \frac{\begin{array}{c}[-A]\\ \vdots\\ \bot\end{array}}{\oplus A}$$

Operation

$$(+\Diamond\mathrm{I.})\ \frac{\oplus A}{+\Diamond A} \qquad (+\Diamond\mathrm{E.})\ \frac{+\Diamond A}{\oplus A} \qquad (\oplus\Diamond\mathrm{I.})\ \frac{\oplus A}{\oplus\Diamond A} \qquad (\oplus\Diamond\mathrm{E.})\ \frac{\oplus\Diamond A}{\oplus A}$$

$$(-\neg\mathrm{I.})\ \frac{+A}{-\neg A} \qquad (-\neg\mathrm{E.})\ \frac{-\neg A}{+A} \qquad (+\neg\mathrm{I.})\ \frac{-A}{+\neg A} \qquad (+\neg\mathrm{E.})\ \frac{+\neg A}{-A}$$

$$(+\wedge\mathrm{I.})\ \frac{+A \quad +B}{+A \wedge B} \qquad (+\wedge\mathrm{I.}_1)\ \frac{+A \wedge B}{+A} \qquad (+\wedge\mathrm{I.}_2)\ \frac{+A \wedge B}{+B}$$

Fig. 4.3. The calculus of Epistemic Multilateral Logic. The notation $\vdots{*}$ denotes a derivation using neither premisses signed with $\ominus$ or $\oplus$ nor $\Diamond$-Elimination rules.

axiom **T** ($\Box p \supset p$) and the Euclidean axiom **5** ($\Diamond p \supset \Box\Diamond p$). We can state this result precisely with the help of some additional notation. Let σ be any mapping from propositional atoms to $\mathcal{L}_\Diamond$-sentences. And if A is a sentence in the language of propositional logic, write $\sigma[A]$ for the $\mathcal{L}_\Diamond$-sentence that results from uniformly replacing every atom p in A with $\sigma(p)$. We have the following theorem.

Theorem 4.6.1. *Let Γ be a set of $\mathcal{L}_\Diamond$ formulae and A an $\mathcal{L}_\Diamond$ formula. If $\Gamma \models^{S5} A$, then $\{+\sigma[B] \mid B \in \Gamma\} \vdash^{\mathsf{EML}} +\sigma[A]$.*

An immediate consequence of this theorem is that EML validates all substitution instances of classically valid arguments in the language of modal logic. Using Theorem 4.6.1, moreover, one can prove that EML is sound and complete with respect to a suitable embedding into **S5**. The precise statement of this soundness and completeness result and its proof are in the appendix to this chapter. The failure of the meta-rules of *reductio*, proof by cases and conditional proof for the material conditional in BML, which we noted in the previous chapter, might suggest that there are similarities between multilateral logic and the supervaluationist logic of vagueness, where these meta-rules also fail (Williamson 1994; Keefe 2000). The soundness and completeness result entails that the similarities run deep: the EML logic of strong assertion axiomatizes the consequence relation of supervaluationist logic (see Incurvati and Schlöder 2022b for an extended discussion of supervaluationism about vagueness in a multilateral setting). We will return to the similarities between multilateral logic and supervaluationism in Chapter 7.

From the soundness result it follows that, since $\Box\Diamond A$ does not entail $\Box A$ in the modal logic **S5**, there is no derivation of $+A$ from $+\Diamond A$ in EML. But if epistemic strengthening preserved evidence, there would be such a derivation, as shown by Schulz's collapse argument. Thus epistemic strengthening does not preserve evidence and any derivation of the rule ($+\Box$I.) must involve a $\Diamond$-Elimination rule. So the collapse argument is invalid in EML. However, we must also consider a related problem pointed out by Jeffrey S. Russell and John Hawthorne (2016). They phrase their result in a dynamic logic. Say that an epistemic state *s rules out* a sentence *A* if updating *s* with *A* results in an absurd state. Then the following are plausible assumptions regarding the dynamics of *might* and *not*.

(MIGHT) Any epistemic state not ruling out *might A* does not rule out *A*.

(NOT) Updating an epistemic state with *not A* yields a state ruling out *A*.

However, Russell and Hawthorne (2016: 326) show, if we accept MIGHT and NOT, we appear to have committed ourselves to *might A* entailing *A*. The proof goes as follows. Suppose for *reductio* that some epistemic state, when updated with *not A*, yields a state *s* that does not rule out *might A*. By MIGHT, whenever *might A* is not ruled out, *A* is not ruled out. Thus, *s* does not rule out *A* either. This contradicts NOT. By *reductio*, updating any state with *not A* yields a state that rules out *might A*. That is, when a state is updated with *not A* and then *might A*, it results in an absurd state. However, the Commutativity of update tells us that updating with *A* and then with *B* has the same effect as updating with *B* and then with *A*. Hence, updating a state with *might A* and then with *not A* results in an absurd state. That is updating any state with *might A* rules out *not A*. Now in dynamic logic, *A* entails *B* just in case updating any state with *A* results in a state that rules out *not B*. Therefore, *might A* entails *A*.

Russell and Hawthorne (2016: 336) conclude that if we want to hold on to MIGHT and NOT, 'the only real option is to go in for some account that violates Commutativity'. Doing so is in fact typical for dynamic accounts (Veltman 1996; Willer 2013). However, these accounts constitute major departures from classical logic.

Although Epistemic Multilateral Logic is, on the face of it, a static logic, one can reproduce in this logic the crucial part of the triviality argument, namely the part showing that if a state is updated with *might A*, it rules out *not A*. In the context of EML, this part of the argument becomes a proof that expressing belief towards *not A* and expressing belief towards *might A* is absurd.

$$\dfrac{+\neg A \qquad \dfrac{\dfrac{+\Diamond A}{\oplus A}\,(+\Diamond\text{E.})}{\ominus\neg A}\,(\ominus\neg\text{I.})}{\bot}\,(\text{Weak Rejection})$$

According to Russell and Hawthorne's argument, one can conclude that expressing belief towards *might A* commits one to believing that *A*. This requires applying *reductio* to derive $+A$ from the fact that $+\neg A$ leads to absurdity. EML does not sanction *reductio*, although it validates a restricted version thereof.

$$(\text{EML } \textit{reductio})\ \dfrac{\begin{array}{c}[+A]\\ \vdots \\ \bot\end{array}}{+\neg A}\ \begin{array}{l}\text{if the derivation of } \bot \text{ does not use}\\ (+\Diamond\text{E.}),\ (\oplus\Diamond\text{E.}) \text{ or premisses signed}\\ \text{with } \ominus \text{ or } \oplus\end{array}$$

However, the derivation above eliminates a $\Diamond$ and hence does not license an application of this restricted version of *reductio*. Indeed, there are models of EML in which $+\Diamond A$ holds but $+A$ does not. So any purported proof of $+A$ from $+\Diamond A$ *has* to fail.

Thus, rejecting Commutativity is not 'the only real option' once we accept that if a state is updated with *not A*, it rules out *might A*. On our account, expressing belief towards *not A* and expressing belief towards *might A* is absurd (so, in dynamic terms, we do have that if a state is updated with *might A*, it rules out *not A*). Yet, triviality is avoided: the culprit is Russell and Hawthorne's tacit endorsement of *reductio*.

4.7 Frege–Geach and its revenge version

We saw in Section 4.1 that the traditional expressivist analysis of *might* is undermined by its embedding behaviour. As already noted, since the following inference appears to be valid, one can run a version of the Frege–Geach argument to establish that the *might* in the second premiss must modify the content of *It is raining*.

(37) a. If it might be raining, we might go to the cinema.
b. It might be raining.
c. We might go to the cinema.

This is analogous to the Frege–Geach argument against traditional expressivism about negation, and the inferential expressivist response is the same. The inferential expressivist agrees with the conclusion of the Frege–Geach argument in question that *might* does not indicate weak assertion. Since *might* is an embeddable operator, the relevant Frege–Geach inference is simply an instance of *modus ponens* for strong assertions. At the same time, the inferential expressivist retains the expressivist insight that the meaning of *might* is to be explained in terms of weak assertion, the speech act expressing the attitude of refraining from disbelieving. Rather than directly indicating weak assertion, the meaning of *might* is inferentially explained in terms of it.

As in the case of strong and weak rejections, the existence of a speech act of weak assertion is not a mere act of faith, but is established on the basis of linguistic evidence: it is the speech act performed by using *perhaps* in otherwise assertoric contexts. However, taking *perhaps* to be a force modifier gives rise to a revenge version of the Frege–Geach argument, just as taking *no* to be a force indicator did. Consider the following Frege–Geach revenge inference, which proceeds as the Frege–Geach inference for *might* above, except that the second premiss and the conclusion are weak assertions instead of epistemic modal assertions.

(38) a. If it might be raining, we might go to the cinema.
b. Perhaps it is raining.
c. Perhaps we will go to the cinema.

This inference appears to be valid. As before, Frege–Geach reasoning shows that the *might* in the antecedent of the first premiss must modify the claim that it is raining. However, on our account *perhaps* is a force modifier, and so its occurrence in the second premiss does not modify the claim that it is raining. It follows that the Frege–Geach revenge inference is not an instance of *modus ponens.*

When confronted with the analogous case involving negation in the previous chapter, we argued that the inferential expressivist should insist that the revenge Frege–Geach inference is not validated by a direct application of *modus ponens.* The inferential expressivist about *might* can give the same response. We can validate the revenge Frege–Geach inference using, besides *modus ponens*, the fact that the weak assertion of a sentence is inferentially equivalent to its epistemic modal strong assertion. In particular, from the weak assertion of *It is raining* (performed by uttering *Perhaps it is raining*), we can infer the strong assertion of *It might be raining.* This, together with the strong assertion of *If it might be raining, we might go to the cinema*, delivers by *modus ponens* the strong assertion of *We might go to the cinema* and hence the weak assertion of *We will go to the cinema* (performed by uttering *Perhaps we will go to the cinema*).

The revenge Frege–Geach inference for *might* involves weak and strong assertion, and is therefore mixed. When considering *no* and rejection in the previous chapter, we presented a mixed inference pattern which can be recognized as valid without theorizing about embeddable operators. One can find analogous cases involving *perhaps.* The following pattern is an example.

(39) a. If *A*, then *B*.
b. Perhaps *A*.
c. Perhaps *B*.

This inference pattern cannot be validated simply by appealing to the inferential equivalence between *perhaps A* and *might A*. To validate this inference pattern, one must coordinate strong and weak assertion. In particular, the inference pattern is validated by the derived rules codifying the subalternity of weak assertion to strong assertion. The situation is analogous to the one in the previous chapter, where we had to coordinate strong assertion and weak rejection to validate the mixed inference pattern discussed by Smiley.

4.8 Modal disagreement and Yalcinean sentences

According to inferential expressivism about epistemic modality, the meaning of *might* is given by the inference rules for $\Diamond$, which tell us to which attitude expressions the speaker is committed in strongly or weakly asserting $\Diamond A$, and when they are committed to these speech acts. We now show that the account retains the advantages of traditional expressivism about epistemic modality in the way it accounts for the embedding behaviour of Yalcinean sentences and the phenomenon of modal disagreement. We begin with the latter. Consider again the following dialogue.

(40) *Alice*: I can't find the keys.
Bob: They might be in the car.
Alice: No, they are not in the car. I just checked.

The problem for the orthodox approach was that it is difficult to locate a single proposition about a single body of knowledge that Bob is warranted in asserting and Alice is warranted in rejecting. Traditional expressivism accounted for the phenomenon of modal disagreement by holding that the attitude expressed by Bob in uttering *The keys might be in the car* is incompatible with the attitude expressed by Alice in asserting that the keys are not in the car. To solve the Frege–Geach Problem, however, and in keeping with minimalism about assertion and belief, we argued that Bob's utterance must be taken to be an assertion, expressing a belief. However, what it is to make an epistemic modal assertion is explained in terms of the speech act of weak assertion, which expresses an attitude other than belief. This suffices to retain the expressivist explanation of the phenomenon of modal disagreement.

On our account, Bob is asserting and thereby expressing belief that the keys might be in the car. It follows from the rules giving the meaning of *might* that he is committed to expressing refraining from disbelieving that the keys are in the car. Alice, on the other hand, is asserting and thereby expressing belief that the keys are not in the car. It follows from the rules giving the meaning of *not* that she is committed to expressing disbelief that the keys are in the car. Thus, Alice and Bob are committed to expressing incompatible attitudes towards the keys being in the car. Alice and Bob are predicted to disagree, and their disagreement is not about some particular body of knowledge, but about the keys' whereabouts, which appears to be the natural reading of the example. This reasoning can be reconstructed formally within our system. Let p be *The keys are in the car*. Then Bob asserts $\Diamond p$, that is $+\Diamond p$, which is inferentially equivalent to $\oplus p$. Alice, by contrast, asserts *not* p, that is $+\neg p$, which is inferentially equivalent to $-p$. Since by the Weak Assertion rule $\oplus p$ and $-p$ are incompatible, Alice and Bob are predicted to disagree. It also follows that Alice is committed to rejecting $\Diamond p$, as she does, since $\ominus\Diamond p$ is derivable from $+\neg p$.

We now turn to the embedding behaviour of Yalcinean sentences. Yalcin observed that sentences of the form *A and it might be that not A* sound bad and, unlike Moorean sentences, continue to do so under supposition and in conditional antecedents. This, we noted, is evidence against a pragmatic explanation of the infelicity of Yalcinean sentences and in favour of the idea that they are semantically contradictory. This idea is vindicated in Epistemic Multilateral Logic, as witnessed by the following derivation of absurdity from $+A \wedge \Diamond\neg A$.

$$\dfrac{\dfrac{+A \wedge \Diamond\neg A}{+A}\,(+\wedge\mathrm{E.}_1) \qquad \dfrac{\dfrac{\dfrac{+A \wedge \Diamond\neg A}{+\Diamond\neg A}\,(+\wedge\mathrm{E.}_2)}{\oplus\neg A}\,(+\Diamond\mathrm{E.})}{\ominus A}\,(+\neg\mathrm{E.})}{\bot}\,(\text{Weak Rejection})$$

This derivation shows that uttering a sentence of the form *A and it might be that not A* is absurd, since this immediately commits one to expressing incompatible attitudes, namely believing A and refraining from believing A. This explains why asserting *A and it might be that not A* is infelicitous. But why would it be infelicitous to suppose it?

To address this question, we must take a closer look at supposition. In English, *supposition* refers both to an attitude and to the speech act used to express this attitude (Green 2000: 377–378). Linguistically, the speech act of supposition may be realized through locutions such as *suppose that A*. Thus, *suppose that* is a force indicator, which we formalize by means of a new primitive force marker 𝕊. Dummett, for one, agrees on the role of locutions such as *suppose that A*.

> In supposition, a thought is expressed but not asserted: 'Suppose . . .' must be taken as a sign of the *force* . . . with which the sentence is uttered. (Certainly, it is not logically an imperative: I could, having said, 'Think of a number', ask 'Have you done so yet?', but it would be a joke if I asked that question having said, 'Suppose the witness is telling the truth'.)
>
> (Dummett 1973a: 309)

In keeping with our methodology, we focus on two dimensions of the speech act of supposition, the attitude it expresses and its essential effect on the conversation. The speech act of supposing that *A* expresses the attitude of supposing that *A*. In supposing that *A*, one is not believing that *A*, but making as if one believes that *A* to determine what follows from it (compare Köhler 2018: 206–207). Thus, in expressing supposition, one is not committing to what one supposes, but is probing what happens if one *were* to commit to it. This understanding of the attitude expressed by the speech act of supposition fits with Stalnaker's (2014: 150–151) characterization of its essential effect. According to him, the essential effect of supposition is a proposal to add the relevant sentence to the common ground, but to do so *temporarily*. One proposes to make as if the sentence is among those mutually taken for granted by the interlocutors, to check what the consequences of this would be.[7]

For this process to work as desired, the internal logic of supposition must be the same as the logic of strong assertion. This sanctions the coordination principle 𝕊-Inference, which states that the suppositional consequences of a suppositional context mirror the strongly assertoric consequences of the corresponding strongly assertoric context.[8]

[7] One may wonder about the relationship between supposition and the proof-theoretic device of adding a dischargeable premiss, for instance [*A*]. The proof-theoretic device can be unconstrained, but may also be restricted, as in the case of the coordination principles*. Certain restrictions on the use of the proof-theoretic device correspond to supposition. We explore the matter in Chapter 8.

[8] Compare with Yalcin (2007: 995), who takes *suppose* to be closed under informational consequence. In particular, his semantics says that *x supposes that A* is true at an information state *s* and a world *w* if *A* is true at the information state S_x^w and all worlds $v \in S_x^w$, where S_x^w is the set of worlds compatible with what *x* supposes in *w*. Thus, for *p and it might be that not p* to be true at all worlds $v \in S_x^w$, it must be the case that *p* is true at all these *v* and false at some such *v*, which cannot be. Hence *Suppose p and it might be that not p* sounds contradictory. It is easy to verify that, if one assumes that $S_x^w \neq \varnothing$, then 𝕊-Inference is sound with respect to Yalcin's semantic entry for *suppose*.

$$\text{(}\mathbb{S}\text{-Inference)}\ \frac{\mathbb{S}A \qquad \begin{matrix}[+A] \\ \vdots \\ +B/\bot\end{matrix}}{\mathbb{S}B/\bot}$$

where the derivation of $+B/\bot$ may only use premisses of the form $+C$ where $\mathbb{S}C$ is derivable in the proof context of $\mathbb{S}A$

This coordination principle immediately implies that an absurdity is derivable from $\mathbb{S}(A \wedge \Diamond\neg A)$ since, as shown above, an absurdity is already derivable from $+p \wedge \Diamond\neg A$. Thus, *Suppose A and it might be that not A* is absurd. Informally, the idea is that just as strongly asserting *A and it might be that not A* commits one to having incompatible attitudes, namely belief and refraining from believing, supposing it still commits one to having incompatible attitudes namely making as if one believes and refrains from believing the same thing. In terms of effects on the conversation, not only is it absurd to propose the addition of a sentence to the common ground while preventing this addition, but it is also absurd to do all of this temporarily.

The absurdity of *Suppose A and it might be that not A* does not quite explain its infelicity, since not all absurd suppositions sound bad. One may felicitously suppose certain logical contradictions. For instance, someone not familiar with the derivability of Peirce's Law in classical logic may felicitously suppose its negation. However, to see that the negation of Peirce's Law is absurd requires a complex argument. By contrast, anyone grasping the meaning of *and* and *not* will immediately recognize the absurdity of, say, *A and not A*, since this absurdity can be inferred by simply applying the elimination rules for *and* and *not*, which give these expressions their meaning. This explains why *A and not A* sounds bad and continues to do so in embedded contexts. The same holds for *A and it might be that not A*: its absurdity can be immediately inferred by applying the meaning-conferring rules of *and*, *not*, and *might*. This absurdity is therefore manifest to anyone who grasps the meaning of these expressions, which explains why *Suppose A and it might be that not A* is infelicitous.

In addition to explaining the infelicity of Yalcinean sentences under *suppose*, our account has the resources to explain why Moorean sentences sound bad in ordinary contexts but cease to do so in suppositional ones. For while strongly asserting *A and I do not believe that A* is *pragmatically incoherent*, it is not *absurd* in the sense of entailing $\bot$. While in strongly asserting *A and it might be that not A* one is committing to both believing and refraining from believing *A*, in strongly asserting *A and I do not believe that A* one is committing to believing *A* and disbelieving that one satisfies this commitment. The latter are not incompatible commitments, since being committed to believing that *A* does not entail that one is committed to believing that one believes that *A*. For example, someone may have overtly expressed some beliefs that jointly entail a commitment to expressing belief towards an obviously absurd sentence *A*. The speaker must concede that they are so committed when this is pointed out, but it would be unwarranted to press them to concede that they actually believe *A*. Thus the inference from $+A$ to $+\mathbf{B}A$ is invalid in that it does not preserve commitment. Accordingly, $+A \wedge \neg\mathbf{B}A$ does not entail $\bot$. From $+A \wedge \neg\mathbf{B}A$ one can derive $+A$ and $-\mathbf{B}A$, but to derive absurdity one would have to derive $+\mathbf{B}A$ from $+A$, and this inference is not valid. Thus, it is not absurd to strongly assert *A and I do not believe that A*.

Nonetheless, it is pragmatically incoherent to assert *A and I do not believe that A*. This, in our view, is because the speaker is undertaking a commitment while announcing that they do not satisfy this commitment (Woods 2018). In uttering the Moorean sentence, the speaker immediately commits both to believing *A* and to not satisfying this commitment. When a speaker immediately commits to an attitude, the listener presumes that they want to be recognized as holding this attitude. But in a Moorean sentence, this presumption is immediately violated by the declaration on the speaker's part that they do not satisfy this commitment. This gives rise to the characteristic reaction to Moorean sentences, namely it being unclear what the speaker is trying to convey by uttering them. By contrast, someone uttering *Suppose A and I don't believe that A* is not presumed to be wanting to be recognized as someone who believes that *A*.

The speech act of strong assertion serves to undertake an explicit commitment towards believing something. One question is what further commitments follow from this. This question is answered by laying down commitment-preserving inference rules. Another question is what further conclusions can be drawn from the performance of the speech act itself. The infelicity of Yalcinean sentences is explained by our answer to the first question, as codified by the calculus of Epistemic Multilateral Logic. By contrast, the infelicity of Moorean sentences is explained, at least in part, by an answer to the second question.[9] This accounts for the divergent embedding behaviour of Yalcinean and Moorean sentences. Inferences

[9] We return to the question of what can be inferred from the performance of attitude expressions in Chapter 6.

preserving commitment also bear on certain embedded uses of sentences, in particular under supposition, whereas conclusions drawn from the performance of a speech act only bear on the speech act itself.

In this chapter, we have presented an inferential expressivist account of epistemic modality. By explaining the meaning of *might* in terms of weak assertion, the account retains the advantages of traditional expressivist accounts, in that it explains the phenomenon of modal disagreement and the infelicity of Yalcinean sentences. By *inferentially* explaining the meaning of *might* in terms of weak assertion, the inferential expressivist account of epistemic modality solves the Frege–Geach Problem and revenge versions thereof. Finally, the account sanctions all classically valid arguments, thus avoiding radical departures from classical logic.

The inferential expressivist accounts we have presented so far explain the meaning of *not* and *might* in terms of the speech acts of strong rejection and weak assertion. Although they do not express belief, both weak assertion and strong rejection express doxastic attitudes, at least on a broad enough understanding of *doxastic attitude*. However, expressivist approaches have traditionally focused on non-doxastic attitudes such as disapproval and their relation to moral language. Indeed, we take it that any expressivist approach worth its name should be able to provide an account of moral vocabulary. We present such an account from an inferential expressivist viewpoint in the next chapter.

4.9 Appendix

All constants under all signs

In the first part of this appendix, we present the multilateral modal calculus, including both the primitive rules of EML and the derived rules that govern the remaining logical constants and their behaviour under all signs. As in the appendix to the previous chapter, we show how to derive a selection of rules, to give a feel for how the calculus works.

Figure 4.4 displays the derived coordination principles of EML, corresponding to the dashed lines in Figure 4.1. The rules for the Boolean connectives under $+$, $-$, and $\ominus$ are as in BML except we can now derive rules for negation that switch between $\ominus$ and $\oplus$, and that $\vdots *$ is now interpreted to denote subderivations using only premisses signed with $+$ or $-$ and using no

$$(\mathrm{SA}_1)\ \frac{+A}{\oplus A} \qquad (\mathrm{SA}_2)\ \frac{-A}{\ominus A} \qquad (\mathrm{SA}_1{}^*)\ \frac{\oplus A \qquad \begin{array}{c}[+A]\\ \vdots^{*}\\ +B/\bot\end{array}}{\oplus B/\bot} \qquad (\mathrm{SA}_2{}^*)\ \frac{\ominus A \qquad \begin{array}{c}[-A]\\ \vdots^{*}\\ -B/\bot\end{array}}{\ominus B/\bot}$$

$$(\mathrm{SC})\ \frac{\begin{array}{c}[\oplus A]\\ \vdots\\ \varphi\end{array} \qquad \begin{array}{c}[\ominus A]\\ \vdots\\ \varphi\end{array}}{\varphi}$$

Fig. 4.4. Subalternity and subcontrariety in Epistemic Multilateral Logic. The notation $\vdots *$ denotes a derivation using only premisses signed with + or − and using no ◇-Elimination rules.

◇-Elimination rules. We illustrate the method for deriving the new negation rules by considering the case of ($\oplus\neg$I.).

$$\cfrac{\cfrac{\ominus A \qquad \cfrac{[-\neg A]^2}{+A}\,(+\neg\mathrm{I.})}{\bot}\,(\text{Weak Rejection})}{\oplus\neg A}\,(\mathrm{SR}_4)$$

The derived rules for the Boolean connectives under ⊕ are as in Figure 4.5. Finally, Figures 4.6 and 4.7 display the rules for the introduction and elimination of ◇ and □ under all signs.

The derivations of (SA_1) and ($\mathrm{SA}_1{}^*$) are as follows.

$$\cfrac{\cfrac{+A \qquad [-A]^1}{\bot}\,(\text{Strong Rejection})}{\oplus A}\,(\mathrm{SR}_4)^1 \qquad\qquad \cfrac{\cfrac{\oplus A \qquad \cfrac{\cfrac{[-B]^1 \qquad \begin{array}{c}[+A]^2\\ \vdots *\\ +B\end{array}}{\bot}\,(\text{Strong Rejection})}{-A}\,(\mathrm{SR}_2{}^*)^1}{\bot}\,(\text{Weak Assertion})}{\oplus B}\,(\mathrm{SR}_4)^2$$

The case of ($\mathrm{SA}_1{}^*$) where one concludes ⊥ is obtained by letting B be any formula such that $-B$ is a theorem. The derivations (SA_2) and ($\mathrm{SA}_2{}^*$) are analogous, and the derivation of (SC) is straightforward.

The derivations of ($\oplus\wedge\mathrm{I.}_1$) and ($\oplus\wedge\mathrm{E.}_1$) are as follows. The derivations of ($\oplus\wedge\mathrm{I.}_2$) and ($\oplus\wedge\mathrm{E.}_2$) are analogous.

$$(\oplus\neg\text{I.})\ \frac{\ominus A}{\oplus\neg A} \qquad (\oplus\neg\text{E.})\ \frac{\oplus\neg A}{\ominus A}$$

$$(\oplus\wedge\text{I.}_1)\ \frac{+A \quad \oplus B}{\oplus A\wedge B} \quad (\oplus\wedge\text{I.}_2)\ \frac{\oplus A \quad +B}{\oplus A\wedge B} \quad (\oplus\wedge\text{E.}_1)\ \frac{\oplus A\wedge B}{\oplus A} \quad (\oplus\wedge\text{E.}_2)\ \frac{\oplus A\wedge B}{\oplus B}$$

$$(\oplus\vee\text{I.}_1)\ \frac{\oplus A}{\oplus A\vee B} \quad (\ominus\vee\text{I.}_2)\ \frac{\oplus B}{\oplus A\vee B} \quad (\oplus\vee\text{E.})\ \frac{\oplus A\vee B \quad \begin{matrix}[\oplus A]\\ \vdots \\ +C\end{matrix} \quad \begin{matrix}[\oplus B]\\ \vdots \\ +C\end{matrix}}{+C}$$

$$(\oplus\supset\text{I.})\ \frac{\begin{matrix}[+A]\\ \vdots \\ \oplus B\end{matrix}}{\oplus A\supset B} \qquad (\oplus\supset\text{E.})\ \frac{\oplus A\supset B \quad +A}{\oplus B}$$

Fig. 4.5. Operations under $\oplus$ in Epistemic Multilateral Logic.

$$(+\Diamond\text{I.})\ \frac{\oplus A}{+\Diamond A} \quad (+\Diamond\text{E.})\ \frac{+\Diamond A}{\oplus A} \quad (-\Diamond\text{I.})\ \frac{-A}{-\Diamond A} \quad (-\Diamond\text{E.})\ \frac{-\Diamond A}{-A}$$

$$(\oplus\Diamond\text{I.})\ \frac{\oplus A}{\oplus\Diamond A} \quad (\oplus\Diamond\text{E.})\ \frac{\oplus\Diamond A}{\oplus A} \quad (\ominus\Diamond\text{I.})\ \frac{-A}{\ominus\Diamond A} \quad (\ominus\Diamond\text{E.})\ \frac{\ominus\Diamond A}{-A}$$

Fig. 4.6. Rules for $\Diamond$ in Epistemic Multilateral Logic.

$$(+\Box\text{I.})\ \frac{+A}{+\Box A} \quad (+\Box\text{E.})\ \frac{+\Box A}{+A} \quad (-\Box\text{I.})\ \frac{\ominus A}{-\Box A} \quad (-\Diamond\text{E.})\ \frac{-\Box A}{\ominus A}$$

$$(\oplus\Box\text{I.})\ \frac{+A}{\oplus\Box A} \quad (\oplus\Box\text{E.})\ \frac{\oplus\Box A}{+A} \quad (\ominus\Box\text{I.})\ \frac{\ominus A}{\ominus\Box A} \quad (\ominus\Box\text{E.})\ \frac{\ominus\Box A}{\ominus A}$$

Fig. 4.7. Rules for $\Box$ in Epistemic Multilateral Logic.

$$\cfrac{+A \qquad \cfrac{\cfrac{\oplus B \qquad \cfrac{\cfrac{\cfrac{[+A]^2\ [+B]^1}{+(A\wedge B)}\,(+\wedge\text{I.}) \qquad [-A\wedge B]^3}{\bot}\,(\text{Strong Rej.})}{-B}\,(\text{SR}_1{}^*)^1}{\bot}\,(\text{Weak Assertion})}{\ominus A}\,(\text{SR}_1)^2}{\cfrac{\bot}{\oplus(A\wedge B)}\,(\text{SR}_4)^3}\,(\text{Weak Rejection})$$

$$\cfrac{\cfrac{\oplus A\wedge B \qquad \cfrac{[-A]^1}{-A\wedge B}\,(-\wedge\text{I.}_1)}{\bot}\,(\text{Weak As.})}{\oplus A}\,(\text{SR}_4)^1$$

We present the derivation of ($\oplus\supset$I.) to show why the subderivation need not be restricted.

$$
\dfrac{\dfrac{\begin{matrix}\dfrac{\dfrac{\dfrac{[-A \supset B]^1}{-\neg(A \wedge \neg B)}\,\text{(Abbreviation)}}{+A \wedge \neg B}\,(-\neg\text{E.})}{+A}\,(+\wedge\text{E.}_1) \\ \vdots \\ \oplus B\end{matrix} \qquad \dfrac{\dfrac{\dfrac{\dfrac{[-A \supset B]^1}{-\neg(A \wedge \neg B)}\,\text{(Abbreviation)}}{+A \wedge \neg B}\,(-\neg\text{E.})}{+\neg B}\,(+\wedge\text{E.}_2)}{-B}\,(+\neg\text{E.})}{\bot}\,\text{(Weak Assertion)}}{\oplus A \supset B}\,(\text{SR}_4)^1
$$

The remaining rules for the constants under $\oplus$ use only methods familiar from the appendix to Chapter 3. Finally, the rules for $\Diamond$ under $-$ and $\ominus$ are derived as follows.

$$
\dfrac{\dfrac{-A \qquad \dfrac{[\oplus\Diamond A]^1}{\oplus A}\,(\oplus\Diamond\text{E.})}{\bot}\,\text{(Weak Assertion)}}{-\Diamond A}\,(\text{SR}_3)^1
\qquad
\dfrac{\dfrac{-\Diamond A \qquad \dfrac{[\oplus A]}{+\Diamond A}\,(+\Diamond\text{I.})}{\bot}\,\text{(Strong Rejection)}}{-A}\,(\text{SR}_3)^1
$$

$$
\dfrac{\dfrac{-A \qquad \dfrac{[+\Diamond A]^1}{\oplus A}\,(+\Diamond\text{E.})}{\bot}\,\text{(Weak Assertion)}}{\ominus\Diamond A}\,(\text{SR}_1)^1
\qquad
\dfrac{\dfrac{\ominus\Diamond A \qquad \dfrac{[\oplus A]}{+\Diamond A}\,(+\Diamond\text{I.})}{\bot}\,\text{(Weak Rejection)}}{-A}\,(\text{SR}_3)^1
$$

Soundness and Completeness

In the second part of this appendix, we prove that EML is sound and complete with respect to **S5** *modulo* the following translation τ.

$$
\tau(\varphi) = \begin{cases} \Box A, \text{ if } \varphi = +A \\ \Box\neg A, \text{ if } \varphi = -A \\ \neg\Box A, \text{ if } \varphi = \ominus A \\ \neg\Box\neg A, \text{ if } \varphi = \oplus A \end{cases}
$$

We begin with the proof of soundness.

Theorem 4.9.1 (Soundness). *Let* Γ *be a set of* $\mathcal{L}_\Diamond$*-formulae. If* $\Gamma \vdash^{\mathsf{EML}} \varphi$ *then* $\tau[\Gamma] \models^{\mathbf{S5}} \tau(\varphi)$.

The main challenge is to show that the restrictions on the coordination principles* are effective in ensuring the soundness of the calculus. To do so, it will be helpful to use the derivable rules for conjunction under weak assertion, that is ($\oplus\wedge$I.$_1$), ($\oplus\wedge$I.$_2$), ($\oplus\wedge$E.$_1$), and ($\oplus\wedge$E.$_2$). Since these rules are derivable in EML, the soundness of EML is equivalent to the soundness of EML$^+$, the calculus of EML plus the derivable rules for conjunction under weak assertion. We use $\vdash^+$ to denote derivability in EML$^+$ and write $\Gamma \vdash^+_D \varphi$ to indicate that the derivation D witnesses the existence of this derivability relation between Γ and φ.

The only difficult step in establishing soundness consists in dealing with the restricted subderivations of Smileian *reductio**. We first tackle the easy cases. To this end, let EML$^-$ be the calculus of EML$^+$ without Smileian *reductio**—that is without (SR$_1$*) and (SR$_2$*)—and write $\vdash^-$ for the resulting derivability relation.

Theorem 4.9.2 (Pre-Soundness). *Let* Γ *be a set of* $\mathcal{L}_\Diamond$*-formulae and* φ *an* $\mathcal{L}_\Diamond$*-formula. If* $\Gamma \vdash^- \varphi$ *then* $\tau[\Gamma] \models^{S5} \tau(\varphi)$.

The proof is a standard induction on the length of derivations and is therefore omitted. Next, we prove the soundness of the full calculus. First, we need an auxiliary definition and a technical lemma. The following definition provides the tools to rewrite a proof D not involving $\Diamond$-Eliminations to a proof where $\Diamond$s only occur in sentences that translate back to **S5**-tautologies.

Definition 4.9.3 Suppose $\Gamma \vdash^+_D \varphi$ where D does not use $\Diamond$-Elimination rules, Γ contains only strongly asserted or strongly rejected formulae and D uses all premisses in Γ. Then construct a mapping π^D as follows: for each sentence Z that occurs anywhere in D pick an unused propositional atom c_Z and let $\pi^D(+Z) = +c_Z$, $\pi^D(-Z) = -c_Z$, $\pi^D(\oplus Z) = \oplus c_Z$, and $\pi^D(\ominus Z) = \ominus c_Z$.

Let Σ^D be the set containing exactly the following formulae. For any sentences X and Y occurring anywhere in D

a. $+(c_{\neg X} \supset \neg c_X)$ and $+(\neg c_X \supset c_{\neg X})$.
b. $+(c_{\neg\neg X} \supset c_X)$ and $+(c_X \supset c_{\neg\neg X})$.
c. $+(c_{X\wedge Y} \supset (c_X \wedge c_Y))$ and $+((c_X \wedge c_Y) \supset c_{X\wedge Y})$.
d. $+(c_X \supset c_{\Diamond X})$.
e. $+(\Diamond c_{\Diamond X} \supset c_{\Diamond X})$.

Note that all formulae in Σ^D substitute to **S5**-tautologies under the map $c_X \mapsto X$ (i.e., the inverse of π^D). The following lemma shows that these added assumptions suffice to rewrite the proof D under the translation π^D.

Lemma 4.9.4. *Suppose* $\Gamma \vdash_D^+ \varphi$ *where* D *does not use* $\Diamond$*-Elimination rules. Then there is a derivation* D^* *such that* $\pi^D[\Gamma] \cup \Sigma^D \vdash_{D^*}^+ \pi^D(\varphi)$ *and* D^* *contains no more applications of Smileian* reductio* *than* D.

Proof. One shows by induction on the length n of D that every derivation D can be rewritten to a derivation D^* as in the Lemma, using the members of Σ^D to replace uses of operational rules. □

Now we are ready to prove the Soundness of the full calculus.

Proof of Theorem 4.9.1. We prove the statement of the theorem for $\vdash^+$, which immediately entails the theorem. The proof proceeds by induction on the number n of times that Smileian *reductio** is applied in a derivation. The base case $n = 0$ is exactly Theorem 4.9.2.

Suppose that Soundness holds for all derivations D in which Smileian *reductio** is applied less than n times. We want to show that derivations with n applications are sound. Let D be a derivation with n applications of Smileian *reductio**. Consider the case in which the nth application of Smileian *reductio** ends with (SR_1*)—the case in which it ends with (SR_2*) is analogous. The local proof context is the following.

$$\dfrac{\begin{array}{c}[+A]^i \\ E \\ \bot\end{array}}{-A}\ (SR_1*)^i$$

In this situation, E does not use $\Diamond$-Elimination rules, and there is a finite subset $\Gamma' \subseteq \Gamma$ such that all formulae in Γ' are signed by $+$ or $-$, such that $\Gamma', +A \vdash_E^+ \bot$. We need to show that $\tau[\Gamma'] \models^{S5} \tau(-A)$.

For readability we will henceforth omit mentioning τ, so that, say, $\Gamma' \models^{S5} +A$ is understood to stand for $\tau[\Gamma'] \models^{S5} \tau(+A)$. Since E contains less than n applications of Smileian *reductio**, by the induction hypothesis we have that $\Gamma', +A \models^{S5} \bot$. The proof that $\Gamma' \models^{S5} -A$ now proceeds in two steps.

i. We show that $\Gamma' \models^{S5} -A$ if A is $\Diamond$-free and Γ' can be split into two sets $\Gamma' = \Delta \cup \Theta$ such that: for all $+C \in \Delta$, C is $\Diamond$-free and for all $+D \in \Theta$, $D = \Diamond X \supset X$ for some $\Diamond$-free X.
ii. Any other application of (SR_1*) can be reduced to (i.).

So, first assume that A is $\Diamond$-free and $\Gamma' = \Delta \cup \Theta$ as above. We need to show that for any model $M = \langle W^M, R^M, V^M \rangle$ of Γ', we have that $M, w \Vdash \Box\neg A$, where $w \in W^M$ and $\Vdash$ is the usual satisfaction relation for worlds in modal logic. Assume for *reductio* that there are an M and a w such that $M, w \Vdash \Gamma'$ and $M, w \Vdash \Diamond A$. Let $v \in W^M$ be a witness for the latter, i.e., $M, v \Vdash A$. Note that for all $+C \in \Delta$ we have that $M, v \Vdash C$, since $M, w \Vdash \Box C$.

Now consider the model M' such that: $W^{M'} = \{v\}$, $V^{M'}(v) = V^M(v)$, $R^{M'} = \{(v, v)\}$. Note that all C with $+C \in \Delta$ are assumed to be $\Diamond$-free. That is, the fact that $M, v \Vdash C$ is dependent *only* on the valuation $V^M(v)$ and not on any other worlds in W^M. Thus it is also the case that $M', v \Vdash C$ for all C with $+C \in \Delta$. For the same reason, $M', v \Vdash A$. Note, further, that, since M' has precisely one world, $M', v \Vdash \Diamond X$ just in case $M', v \Vdash X$. So $M', v \Vdash \Diamond X \supset X$ for any X. Thus $M' \Vdash \Theta$.

Hence $M', v \Vdash \Gamma'$. By construction, $M', v \Vdash \Box A$. So M' is a countermodel to $\Gamma' \cup \{+A\} \models^{S5} \bot$, which however followed by induction. Contradiction. Thus, there is no such M. This establishes (i.).

For (ii.), we relax our assumption so that A and the formulae in Γ' may be arbitrary. By Lemma 4.9.4 and writing $+c_A$ for $\pi^E(+A)$, we have a derivation E^* such that:

$$\pi^E[\Gamma'] \cup \Sigma^E, +c_A \vdash^+_{E^*} \bot$$

Note that since π^E maps everything to $\Diamond$-free formulae, the elements of $\pi^E[\Gamma'] \cup \Sigma^E$ are as described in (i): $\pi^E[\Gamma'] \cup \Sigma^E = \Delta \cup \Theta$ with Θ being exactly all formulae added in clause (e.) in the construction of Σ^E (Definition 4.9.3). Thus we obtain $\pi^E[\Gamma'] \cup \Sigma^E \models^{S5} \Box\neg A$ by the argument of (i).

Now let $\Sigma = (\pi^D)^{-1}[\Sigma^E]$. Since **S5** is closed under uniform substitution, $\Gamma' \cup \Sigma \models^{S5} \Box\neg A$. But Σ contains only **S5**-tautologies (by inspection of Definition 4.9.3). Hence $\Gamma' \models^{S5} \Box\neg c_A$. □

We now show that, *modulo* the translation τ, EML is also complete with respect to **S5**.

Theorem 4.9.5 (Completeness). *Let Γ be a set of $\mathcal{L}_\Diamond$-formulae and φ an $\mathcal{L}_\Diamond$-formula. If $\tau[\Gamma] \models^{S5} \tau(\varphi)$ then $\Gamma \vdash^{EML} \varphi$.*

This is shown by a model existence theorem. The construction of a canonical term model has to respect the difference between derivations that use $\Diamond$-Eliminations and those that do not. To this end, we need some additional definitions. We write $\vdash^*$ to denote the derivability relation of the evidence-preserving fragment of EML. That is, $\Gamma \vdash^* \varphi$ if one can derive φ from Γ using only premisses signed with $+$ or $-$ and no $\Diamond$-Elimination rules. In the typical canonical construction, one takes maximally consistent sets of formulae to be the worlds. We will instead take maximally S-consistent sets of formulae, where a set Γ is *S-consistent* if $\Gamma \nvdash^* \bot$ and *S-inconsistent* otherwise. Note that there are inconsistent S-consistent sets, such as $\{+p, +\Diamond\neg p\}$ and $\{+A, \ominus A\}$.

Theorem 4.9.6 (Model Existence). *Let Γ be a set of $\mathcal{L}_\Diamond$-formulae. If Γ is consistent, then there is an* **S5**-*model M such that $M \models \tau[\Gamma]$.*

Proof. Let $Cl^+(\Gamma)$ consist of all strongly asserted formulae in the closure of Γ under derivability in EML. Concisely, $Cl^+(\Gamma) = \{+A \mid \Gamma \vdash^{EML} +A\}$. Moreover, let $\mathcal{E} = \{\Delta \mid \Delta$ is a maximal S-consistent extension of $Cl^+(\Gamma)\}$. Finally, define a model $M = \langle W, R, V\rangle$ where (i) $W = \mathcal{E}$, (ii) wRv just in case, for all $+A \in v$, $+\Diamond A \in w$, and (iii) $V(w) = \{p \mid +p \in w\}$. We show by induction on the complexity of sentences A that $+A \in w$ if and only if $M, w \Vdash A$. The cases for atomic A and $A = B \wedge C$ are straightforward, so we only cover negation and the modal.

Let us start from the negation case. If $+\neg A \in w$, then $+A \notin w$. By the induction hypothesis, $M, w \nVdash A$. Thus $M, w \Vdash \neg A$. Conversely, if $M, w \Vdash \neg A$, then $M, w \nVdash A$, so $+A \notin w$ by the induction hypothesis. Since w is a maximally S-consistent set, this means that $+A$ is S-inconsistent with w. By Signed *reductio*, $+\neg A \in w$.

Now let us turn to the case of the modal. Suppose $+\Diamond A \in w$. Let $\{B_i \mid i \in \omega\}$ be the set of sentences such that $+\Diamond B_i \notin w$. We need a v such that $+A \in v$ and for all i, $+B_i \notin v$. Note that if $+\Diamond B_i \notin w$, then $+\Diamond B_i$ is S-inconsistent with w, so $+\neg\Diamond B_i \in w$. Hence, $+\neg B_i \in w$ for all i since $Cl^+(\Gamma)$ contains Axiom **T**. Thus, $Cl^+(\Gamma) \cup \{+\neg B_i \mid i \in \omega\}$ is S-consistent, since it is a subset of w. Now, if $Cl^+(\Gamma) \cup \{+\neg B_i \mid i \in \omega\} \cup \{+A\}$ is S-consistent, there is a v as needed. Towards a contradiction, assume this set is S-inconsistent. By Signed *reductio*, this means that $Cl^+(\Gamma) \cup \{+\neg B_i \mid i \in \omega\} \vdash^* +\neg A$. It follows by $(+\supset\text{I.})$ that there is a finite set of B_is (say, all $i < n$) such that $Cl^+(\Gamma) \vdash^{\mathsf{EML}}$ $+((\bigwedge_{i<n} \neg B_i) \supset \neg A)$. Since EML extends **S5**, it follows that $\Gamma \vdash^{\mathsf{EML}} +\Diamond A \supset (\bigvee_{i<n} \Diamond B_i)$. Since $\Diamond A \in w$ this means that $+(\bigvee_{i<n} \Diamond B_i) \in w$. But we saw that for any $i < n$, $+\neg\Diamond B_i \in w$. Hence w is S-inconsistent. Contradiction. Thus, there is a $v \in W$ with $+A \in v$ and wRv. By the induction hypothesis, $M, v \Vdash A$. Thus, $M, w \Vdash \Diamond A$.

Conversely, suppose $M, w \Vdash \Diamond A$. Then there is a v, wRv such that $M, v \Vdash A$. By the induction hypothesis, $+A \in v$. By definition of R, $+\Diamond A \in w$.

This completes the proof that $+A \in w$ if and only if $M, w \Vdash A$. Now, take an arbitrary $w \in W$. Without loss of generality, we can assume that all formulae in Γ are signed by $+$, since $-A$ can be replaced by $+\neg A$, $\oplus A$ by $+\Diamond A$, and $\ominus A$ by $+\Diamond\neg A$. Since $\Gamma \subseteq w$, it follows that $M, w \Vdash \varphi$ for each $\varphi \in \tau[\Gamma]$. It remains to show that $\langle W, R, V\rangle$ is an **S5** model. This follows from the fact that the **S5** axioms are contained in $Cl^+(\Gamma)$ (see Incurvati and Schlöder 2022a). □

5
Moral Vocabulary

We now come to the inferential expressivist treatment of moral vocabulary. The expressivist analysis of moral vocabulary has several attractive features. It replaces referentialist (and hence metaphysically loaded) explanations of meaning with non-referentialist ones by linking meaning with our psychological lives. In addition, ethical expressivism has the capacity to explain disagreement phenomena and the motivating power of moral judgements. But, once more, an appealing expressivist story about the meaning of some terms must contend with the Frege–Geach Problem.

In its original, *conditional* version, the Frege–Geach Problem challenges the expressivist to explain how a term that putatively expresses an attitude can embed in conditional antecedents. While traditional ethical expressivism succumbs to this version of the problem, inferential expressivism about moral vocabulary has the resources to solve it, much in the way in which it solved the corresponding versions of the problem for expressivism about negation and epistemic modality.

However, current debate in meta-ethics centres on the *negation* version of the Frege–Geach Problem (Unwin 1999, 2001; Schroeder 2008a). In brief, it goes as follows. The negation of a descriptive term appears to be different from the negation of a term treated in expressivist terms. The two moral claims *Lying is wrong* and *Lying is not wrong*, if receiving expressivist treatment, must be read as expressing *different attitudes* towards the *same content*. Thus, it seems that *not* in *Lying is not wrong* modifies attitudes. However, in the descriptive claim *Grass is not green*, the term *not* modifies the *content* of what is claimed. The expressivist is challenged to explain these facts without any ad hoc stipulations. The problem permeates semantic analysis at different levels; we discuss its different aspects below. The negation version of the Frege–Geach Problem does not only affect traditional forms of expressivism, but seems to undermine also sophisticated forms thereof. Indeed, Mark Schroeder concludes:

Reasoning with Attitude: Foundations and Applications of Inferential Expressivism. Luca Incurvati and Julian J. Schlöder, Oxford University Press. © Oxford University Press 2023.
DOI: 10.1093/oso/9780197620984.003.0005

> [t]o date... expressivists have not yet managed even to supply a semantics for 'not' that explains why atomic moral sentences are inconsistent with their negations. So the problem is about as far from being discharged as problems come. (Schroeder 2009: 265)

In this chapter, we show that an inferential expressivist treatment of moral vocabulary has the resources to solve the negation version of the Frege–Geach Problem, in addition to its conditional version. Our proposal builds on the inferential expressivist accounts of negation and epistemic modality developed in the previous chapters. Our account of the meaning of negation is central to our solution to the negation version of the Frege–Geach Problem. And our account of the meaning of epistemic modals enables us to explain how epistemic modals interact with moral vocabulary, improving on Schroeder's (2008a) analysis, by solving what he calls the *New New Negation Problem* concerning sentences such as *Lying might be wrong.* We conclude the chapter by discussing how our approach deals with the Wishful Thinking Problem and how it can account for the motivating power of moral judgements.

5.1 The Negation Problem

Traditional expressivists claim that a sentence such as *Lying is wrong* (1a) is used to express the attitude of *disapproval* towards the act of lying. The expressivist problem with negation is to explain what is expressed by *Lying is not wrong* (1b).

(1) a. Lying is wrong.
    b. Lying is not wrong.

There is one clear constraint on any such explanation: (1a) and (1b) are *inconsistent* in prima facie the same way that *Grass is green* and *Grass is not green* are inconsistent. This ought to be explained.

Which attitude is expressed by (1b)? It cannot be *disapproval of not lying*, since that is what one expresses by uttering (2a). It is a mistake to equate (1b) with (2a), since (2a) is inconsistent with (2b), whereas (1b) is not.

(2) a. Not lying is wrong.
    b. Not lying is not wrong.

Moreover, (1b) cannot express the *absence* of *disapproval of lying*, as this would wrongly conflate (3a) and (3b).

(3) a. Jane does not think that lying is wrong
       = Jane does not disapprove of lying
    b. Jane thinks that lying is not wrong
       = ?
    c. Jane thinks that lying is wrong
       = Jane disapproves of lying
    d. Jane thinks that not lying is wrong
       = Jane disapproves of not lying

As Nicholas Unwin (1999, 2001) notes, the problem is really quite simple: in *Jane does (not) think that (not) lying is (not) wrong* one can insert up to three negations, and no two choices should be equivocated. But the traditional expressivist paraphrase *Jane does (not) disapprove of (not) lying* has only space for two, so the expressivist requires an attitude other than disapproval to describe what attitude is expressed by (1b). One may—by stipulation—call this attitude *tolerance* and describe (1b) as expressing tolerance of lying. However, according to Schroeder (2008a), this leaves an explanatory gap. Expressivists must explain the inconsistency of *Lying is wrong* and *Lying is not wrong*. They may attempt to do so by claiming that it is inconsistent to simultaneously tolerate and disapprove of the same act, in much the same way that it is inconsistent to hold the attitude of belief towards two inconsistent sentences or the attitude of intent towards two incompatible goals.

Schroeder, however, argues that this does not suffice. Any such putative explanation appears to entail that the expressivist is committed to there being two ways for attitude expressions to be inconsistent. The first consists in *A-type* inconsistency, which arises from holding a single kind of *inconsistency-transmitting* attitude, such as belief, towards truth-conditionally incompatible sentences, such as *A* and *not A*. The second consists in *B-type* inconsistency, which arises from holding different kinds of attitude that are in some sense mutually exclusive towards the *same* sentence. The supposed explanation of the inconsistency of (1a) and (1b)—that they express, respectively, disapproval and tolerance of the same act—is

B-type. A-type inconsistencies arise in descriptive language: the sentences in (4a) and (4b) are incompatible and hence it is inconsistent to express belief towards both.

(4) a. Grass is green.
 b. Grass is not green.

A-type inconsistency feels familiar, whereas B-type inconsistency may appear to be an artefact of expressivism. Expressivists cannot, Schroeder contends, create the attitude of tolerance and its B-type properties by *fiat*. Rather, they need to explain *why* there is an attitude that behaves in the same manner as what they call *tolerance*, by appealing to more fundamental concepts. To find these concepts or to explain away the complaint is the *Negation Problem*.

The expressivist may point to natural cases of B-type inconsistencies. For example, *believing* and *wondering whether* seem to be B-type inconsistent attitudes, regardless of whether expressivism about any particular class of terms is correct. Baker and Woods (2015) discuss further natural examples of B-type inconsistencies and conclude that they are no less familiar than A-type inconsistencies. May the expressivist, therefore, turn the Negation Problem on its head? We are inclined to take the inconsistency of *Grass is green* and *Grass is not green* to display the inconsistency-transmitting property of belief. The expressivist might claim that the very data constituting the Negation Problem show that *Lying is wrong* and *Lying is not wrong* express different types of attitude, and that these attitudes are B-type inconsistent. Unfortunately for the expressivist, this will not suffice. At least the following problems remain.

The problem of compositionality. The contrast between (1) and (4) brings to light the *compositional* version of the Negation Problem. For these examples appear to show that expressivism requires two types of sentential negation: an A-type negation operating on *sentences*, such as the *not* in (4b), and a B-type negation operating on *attitudes*, such as the *not* in (1b). This division however is not reflected in language—there are no *A-not* or *B-not* realized as distinct lexical items. Thus, one should give a uniform analysis of the particle *not* as a single compositional operator. It is unclear at this point how the expressivist could do this. It is also a mistake to take *not* to be *ambiguous* between two distinct operators not_A and not_B. In *Lying is not wrong*, the B-reading is mandatory and the A-reading is not available.

If an ambiguity is postulated, the unavailability of the A-type reading is still in need of explanation.

The problem of logicality. We have observed that *Lying is wrong* and *Lying is not wrong* sound inconsistent, but we have been using the term 'inconsistency' quite loosely. But *Lying is wrong* and *Lying is not wrong* are inconsistent in a precise sense: they are *logically* inconsistent, meaning that it is sufficient to understand the meaning of *not* to recognize them as inconsistent; one may be entirely ignorant about the meaning of *lying* or *wrong*. This is in contrast with *factual* inconsistencies, like *Grass is green and grass is blue*. To recognize these as inconsistent, one needs knowledge of the non-logical vocabulary featuring in them. However, the rough expressivist explanation of the relevant inconsistency rests on the recognition that *not* modifies the attitude expressed by *wrong* and that *not* does not modify attitude in every context. So it seems that some understanding of *wrong*—at least that it *is* an attitude expression—is required to recognize the inconsistency of *Lying is wrong* and *Lying is not wrong*, contravening logicality.

The explanatory problem. The expressivist has made the following assumptions: there are B-type inconsistencies; for each attitude whose expression is linguistically realized (for instance *disapproval*, whose expression is realized by *wrong*) there is another attitude (for instance *tolerance*) that may not be linguistically realized; all these attitudes stand in appropriate logical relations to one another (for instance, disapproval and tolerance towards the same act are B-type inconsistent). Taken together, these assumptions commit the expressivist to the existence of a set of attitudes standing in a reasonably complex web of logical relations. The expressivist cannot just assume that this web exists but must explain where it comes from.

An account that must contend with the explanatory problem is due to Terence Horgan and Mark Timmons (2006). They construct a language for *logically complex* attitude expressions, starting with two basic sentence-forming operators to express, respectively, descriptive is-commitment and normative ought-commitment. They then recursively define logical connectives on these operators: for instance, we have that for each operator there is another one that negates it. The meanings of these complex operators are given by their inferential role. However, Schroeder (2008a) incisively points out, commitments (or attitudes) cannot themselves be logically complex—only the language that Horgan and Timmons use to describe these commitments is logically complex. But a logically complex language might be overwrought and does not in itself guarantee that its expressions denote anything, so Horgan and Timmons have yet to explain why we should suppose the attitudes their language requires exist. To suppose without further argument that all attitudes expressible in this language exist is, as Schroeder (2008a: 51) puts it, 'sheer optimism'.

Gibbard's (1990, 2003) sophisticated version of expressivism, which we outlined in Chapter 1, may at first glance appear to fare somewhat better. Recall that Gibbard defines semantic values with respect to fact-prac worlds that settle all descriptive facts (such as whether grass is green) and settle all plans on what to do (such as whether to blame for lying). The content of *Lying is wrong* is then the set of all fact-prac worlds in which one plans to blame for lying. Gibbard can help himself to straightforward and familiar definitions of inconsistency and negation: two contents are inconsistent if their intersection is empty and the compositional semantics of negation is set-theoretic complementation. It follows immediately that *Lying is wrong* and *Lying is not wrong* are incompatible and to recognize this, one must only understand the meaning of *not*.

As appealing as this approach seems, it delivers the wrong result. On Gibbard's account, *Lying is wrong* expresses a mental state that can be represented by the set of fact-prac worlds in which, at each practical world, one plans to blame for lying. Since negation denotes complementation, *Lying is not wrong* expresses a mental state that can be represented by its complement, the set of fact-prac worlds in which, at each practical world, one does not plan to blame for lying. But someone who is neutral about whether to blame for lying could be in this mental state, whereas *Lying is not wrong* expresses a moral conviction stronger than neutrality. The problem, to stress, is not that the structure of fact-prac worlds does not afford the resources to represent the mental state expressed by *Lying is not wrong*. For we can represent this mental state by a set of fact-prac worlds in which, at each practical world, one plans not to blame for lying. The problem is that Gibbard's semantics for negation does not assign the required set of fact-prac worlds to *Lying is not wrong*.

Gibbard could appeal to the fact that practical worlds represent hyperplans and that hyperplanners are maximally decided. For this means that if at a practical world one is not planning to blame, then at that world one is planning not to blame. It then follows that the complement of the set representing the state expressed by *Lying is wrong* is, correctly, the set of fact-prac worlds in which, at each practical world, one plans not to blame for lying. To make room for neutrality about lying, Gibbard could then resort to standard patterns of possible world semantics, for instance by representing the mental state of someone who does not think that lying is wrong by a set of fact-prac worlds in which, at *some* practical world, one plans not to blame for lying. However, it is unclear whether hyperplanners who do not plan to do something are thereby planning not to do it.[1] If this is the case, Gibbard's semantics equivocates *I ought not to lie* with *It is not the case that I ought to lie* (Schroeder 2008a: 53). Gibbard (2003) holds that *I ought to lie* expresses a mental state that is represented by a set of fact-prac worlds in which, at each practical world, one plans to lie. The state expressed by *It is not the case that I ought to lie* is its complement, the set of fact-prac worlds in which, at each practical world, one does not plan to lie. If at each practical world, planning not to lie is the same as not planning to lie, then this is the same as the set in which, at each practical world, one plans not to lie. But this set represents the state expressed by *I ought not to lie.*

Thus, Gibbard faces a dilemma. If hyperplanners who do not plan to do something are not thereby planning not to do it, complement negation assigns the wrong meaning to *Lying is not wrong*. If hyperplanners who do not plan to do something are thereby planning not to do it, complement negation assigns the wrong meaning to *It is not the case that I ought to lie.* Gibbard's only hope, Schroeder concludes, is to add additional structure to the hyperplans themselves, perhaps by adopting Dreier's (2006) distinction between indifference and indecision.[2] But again, this distinction cannot simply be stipulated, it must be explained. The challenge for the expressivist was to explain what it is to not-disapprove. The challenge for Gibbard is to explain what it is to not-plan. No headway has been made.

[1] Gibbard (2003: 56) initially says that a hyperplan specifies, for each action, whether it is permitted or forbidden. This seems to leave room for a hyperplanner deeming both blaming and not blaming as being permitted without planning to do either. Yalcin (2018) stresses this point in response to Schroeder (2008a), who attributes to Gibbard the view that a hyperplanner who is not planning to do something is planning not to do it. Schroeder's reading is supported by Gibbard's elaboration that 'hyperplans leave nothing indeterminate as to what to do' (p. 57), which suggests that of the alternatives of blaming and not blaming, one must be selected by any hyperplan as the thing to do. We need not take a stand on this exegetical matter, since our arguments show that there is a negation problem either way. On Yalcin's reading, there is a problem with *wrong*, and on Schroeder's reading there is a problem with *ought*.

[2] By taking *wrong* to express *planning to blame* Gibbard has enough structure to avoid the Negation Problem for *wrong*. (This, incidentally, is the same amount of structure Schroeder obtains by taking *wrong* to express *being for blaming*, as we shall see below.) By adopting the view that a hyperplanner who does not plan to do something is thereby planning not to do it, Gibbard avoids the problem with complement negation we pointed out. Neutrality can be captured in terms of standard patterns in possible-world semantics. The most promising avenue, it therefore seems to us, would be to adjust the meaning of *ought* to add further structure, as in the case of *wrong*. Rather than taking *I ought to lie* to express a plan to lie, one could take it to express a plan to have a certain attitude towards lying, just as *wrong* is taken to express a plan *to blame* for lying.

5.2 Bilateralism and the Negation Problem

In Chapter 3 we argued that bilateralism is best seen as a form of inferential expressivism about negation: the meaning of *not* is inferentially explained in terms of rejection, the speech act expressing the attitude of disbelief. As a version of inferential expressivism about negation, bilateralism meets the original Frege–Geach challenge head on. Bilateralists agree that attitude expressions do not embed: *yes* and *no* answers to self-posed polar questions are linguistically unembeddable and the signs + and − *decorate* formulae, so cannot embed either. Bilateralists also accept Frege's conclusion that *not* cannot be an attitude expression, since it embeds. But, they continue, this does not mean that the meaning of *not* cannot be explained in terms of the attitude expression of disbelief. The general semantic framework of inferential expressivism makes clear what kind of meaning bilateralists are proposing for negation: they explain *not* by stating which attitudes one can infer from its use.

However, the bilateralist strategy to explain the meaning of negation raises a question. The Negation Problem has to do with the explanation of certain inconsistencies. Consider how the bilateralist accounts for the inconsistency of *Grass is green* and *Grass is not green*. Uttering the sentence *Grass is not green* expresses belief towards it. The meaning of this expression is given by the inferential roles of the sentence's constituents, in particular the role of *not*. Part of this meaning is that expressing belief towards *Grass is not green* commits one to expressing disbelief towards *Grass is green*, as per (+¬E.). This is in contrast with traditional expressivism about *not*, which holds that uttering *Grass is not green* would *directly* express disbelief towards grass being green. So, according to the bilateralist, uttering *Grass is not green* entails a commitment to expressing disbelief towards *Grass is green*; and uttering *Grass is green* expresses belief towards it. Thus, using the coordination principle of Strong Rejection, we can show that *Grass is green and grass is not green* entails ⊥. But, in Schroeder's terminology, this derivation of inconsistency is *B-type*. It arises from being committed to expressing two incompatible attitudes—belief and disbelief—towards the same sentence, namely *Grass is green*. Does bilateralism therefore fall prey to the negation version of the Frege–Geach Problem?

Bilateralists take it as primitive that it is inconsistent to simultaneously express belief and disbelief towards the same sentence. One may object that this already falls prey to the Negation Problem: the bilateralist owes an explanation of *why* these attitude expressions are inconsistent. But this objection is unfair. It should be uncontroversial that one needs to take *some* basic type of inconsistency as primitive. Schroeder (2008a) posits as basic (i) the truth-conditional inconsistency between a sentence and its negation and (ii) that some attitudes, such as belief and intent, are *inconsistency transmitting*. The A-type inconsistency of *believing that A and believing that not A* follows from *A and not A* being truth-conditionally inconsistent and from this inconsistency being *transmitted* to the level of belief. The bilateralist, by contrast, posits as basic (i) the B-type inconsistency of simultaneously expressing belief and disbelief and (ii) that negation is explained by its inferential relation to the expression of disbelief. This explains the purportedly A-type inconsistency of expressing belief towards *A* and towards *not A* by noting that someone who expresses belief towards *not A* is committed to expressing disbelief towards *A*—and it is inconsistent to express both belief and disbelief towards *A*.

Abstracting from there, we can say that there are *A-type expressivists* such as Schroeder who posit A-type inconsistency as basic and attempt to reduce apparent B-type inconsistencies to A-type. Schroeder (2008a) explains the apparent B-type inconsistency between *Lying is wrong* and *Lying is not wrong* as the A-type inconsistency between *being for blaming for lying* and *being for not blaming for lying*, stipulating that *being for* is an inconsistency-transmitting attitude. We discuss Schroeder's positive view in Section 5.4.

Bilateralists, by contrast, take B-type inconsistency as basic and attempt to reduce apparent A-type inconsistencies to B-type. Both A-type expressivists and bilateralists need to take some kind of inconsistency as basic and use it to explain the other one. An A-type expressivist might demand an explanation from the bilateralist of *why* expressions of belief and disbelief are inconsistent—supposedly, asking the bilateralist to provide some fundamental property of the mind that would entail this inconsistency. But the bilateralist might equally demand an explanation of why some attitudes are inconsistency transmitting—asking the A-type expressivist to name the underlying reason that explains why *belief* transmits inconsistencies, but *wonder* does not. So in terms of explanatory credentials, A-type expressivism and bilateralism are on equal footing.

To be sure, the Negation Problem we discussed in Section 5.1 is not thereby solved. Both A-type expressivists and bilateralists still need to explain compositionality and logicality. In addition, bilateralism does not provide a semantics for *wrong*—though, as we shall shortly see, inferential expressivism has the resources to do so. At this stage, we are merely defending the starting point of bilateralism—that there are two attitudes we may call *belief* and *disbelief* whose expression stands in a B-type inconsistency relation—against the charge of being unexplanatory.

We should also stress that our claim is not that the entire explanatory challenge is unfair. Expressivists such as Horgan and Timmons (2006), who take logical operators such as negation to *generate new* operators for attitude expressions, are fairly challenged to explain what these attitudes are and why they have the required logical properties. But bilateralism, as a form of inferential expressivism, does nothing of the sort. Horgan and Timmons claim that a logically complex sentence expresses a logically complex attitude; the meaning of this attitude, they continue, is defined by its inferential role. Inferential expressivism agrees with Schroeder (2008a) that there are no logically complex attitudes. It states the meanings of sentences under various attitude expressions in terms of their inferential roles, but the complexity lies with the sentences towards which those attitudes are expressed: inferential expressivism deals in *basic* attitude expressions (such as the expressions of belief and disbelief), *towards* complex sentences. Furthermore, the multilateral methodology we outlined in Chapter 4 guarantees that the basic attitudes of inferential expressivism are linguistically realized and their logical properties are linguistically observable. While Horgan and Timmons do have an explanatory problem, inferential expressivism does not.

Schroeder nevertheless considers A-type expressivism superior, since 'B-type inconsistency is not something that expressivists can take for granted, because there are no good examples of it' (2008a: 48). However, there *are* natural examples of attitudes that stand in a relation of B-type inconsistency, e.g., *belief* and *wonder*, or in other logical relations, e.g., *certainty* entails *belief*. The expression of such attitudes and their logical relations are linguistically observable. In terms of what there is to explain, such observations are on a par with Schroeder's observation that certain attitudes, such as *belief*, are inconsistency transmitting (Baker and Woods 2015).

In sum, either there is no explanatory problem for the bilateralist or there is an analogous problem for the A-type expressivist. However, there is a serious explanatory challenge for expressivists who take attitudes and logical properties that are not observable as basic, and for expressivists who take *both* A-type and B-type inconsistency as basic. Bilateralists fall in neither camp. Expressions of belief and disbelief, as well as their logical properties, are linguistically observable by examining answers to polar questions.

And A-type inconsistency is explained in terms of B-type inconsistency. If inferential expressivists can provide analogous justifications for all primitive attitude expressions in their system, then they similarly face no explanatory problem.

Nonetheless, we must still address the aspects of the Negation Problem related to compositionality and logicality and provide an inferential expressivist account of the meaning of moral vocabulary. To this end, we extend our multilateral framework by taking further linguistically observable attitudes as primitive.

5.3 Disapproval and moral vocabulary

According to traditional ethical expressivism, *wrong* has no referential semantic value but instead indicates the expression of disapproval. According to inferential ethical expressivism, the semantic value of *wrong* is given by the fact that in asserting that, say, lying is wrong one incurs a commitment to expressing disapproval of lying, and that in expressing disapproval of lying one commits to expressing belief towards lying being wrong. Formally, this is captured by the following inference rules, where ☹ indicates the expression of disapproval.

$$(+WI.)\ \frac{\text{☹}a}{+wrong\ (a)} \qquad (+WE.)\ \frac{+wrong\ (a)}{\text{☹}a}$$

Thus, as in the case of *not* and *might*, the meaning of *wrong* is given by how expressions of attitudes towards sentences containing *wrong* inferentially interact with other expressions of attitudes. The inferential expressivist semantics for *wrong* provides enough places to insert a negation, since *wrong* is treated as an embeddable predicate. The attitude expressed by uttering *Lying is not wrong* is belief towards $\neg wrong(l)$, the attitude expressed by *Not lying is wrong* is belief towards $wrong(\neg l)$.[3] This, to stress, markedly

[3] Little is said in the literature on the Negation Problem on the logical form of negated subject gerunds like *not lying*. Notably, Schroeder (2008a: 73ff.) relegates this question to a logic of blaming that he leaves open. We assume that $wrong(\neg m)$ is a placeholder for a more complex first-order formula. Linguistic evidence suggests that subject gerunds are quantificational (Portner 1995), so $wrong(l)$ stands for $\forall x\, act\text{-}of\text{-}lying(x) \supset wrong(x)$. Care is needed when negating this, however. *Not lying is wrong* should not entail that any non-lying action is wrong, since this would mean that everything except lying is wrong. One could let $wrong(\neg l)$ stand for $\forall^D x\, \neg act\text{-}of\text{-}lying(x) \supset wrong(x)$

distinguishes inferential expressivism from Horgan and Timmons's (2006) proposal: they describe the meaning of *Lying is not wrong* as a logically complex attitude; we describe *Lying is not wrong* as a logically complex sentence towards which one may express attitudes.

What does ¬*wrong*(*l*) mean? The glib answer is that the meaning of this sentence is given by the inference rules defining the meaning of ¬ and those defining the meaning of *wrong*. This answer may be too glib to be satisfactory, so let us explore this inferential meaning.

The meaning conferred on ¬*wrong*(*l*) by the inference rules for negation and *wrong* will license different inferences depending on which attitude one expresses towards this sentence. Let us consider its assertion, +¬*wrong*(*l*). This entails −*wrong*(*l*), formalizing the strong rejection of *Lying is wrong*, which is inconsistent with +*wrong*(*l*) and hence also with ☹*m*. It follows that in asserting *Lying is not wrong* the speaker is committed to proposing to modify the common ground and negative common ground so that disapproval of lying is incompatible with the shared presuppositions of the conversation participants. Note the difference with the assertion of *Lying might not be wrong*, which only entails the *weak* rejection of *Lying is wrong*. A strong rejection is a proposal to update the negative common ground, or the acceptance of such a proposal. Thus, someone uttering *Lying is not wrong* is contributing to making it so that henceforth in the conversation, it is taken for granted that the interlocutors do not disapprove of lying. By contrast, a weak rejection merely prevents a common ground update, but does not contribute to making changes to the common grounds.

It bears emphasis that we did not introduce an auxiliary attitude expression of tolerance to account for what is expressed by uttering *Lying is not wrong*: we only need the basic attitude expressions of inferential expressivism together with the expression of disapproval. If someone wishes to call *tolerance of lying* the attitude expressed by −*wrong*(*l*), they can do so—this may be a useful and illuminating shorthand. But *only* a shorthand. The inferential expressivist does not need to take the expression of tolerance as a primitive.

where the universal quantifier is restricted to a domain that stems from a presupposed set of alternatives. Alternatively, *not lying* could be interpreted as a term negation, *viz.* ∀*x act-of-un-lying*(*x*) ⊃ *wrong*(*x*). The former approach appears to be supported by linguistic evidence (Portner 1995). We leave this matter open, but note that our semantics for negation and moral vocabulary would work on either approach.

However, we do take the expression of disapproval as a primitive and, in line with our multilateral methodology, we must ensure that this attitude expression may be linguistically realized. It suffices to combine Smiley's insight about polar questions with Ayer's vivid exposition of his traditional expressivism. We can express our disapproval of some act type such as lying by *booing* it.

(5) Lying? Boo!

In addition to indicating how expression of disapproval may be linguistically realized, we should also say what its essential effect on the conversation is. We explained the essential effect of the four speech acts of weak and strong assertion and rejection in terms of their role in managing the common ground and the negative common ground, the collections of beliefs and disbeliefs mutually taken for granted by the interlocutors. Similarly, we explain the essential effect of non-cognitive attitude expressions such as *Lying? Boo!* in terms of their role in managing non-cognitive analogues of the positive and negative common grounds: the collections of non-cognitive attitudes mutually taken for granted by the interlocutors. Call the collection of acts of which the conversation participants mutually suppose to disapprove the *negative moral common ground*. The essential effect of the expression of disapproval of some act is then a proposal to add that act to the negative moral common ground, or to accept such a proposal. Analogously, the essential effect of the expression of approval of some act is a proposal to add that act to the *moral common ground*, the collection of acts of which the conversation participants mutually suppose to approve. This dimension of expressions of moral attitudes accounts for their social function, stressed by Stevenson (1937). One expresses a moral attitude to coordinate one's non-cognitive attitudes with others' and affect other people's non-cognitive attitudes. This is analogous to the case of assertions, which are used to coordinate beliefs and influence others' beliefs.

One may now think that, in analogy with the four speech acts managing the common ground and negative common ground, there should be an attitude expression whose essential effect is to prevent some attitude being added to the negative moral common ground. And wouldn't this be the maligned expression of tolerance? The absence of such an attitude expression may be a simple matter of economy. One can prevent the addition of lying to the negative moral common ground by uttering *Lying is not wrong* (or, in fact, *Perhaps lying is not wrong* to prevent such an addition without changing the common ground). The various common grounds constrain each other. As mentioned in the previous chapter, it follows from the meaning of *not* that the common ground and the negative common ground are constrained so that if some sentence is a member of the one, its negation is a member of the other. Likewise, it follows from the meaning of *not* and *wrong* that if *Lying is not wrong* is in the common ground, then it cannot be the case that *lying* is in the negative moral common ground. The original Negation Problem showed that negations of moral vocabulary burden the traditional expressivist with the unexplained, putative attitude of tolerance. The inferential expressivist can explain why there is no need for a primitive expression of tolerance: we can negate moral vocabulary.

To sum up, inferential expressivism solves the Negation Problem for expressivism about *wrong*. It deals with the problems of compositionality and logicality, since it posits the same negation operator for descriptive and moral language. And it avoids the explanatory problem, since it explains the meaning of *Lying is not wrong* not by stipulating auxiliary or complex attitudes, but by appealing to the joint inferential meaning of *not* and *wrong*.

The Negation Problem is the instance concerning embeddings under negation of the general form of the Frege–Geach Problem understood as a problem about any kind of embedding. Similarly, the original Frege–Geach Problem is the instance concerning embeddings in conditional antecedents. It is therefore natural to consider embeddings in other linguistic contexts. For instance, one may ask what is expressed by the sentence *Lying is wrong and grass is green*. Does it express two attitudes at the same time? Or one attitude that is somehow a fusion of disapproval and belief? In the latter case, expressivists would be once again committed to the existence of complex attitudes that only seem needed to make their semantics work. Schroeder (2008a: 124–127) takes embeddings under disjunction to give rise to instances of the general form of the Frege–Geach Problem that are particularly troublesome for expressivists. In uttering the sentence *Lying is wrong or grass is green*, one expresses neither disapproval of lying nor the belief that grass is green. But the logical behaviour of *Lying is wrong or grass is green* reveals that in uttering this sentence, one performs an attitude expression that does stand in some appropriate relation to expressions of disapproval and belief. The disjunctive syllogism allowing us to conclude that lying is wrong from *Lying is wrong or grass is green* and *Grass is not green* is valid. Does this mean that we need to take *Lying is wrong or grass is green* to express a disjunctive attitude, thereby falling prey to the explanatory problem?

While these may be pressing issues for the traditional expressivist, there is no problem here for the inferential expressivist. In uttering *Lying is wrong or grass is green*, one expresses a single attitude, namely belief, towards that very sentence. The logical behaviour of this attitude expression—including the fact that it licenses reasoning by disjunctive syllogism—is governed by the multilateral rules for *or*. And there is no need for a disjunction operator on attitudes, since the linguistic realizations of attitude expressions do not embed: *Lying? Boo! Or is grass green? Yes!* is gibberish. There is no embedding problem for inferential expressivism, at least not as far as embeddings under logical vocabulary are concerned. We discuss embeddings under propositional attitude verbs in the next chapter, when addressing the Many Attitudes Problem.

The inferential expressivist account of *wrong* straightforwardly generalizes to further pieces of moral vocabulary. For instance, according to traditional expressivism, *right* indicates the expression of approval. According to inferential expressivism, the expression of approval is realized not by predicating *right* of some act type, but by *hooraying* it.

(6) Giving to charity? Hurrah!

The essential effect of hooraying some act type is a proposal to update the moral common ground with that act, or the acceptance of such a proposal. The meaning of *right* can then be inferentially explained in terms of the expression of approval. In particular, we can take it to be given by the following rules, where ☺ indicates the expression of approval.

$$(+RI.)\ \frac{\text{☺}a}{+right\ (a)} \qquad (+RE.)\ \frac{+right\ (a)}{\text{☺}a}$$

Since the essential effect of an expression of approval is a proposal to update the moral common ground, the meaning of *Lying is not right* can be explained analogously to how we explained the meaning of *Lying is not wrong*.

We now turn to the coordination principles involving expressions of approval and disapproval. We do not need to lay down rules governing the interaction of expressions of cognitive attitudes (such as belief and disbelief) and non-cognitive attitudes (such as approval and disapproval). For we can account for the validity of inferences involving both cognitive and non-cognitive attitude expressions by using the rules for embeddable items such as *wrong*, which allow us to move between these attitude expressions. Consider, for instance, the following inference involving both the expressions of belief and the expression of disapproval.

(7) a. If lying is wrong, liars go to hell.
b. Lying? Boo!
———————————
c. Do liars go to hell? Yes!

We can account for the validity of this inference by using the rules giving the meaning of *wrong* and of the conditional, without needing to appeal to principles coordinating expressions of belief and expressions of disapproval. In particular, from *Lying? Boo!* we can infer *Lying is wrong*, which together with the conditional premiss of the inference yields *Do liars go to hell? Yes!* by *modus ponens*.

However, we do need to explicitly coordinate expressions of approval and disapproval to account for their linguistic behaviour. In particular, it sounds morally incoherent to express approval of some act immediately after expressing disapproval of the same act.

(8) # Lying? Boo! Lying? Hurrah!

To account for this fact, we lay down the following rule, which states that it is inconsistent to simultaneously express approval and disapproval of the same act, just as it is inconsistent to assert and reject the same sentence.

$$\text{(AD-Inconsistency)}\ \frac{☺A \qquad ☹A}{\bot}$$

Thus, the moral common ground and the negative moral common ground are mutually constraining: the same act type cannot be a member of both. We are not aware of data motivating a *reductio*-like principle such as the one governing the interaction of strong assertion and strong rejection. Thus, to govern the interaction of expressions of approval and disapproval, we only lay down the AD-Inconsistency rule. This rule also features in the explanation of why *Lying is wrong and lying is right* sounds inconsistent. Using the elimination rules for *right* and *wrong*, the AD-Inconsistency rule immediately allows one to infer an absurdity.

5.4 Schroeder and the Negation Problem

Schroeder (2008a) defends an A-type account of *wrong*. Instead of analysing the attitude expressed by *wrong* as *disapproval*, he analyses it as *being for blaming for*, stipulating *being for* as his only primitive attitude. This allows him to insert a negation in the right place: *Lying is not wrong* is analysed as *being for not blaming for lying*. Then Schroeder claims that *being for* is an inconsistency-transmitting attitude like belief, thereby accounting for all data involved in the Negation Problem. This is remarkably elegant, but Schroeder himself points out some issues.

Notably, he acknowledges the *New Negation Problem*, a revenge version of the compositionality problem. It goes as follows. To give a uniform semantics—covering both descriptive and moral vocabulary—that is grounded in the basic attitude of *being for*, Schroeder analyses descriptive claims as expressions of *being for proceeding as if* (*being for pai* for short). Thus, for instance, an utterance of *Grass is green* expresses the attitude *being for pai grass is green*. But then an utterance of *Lying is wrong* expresses *being for blaming for lying* and an utterance of *Grass is green* expresses *being for pai grass is green*. Thus, if an utterance of *Lying is not wrong* expresses *being for not blaming for lying* and the compositional semantics of *not* is uniform—as it should be—an utterance of *Grass is not green* expresses *being for not pai grass is green*. But this is a mistake: the attitude expressed by uttering *Grass is not green* should be *being for pai grass is not green*. In essence, the Negation Problem is that, for moral language, the expressivist analysis has one space too few for negation. The New Negation Problem is that, for descriptive language, Schroeder's analysis has one space *too many* for negation.

To solve the New Negation Problem, Schroeder introduces the notion of a *biforcated* attitude. A biforcation is a pair of two *being for* attitudes such that the *major attitude* entails the *minor* one.[4] We write biforcations as tuples and treat the left-hand side of a tuple to be the major attitude. What exactly the minor attitude is is determined by the lexical semantics of the relevant

[4] The term *biforcated* is Schroeder's, a pun on *bifurcated* and *being for*.

predicates; for descriptive vocabulary, one obtains the minor attitude from the major one by inserting two negations so that, for instance, the biforcated attitude expressed by uttering *Grass is green* is ⟨*being for pai grass is green, being for not pai grass is not green*⟩, where *being for pai grass is green* is the major attitude (Schroeder 2008a: 98).

Using biforcations, Schroeder defines the meaning of negation as follows: the negation of a biforcation inserts in both the minor and major attitude a *not* between *being for* and the embedded action term and then switches major and minor attitude. The following is Schroeder's (2008a: 106) compositional semantics for *not*.

$$[\![not]\!](\langle \textit{being for } a,\ \textit{being for } b\rangle) = \langle \textit{being for } \neg b,\ \textit{being for } \neg a\rangle$$

Thus, if an utterance of a descriptive sentence *A* expresses the biforcation ⟨*being for pai A, being for ¬ pai ¬ A*⟩, then an utterance of *not A* expresses ⟨*being for ¬¬ pai ¬A, being for ¬ pai A*⟩. Deleting double negations, one obtains the desired major attitude *being for pai ¬A*.

However, Schroeder now needs to insist that for *moral* sentences, the minor and major attitudes in a biforcation are identical, lest he run into the original Negation Problem again. For descriptive language, by contrast, the minor attitude is strictly weaker than the major one. This is not an ad hoc move, but a sound strategy: by encoding a difference between the attitudes expressed by descriptive and moral terms at the level of lexical semantics, Schroeder can maintain uniform compositional semantics for negation while solving both the Negation Problem and the New Negation Problem. As a matter of fact, we agree that lexical semantics is the right place to locate this difference.

Unfortunately, this move leads to the *New New Negation problem* (Schroeder 2008a: 113). Schroeder justifies his introduction of minor attitudes by appealing to a notion he calls *disacceptance*. Informally, to disaccept a sentence is to not agree with it without agreeing with its negation. Formally, Schroeder characterizes disacceptance by saying that to disaccept *A* is to accept the minor attitude in the biforcation expressed by *not A*. But since moral vocabulary sentences express biforcations with identical major and minor attitudes, the distinction between disacceptance and acceptance of a negative collapses for moral sentences. This entails that one cannot withhold judgement about an act *a* by disaccepting both *a is wrong* and *a is not wrong*. So the only way to withhold judgement is to *not express* an attitude. Schroeder argues that this is not an outright bad result, concluding that it might be 'a bullet to bite' (2008a: 115).

However, this bullet becomes harder to bite when also considering epistemic *might*. Schroeder (2008a: 181) suggests the following semantics for epistemic modals: *might* deletes the major attitude and replaces it with the minor attitude.

$$[\![might]\!](\langle being\ for\ a,\ being\ for\ b\rangle) = \langle being\ for\ b,\ being\ for\ b\rangle$$

But given Schroeder's solution to the New Negation Problem, this means that *Lying might be wrong* expresses the same attitude as *Lying is wrong*, since *Lying is wrong* expresses a biforcation with identical minor and major attitudes. Schroeder (2008a: 182) concludes that either his semantics for *might* or his identification of minor and major attitudes for moral vocabulary must be revised. But no suitable such revision is in sight. The semantics for *might* has compelling features, as Schroeder himself notes. To preserve these features while making sense of moral vocabulary under epistemic modality, Schroeder would likely need to revise the foundations of the biforcated attitude framework. However, it is unclear whether this can be done so as to solve both the original Negation Problem and the New Negation Problem.

Now, according to inferential expressivism, the essential effect of expressing the belief that grass is green is a proposal to add *Grass is green* to the common ground. Since the common ground records the tacit assumptions in a conversation, this can be roughly paraphrased as *being for pai grass is green*. Thus, *being for pai* is analogous, in our framework, to belief, the attitude expressed by assertion. And, as is highlighted by Schroeder's proposed semantics for *might*, his minor attitudes are analogous to refraining from disbelieving, the attitude expressed by weak assertion. His notion of disacceptance similarly corresponds to our refraining from believing, the attitude expressed by weak rejection. Likewise, the essential effect of expressing the belief that lying is wrong is a proposal to add *Lying is wrong* to the common ground, which also modifies the negative moral common ground so as to include disapproval of lying. This can be roughly paraphrased as *being for blaming for lying*.

Thus, inferential expressivism, as we have developed it so far, can be examined in Schroeder's terms. Doing so shows no trace of the New and New New Negation Problems. The inferential expressivist account differs from Schroeder's in that it does not make fundamental distinctions among its basic attitude expressions in the way they operate on moral vis-à-vis descriptive vocabulary: we do not collapse strong and weak assertion anywhere, whereas Schroeder collapses major and minor attitudes towards moral content. But how does inferential expressivism distinguish between moral and descriptive vocabulary if it does not, like Schroeder, locate their difference in the minor attitudes?

Inferential expressivism locates this difference in whether the inference rules representing the semantic value of an item *change attitude expression*. The semantics of descriptive vocabulary is given by rules which do not change attitude expression. For instance, a rule allowing us to infer the assertion of *a is unmarried* from the assertion of *a is a bachelor* does not change attitude expression because it only allows us to pass from one expression of belief to another. By contrast, the semantics of moral vocabulary is given by rules which *do* change attitude expression. For instance, the rules for *wrong* allow one to switch between the expression of belief and the expression of disapproval. This is why when one negates *wrong*, it may appear as if one is negating an attitude (namely, disapproval, the attitude *wrong* switches with), whereas this is not so when one negates a descriptive term. Like Schroeder, we locate the difference between moral and descriptive vocabulary at the level of lexical semantics: in the way the inference rules constituting lexical meaning characterize the interaction of a term with attitudes. This allows us to have uniform semantics for operators such as negation, without collapsing any of our notions.

5.5 Wishful thinking and evidence

We have arrived at an inferential expressivist account of the meaning of *wrong* that avoids the Frege–Geach Problem, including its negation version and iterations thereof. According to this account, an utterance of *Lying is wrong* expresses a cognitive attitude, namely the belief that lying is wrong. Nevertheless, the account is expressivist since it takes the meaning of *wrong* to be explained in terms of the non-cognitive attitude of disapproval. However, the Wishful Thinking Problem we encountered in Chapter 1 purports to undermine expressivism even granting that it can meet the Frege–Geach challenge. Thus, we must now examine how the inferential expressivist account fares with respect to this problem.

Recall Edgar, who accepts the premiss that if lying is wrong, the souls of liars will be punished in the after life. He comes to accept the further premiss that lying is wrong by reading a philosophy book. On the basis of these two premisses, he comes to believe that the souls of liars will be punished in the afterlife by performing the following inference.

(9) a. If lying is wrong, the souls of liars will be punished in the afterlife.
    b. Lying is wrong.
    ____
    c. The souls of liars will be punished in the afterlife.

The inferential expressivist account of *wrong* immediately validates this inference. However, according to Dorr (2002), even if they succeed in validating the inference, expressivists make Edgar's reasoning irrational, since they take the second premiss of the inference to express a non-cognitive attitude, and it is wishful thinking to form beliefs about the world on the basis of a change in one's non-cognitive attitudes. Now, the inferential expressivist account takes the second premiss to express a cognitive attitude. Nonetheless, it also takes the second premiss to be inferentially equivalent to the expression of disapproval. Does the account thereby make Edgar's reasoning a case of wishful thinking?

The inferential expressivist can simply endorse the cognitivist's reconstruction of Edgar's reasoning. Edgar acquires new evidence relevant to the question of the fate of liars in the afterlife. He reads and considers relevant arguments related to the wrongness of lying and forms the belief that lying is wrong. The arguments in the philosophy book are evidence bearing on the fate of liars in the afterlife, given Edgar's acceptance of the conditional linking the wrongness of lying to the fate of liars. To be sure, it follows from the inferential expressivist meaning of *wrong* that when he expresses his belief that lying is wrong, Edgar undertakes a commitment to disapproving of lying. But this commitment is not what explains the validity of Edgar's reasoning. On our approach, Edgar has a prior belief that if lying is wrong, the souls of liar will be punished. He then acquires the further belief that lying is wrong. On the basis of these two beliefs, he comes to believe that the souls of liars will be punished in the afterlife via the reasoning above. In this respect, our reconstruction of Edgar's reasoning is similar to the one offered by hybrid expressivist accounts.

As noted in Chapter 1, however, the hybrid solution to the Wishful Thinking Problem makes the non-cognitive aspects of moral statements utterly irrelevant to the use of moral vocabulary in inferences. Our framework has the resources to do better on this score, since the expression of disapproval of lying licenses the expression of the belief that lying is wrong. However, this also means that our solution to the Wishful Thinking Problem cannot simply be the one offered by hybrid expressivist accounts. To arrive at our proposed solution, it will be helpful to compare Edgar's case with Herman's. Like Edgar, Herman accepts the premiss that, if lying is wrong, the souls of liars will be punished in the afterlife. Herman, who never had any moral qualms about lying, wakes up one morning and spontaneously declares that lying is wrong on the basis of a groundless change of non-cognitive attitude towards lying. He then reasons like Edgar: he performs an inference from *If lying is wrong, the souls of liars will be punished in the afterlife* and *Lying is wrong* to *The souls of liars will be punished in the afterlife.*

Clearly, if the inference is valid when performed by Edgar, it remains valid when performed by Herman. But it *would* be wishful thinking on Herman's part to come to believe that the souls of liars will be punished in the afterlife on the basis of his belief that if lying is wrong, the souls of liars will be punished in the afterlife and his disapproval of lying. The difference between Edgar's reasoning and Herman's reasoning is clear. Edgar has evidence for the premisses of the inference. Since the inference preserves evidence, he can use it to form a new belief. By contrast, Herman does not have evidence for one of the premisses, since it was based on a groundless change of non-cognitive attitude. So it is a mistake for Herman to use the inference to form the belief that the souls of liars will be punished in the afterlife. Herman has acquired no new evidence relevant to this belief.

There is nothing special about moral beliefs here. In general, it is mistaken to form new beliefs on the basis of one's extant conditional beliefs and groundless changes of mind. Suppose, for instance, that one believes that if it does not rain at the weekend, the picnic will not be cancelled. Without considering any evidence, one may spontaneously come to believe that it will not rain at the weekend and thus conclude that the picnic will not be cancelled. But in so doing, one would be just as guilty of wishful thinking as Herman. To insist, without further argument, that expressions of moral belief cannot occur in evidence-preserving inferences but expressions of non-moral belief can is to beg the question against expressivism.

What about direct expressions of attitude such as *Lying? Boo!*? These have the same status as expressions of belief towards *Lying is wrong* in that they can be made on the basis of evidence or of groundless changes of mind.

One may have reasons for one's non-cognitive attitudes and use them to justify expressions of non-cognitive attitudes. Moreover, these reasons need not necessarily be beliefs. It may be that some of our moral attitudes bottom out at self-justifying non-cognitive attitude expressions or that there is a distinctive kind of moral evidence. Suppose, for instance, that expressing disapproval of treating people as mere means is self-justifying and one has reasons to believe that lying is treating people as mere means. Or suppose, alternatively, that there is a distinctive kind of moral evidence for *Lying is wrong*. Then it is not wishful thinking to conclude that liars will be punished in the afterlife, provided one believes the conditional linking the wrongness of lying to the fate of liars in the afterlife.

Edgar and Herman perform the same inference. The inference preserves evidence. Thus, having obtained evidence for the premiss that lying is wrong, Edgar rationally comes to believe its conclusion. Herman does not rationally come to believe the conclusion, since his reason for coming to accept the premiss is a mere change in his non-cognitive attitudes. To come to accept a moral premiss on the basis of a groundless change in one's non-cognitive attitudes is bad reasoning—indeed, it is wishful thinking. But it is similarly bad reasoning to come to accept a non-moral premiss on the basis of a groundless change in one's cognitive attitudes. Thus, there is no relevant difference between the moral and the non-moral case here. Indeed, if Herman had reasons for his disapproval of lying or his disapproval had been based on self-justifying expressions of attitudes, he would not be guilty of wishful thinking. The inferential expressivist can therefore account for the rationality of Edgar's reasoning while preserving a role for non-cognitive expressions of attitudes in moral reasoning.

5.6 Moral Moorean sentences

According to inferential expressivism about moral vocabulary, there is an inferential connection between expressions of belief towards *Lying is wrong* and expressions of disapproval towards lying, in virtue of the meaning of *wrong*. Thus, like its traditional counterpart, inferential expressivism holds there to be a strong connection between the meaning of *wrong* and the expression of disapproval. This connection, however, has been challenged by Jack Woods (2014) on the basis of *moral Moorean sentences* such as the following.

(10) Lying is wrong, but I don't disapprove of it.

Woods argues that utterances of this form sound coherent or, at any rate, not nearly as bad as traditional Moorean sentences such as the following.

(11) It is raining, but I don't believe that it is.

This, Wood continues, is a problem for expressivism. For according to traditional expressivists 'moral assertions express non-cognitive attitudes like disapproval in exactly the same way that non-moral assertions express cognitive attitudes like belief' (Woods 2014: 4). Thus, according to traditional expressivists, in uttering both (10) and (11), the speaker expresses an attitude and then reports that they do not have that attitude. But then, Woods notes, (10) should be as incoherent as (11), which it is not.

We do not agree with Woods that utterances of (10) are, in general, entirely felicitous.[5] Woods's point is that the reaction to a traditional Moorean sentence is typically befuddlement about what is even said, whereas we are able to assign a coherent meaning to (10). But felicitous utterances of (10) seem to be confined to special circumstances (see Toppinen 2014 for a related point). For instance, someone who is unable to feel the proper non-cognitive moral attitudes but nevertheless aims to have the proper moral beliefs may utter (10), and we understand what they are conveying. But, similarly, someone who is hallucinating but nevertheless aims to have only true beliefs may utter (11), and we understand what they are conveying. Either sentence is interpretable in the right, somewhat fringe, circumstances.

Nevertheless, we do agree with Woods that there is a difference between (10) and (11) in that the former does not sound as infelicitous as the latter. Our inferential expressivist semantics for *wrong* can account for this difference, since it does not render (10) analogous to (11). An utterance of a Moorean sentence expresses an attitude, namely belief, followed by the claim not to have this attitude. But *lying is wrong* in (10) does not express the same attitude that the speaker then claims not to have. In uttering *Lying is wrong*, one expresses belief, albeit towards a sentence whose meaning is inferentially

[5] Franzén and Soria-Ruiz (Forthcoming) provide experimental evidence that moral Moorean sentences are largely infelicitous, although not as infelicitous as their non-moral counterparts, in line with the discussion to follow.

explained in terms of the attitude of disapproval. To formulate a proper moral analogue to (11), we need to directly express disapproval.

(12) Boo lying! But I do not disapprove of lying.

This utterance is analogous to a traditional Moorean sentence. But it also sounds exactly as incoherent in typical circumstances, as Woods himself (2014: 5) notes. The reason why it sounds incoherent is that someone uttering (12) is presumed to want to be recognized as somebody who disapproves of lying, but this presumption is violated by the second sentence. This is analogous to the explanation of the traditional Moorean sentence we gave in Chapter 4, and likewise leads to the characteristic interpretation of Moorean sentences: it is unclear what the speaker is trying to achieve or convey with their utterance.

Like a traditional Moorean sentence, the moral Moorean sequence (12) is incoherent because the speaker is expressing an attitude, namely disapproval, which they then claim not to have. This is in contrast with Woods's moral Moorean sentence (10), which does not feature an expression of disapproval and so is not analogous to the traditional Moorean sentence. But why does Woods's moral Moorean sentence appear to sound somewhat better than a traditional Moorean sentence? To answer this question, let us consider a non-moral analogue of Woods's sentence. We need an expression other than *wrong* whose meaning depends on an attitude other than belief. The epistemic modal *might* is such an expression. In the previous chapter, we inferentially explained the meaning of *might* in terms of the attitude reported by *I refrain from disbelieving*. And in this chapter we inferentially explained the meaning of *wrong* in terms of the attitude reported by *I disapprove*. Thus, on our account, *might* stands to *I refrain from disbelieving* as *wrong* stands to *I disapprove*. Thus, the following is analogous to Woods's sentence.

(13) This might be a sparrow, but I don't refrain from disbelieving it is.

This sentence is not especially idiomatic, but, for current purposes, we can eliminate the double negative *don't refrain* and replace *disbelieve* with the more natural *believe that not* to obtain the following more natural sentence.

(14) This might be a sparrow, but I believe it is not.

This sentence is not incoherent in exactly the same way as the traditional Moorean sentence *This is a sparrow, but I don't believe that it is.* The only way to hear the traditional Moorean sentence as coherent is if one assumes that the speaker is being coerced, is speaking tongue-in-cheek, or is self-aware of an impairment of their faculties, such as in cases of hallucination. By contrast, sentences such as (14) can be coherently and sincerely uttered in normal circumstances when the meaning of *might* is understood to be inferentially related to an attitude expression managing a common ground other than the one among the speaker and their audience. Suppose, for instance, that the speaker is invigilating an exam on the identification of birds and is showing a picture of a wren. They turn to their colleague and whisper *This might be a sparrow, but I believe it is not.*[6] In this case, the meaning of *might* is inferentially related to an attitude expression managing the common ground of the classroom. In Chapter 4 we outlined how to account for such cases by relativizing *might* to different common grounds.

We submit that Woods's *Lying is wrong, but I don't disapprove of it* is similar to (14) in this respect. One can hear this sequence as felicitous because one can read *wrong* as inferentially related to an attitude expression managing a common ground other than the one among the speaker and their audience. In analogy with (14), we may imagine a prosecutor who just won a trial against a perjurer, but who was just doing their job, without having any personal moral conviction that the defendant ought to be punished. When asked privately by a colleague about why they pushed for a conviction, they answer *Lying is wrong, but I don't disapprove of it*, where *wrong* is understood with reference to the negative moral common ground of the courtroom, where everybody is supposed to assent to the letter of the law.

Woods might insist that this is equally problematic for the expressivist, since we are admitting a reading of *wrong* on which its meaning is not explained in terms of the expression of the speaker's attitude. But this would not be the correct account of what is going on. The meaning of *wrong* is still inferentially explained in terms of the expression of disapproval. Thus, its meaning remains constant across all uses. What *may* change across uses is the negative moral common ground with reference to which *wrong* is

[6] To our knowledge, such examples were first brought to wider attention by Angelika Kratzer, notably in her David Lewis lecture (2017). In Kratzer's original example, the teacher uses the sentence *This might be a sparrow, but I know it is not.*

being understood. In our example, for instance, the prosecutor commits to accepting lying as part of the courtroom's negative moral common ground, so that disapproval of lying is mutually taken for granted by the people party to this common ground. The prosecutor goes on to propose to make it part of the common ground between them and their colleague that they do not disapprove of lying. Similar considerations apply to the cases involving epistemic vocabulary and epistemic attitudes with respect to different common grounds.

Woods (2014: 7–8) considers the objection that *Lying is wrong, but I don't disapprove of it* admits an inverted-comma reading and that it is the availability of this reading that renders the sentence more felicitous than a traditional Moorean sentence. He replies that inverted-comma readings are typically indicated by clues such as stress, which are not needed to render his moral Moorean sentence more felicitous than a traditional Moorean sentence. Moreover, Woods continues, a traditional Moorean sentence also admits inverted-comma readings and is nevertheless infelicitous. These observations do not undermine our explanation of the difference between moral Moorean sentences and traditional Moorean sentences. The readings which, on our explanation, give rise to the difference are not inverted-comma readings, but result from the semantics of *might* and *wrong* being sensitive to a contextual parameter. Thus, while Woods is correct that there is a difference between moral Moorean sentences and traditional Moorean sentences, this difference can be explained from an inferential expressivist point of view. Vocabulary such as *wrong* and *might* can be understood with reference to attitude expressions managing a common ground other than the one among the speaker and their audience, due to the availability of a contextual parameter for the targeted common ground. Such a contextual parameter is not present in direct attitude expressions.

5.7 Moral motivation

Inferential expressivism about moral vocabulary retains certain key advantages of its traditional counterpart. It can account for the phenomenon of moral disagreement, since if someone utters *Lying is wrong* and someone else utters *Lying is not wrong*, they are immediately committed to expressing incompatible attitudes. Moreover, inferential expressivism about moral vocabulary offers non-referentialist explanations of meaning, since, for instance, it takes the meaning of *wrong* to be completely explained in terms of the inferential connections between expressions of beliefs towards sentences containing *wrong* and expressions of disapproval. In keeping with the understanding of the expressivist and inferentialist projects that we have been defending, this does not mean that inferential expressivism is an anti-realist approach to the status of moral properties. Rather, inferential expressivism is neutral with regards to the question of whether there are moral properties beyond what is entailed by a suitable minimalism about *property*. There remains nonetheless the question of whether we can make sense of a distinction between realist and quasi-realist forms of ethical inferential expressivism, and we shall address this question in Chapter 7. Does inferential expressivism also enjoy another key advantage of traditional ethical expressivism, namely its capacity to explain the motivating power of moral judgements? We are now going to argue that it does.

Inferential expressivism explains the meaning of *wrong* in terms of its inferential connections to the expression of disapproval. As noted in Chapter 1, expressivists are concerned with a non-achievement reading of attitude expression, according to which one can express an attitude without having that attitude. In the case at hand, one can express disapproval of lying without disapproving of lying. But this does not mean that attitude expression and attitude possession are not suitably related. Indeed, as we saw in Chapter 1, sophisticated expressivists are wont to endorse Wright's (1992) minimalism about belief. According to minimalism about belief, someone who sincerely asserts a sentence with a certain content has a belief whose content may be captured by that sentence. However, there seems to be nothing special here about assertion and belief. Thus, it makes sense for expressivists to endorse a more general form of minimalism about the attitudes.

According to minimalism about the attitudes, someone who sincerely expresses an attitude towards some sentence has the same attitude towards the content expressed by that sentence, and someone who sincerely expresses an attitude towards an act has the same attitude towards that act. Minimalism about the attitudes provides the required general connection between attitude expression and attitude possession. It entails that someone who sincerely expresses disapproval of lying, for instance by sincerely uttering *Lying? Boo!*, disapproves of lying. On the assumption, familiar from the standard expressivist story about moral motivation, that disapproval is intrinsically motivating, it follows that someone who sincerely expresses disapproval of lying is motivated not to lie.

However, it does not yet follow that someone who sincerely asserts that lying is wrong, for instance by sincerely uttering *Lying is wrong*, is motivated not to lie. For, according to inferential expressivism, sincere moral assertions express beliefs, and only beliefs. Thus, the standard expressivist story about moral motivation does not carry over to the case of moral assertions. What is more, the problem of moral motivation primarily concerns the motivating power of moral *judgements*, so whatever story the inferential expressivist is going to give about moral motivation, it must extend to the case in which someone merely has the belief that lying is wrong, without expressing this belief.

Not all is lost, however. For although inferential expressivism holds that moral assertions only express beliefs, it also holds that they immediately commit the speaker to expressing conative attitudes of approval or disapproval. For instance, the assertion of *Lying is wrong* immediately commits the speaker to expressing disapproval of lying. This fact, we submit, suffices to account for the motivating power of moral judgements.

What is needed is a principle linking, under certain conditions, entailment between attitude expressions and the agent's attitudes. In the debate on the normativity of logic, principles linking facts about what follows from what and norms governing an agent's doxastic attitudes are known as *bridge principles* (MacFarlane 2004). Bridge principles have been formulated in response to Harman's (1986) challenge to the idea that logic is normative. Harman argues that there is a gap between logic and norms of reasoning. Bridge principles are intended to bridge this gap. The problem is that the naïve bridge principle that one ought to hold all beliefs that logically follow from one's beliefs will not do. Building on Harman, Steinberger (2019) identifies four problems with the naïve bridge principle, two of which are relevant to us here. First, one's beliefs have infinitely many logical consequences, most of which are irrelevant to one's concerns. It is therefore cognitively implausible as well as inefficient that one ought to believe all logical consequences of one's beliefs. Call this the *Problem of Clutter Avoidance*. Second, among the logical consequences of one's beliefs are some that, albeit relevant to one's concerns, require infeasible mental effort to compute—for instance because the proof of these consequences would require more lines than there are particles in the universe. It is therefore also unreasonable to demand that one ought to believe such consequences. Call this the *Problem of Excessive Demands*.

These problems carry over, *mutatis mutandis*, to our present concerns. We are looking for a principle that bridges the gap between entailment among attitude expressions and the agent's attitudes, so that from the fact that the assertion of *Lying is wrong* entails disapproval of lying it follows that someone who judges that lying is wrong disapproves of lying. The following naïve bridge principle delivers this result: if expressing attitudes $\mathcal{A}_1, \ldots, \mathcal{A}_n$ entails expressing an attitude $\mathcal{C}$ and one has $\mathcal{A}_1, \ldots, \mathcal{A}_n$, then one has $\mathcal{C}$. However, this naïve bridge principle suffers from the same problems that we described for its counterpart linking facts about what follows from what and norms of reasoning. Indeed, even more starkly so, on the face of it, since it links facts about entailments with facts about the attitudes one actually has, rather than the attitudes one ought to have. Nonetheless, examining the Problems of Clutter Avoidance and Excessive Demands as applied to the naïve bridge principle will serve to work our way to an appropriate bridge principle.

Let us start from the Problem of Excessive Demands. Among the entailments between attitude expressions, there are clearly many that are unfeasible to compute, so the problem affects the naïve bridge principle between entailment among attitude expressions and the agent's attitudes. Now, we have been understanding inference between attitude expressions in terms of commitment. But recall that in our explanation of the infelicity of Yalcinean sentences in Chapter 4 we appealed to the fact that some commitments are *immediate*. The reason why *It is raining and it might not be raining* is infelicitous is that the expression of belief towards *It might not be raining* immediately commits one to expressing incompatible attitudes, namely belief and refraining from believing towards it being raining. By contrast, although the expression of belief towards the negation of Peirce's Law also commits one to expressing incompatible attitudes, this commitment is not immediate—to see that it holds requires a sophisticated logical argument.[7] Immediate commitments are a special case of the commitments codified by the entailments that the speaker is implicitly or explicitly aware of. Thus, to address the Problem of Excessive Demands, we require that the entailments between attitude expressions at stake in our bridge principle be ones that the speaker is implicitly or explicitly aware of.

Next, let us move to the Problem of Clutter Avoidance. Just as it is cognitively implausible and inefficient that one ought to have all beliefs that

[7] Immediacy is a vague concept and we cannot say where precisely the boundary between immediate and non-immediate commitments lies. But as with most vague concepts, we can identify definite cases of immediate and non-immediate commitments. Direct application of one or two meaning-conferring rules give rise to commitments that are definitively immediate; complex logical arguments such as the derivation of Peirce's Law give rise to commitments that are definitely not immediate.

follow from one's beliefs, so is it cognitively implausible and inefficient that one has all attitudes that follow from one's attitudes via the logic of attitude expressions. This is the case even if one restricts attention to immediate entailments among attitude expressions. The belief that it is raining and it is raining is presumably an immediate consequence of the belief that it is raining, but it is implausible and inefficient that one has the former belief whenever one has the latter.

To address this problem, we must focus on the attitudes one is considering whether to hold. Suppose that one is considering whether to believe that there are eggs in the fridge. In that case, one should attend to the fact that the expression of belief that one bought eggs yesterday and the expression of belief that if one bought eggs yesterday, then one has eggs in the fridge jointly entail the expression of belief that there are eggs in the fridge. Nonetheless, in the same circumstances, one should not attend to what follows from the expression of belief that Fermat's Last Theorem is true. However, as Steinberger (2019: 321) points out, it is too restrictive to say that the speaker must consider whether to hold a particular attitude for an immediate entailment to be relevant to which attitudes one has (or ought to have, in Steinberger's case). For one might fail to consider consequences of one's attitudes that one has good reasons to consider. For example, one might not consider an implausible consequence of one's position out of intellectual dishonesty. Thus, as Steinberger suggests, we must restrict our bridge principle to commitments to attitudes that the speaker considers or has reasons to consider.

We therefore arrive at the following provisional bridge principle: if one is implicitly or explicitly aware that expressing attitudes $\mathcal{A}_1, \ldots, \mathcal{A}_n$ commits one to expressing attitude $\mathcal{C}$, one considers or has reasons to consider whether to hold $\mathcal{C}$, and one has $\mathcal{A}_1, \ldots, \mathcal{A}_n$, then one has $\mathcal{C}$. However, this principle is clearly still too strong. Plausibly, expressing belief that if it is raining, one must take an umbrella and expressing belief that it is raining immediately commits one to expressing belief that one must take an umbrella. However, it is not *impossible* for someone who is considering whether to take an umbrella to believe that if it is raining one must take an umbrella and to believe that it is raining while failing to believe that they must take an umbrella. Nonetheless, it would be *incoherent* for them to fail to form the belief that they must take an umbrella. We therefore obtain the following bridge principle linking entailment between attitude expressions and an agent's attitudes.

Commitment-to-Incoherence. If one is implicitly or explicitly aware that expressing attitudes $\mathcal{A}_1, \ldots, \mathcal{A}_n$ commits one to expressing attitude $\mathcal{C}$, and one considers or has subjective reasons to consider whether to hold $\mathcal{C}$, then it is incoherent for one to hold $\mathcal{A}_1, \ldots, \mathcal{A}_n$ but not $\mathcal{C}$.

This bridge principle entails that if someone believes that lying is wrong and considers or has reasons to consider whether to disapprove of lying—for instance because they are in a position in which it would be useful for them to lie—then it is incoherent for them not to disapprove of lying. However, it is not impossible for them not to disapprove of lying. Moreover, when combined with minimalism about the attitudes, Commitment-To-Incoherence entails that if someone sincerely asserts that lying is wrong and considers or has reasons to consider whether to disapprove of lying, then they are either incoherent or disapprove of lying.

As we saw in Chapter 1, motivational internalists hold that if someone sincerely judges that lying is wrong, they are, at least to some degree, motivated not to lie. Motivational internalism makes sense of the intimate connection that there appears to be between moral judgement and motivation. However, it also makes the existence of amoralists a conceptual impossibility. Our account can make sense of the connection between moral judgement and motivation without ruling out the existence of amoralists. An amoralist is someone who, say, sincerely judges that lying is wrong without thereby being motivated not to lie. So consider Joyce, who indeed sincerely judges that lying is wrong but does not disapprove of lying. According to Commitment-to-Incoherence, when Joyce considers or has subjective reasons to consider whether to disapprove of lying, her attitudes are incoherent. Thus, Joyce's existence is not ruled out as a conceptual impossibility. Nonetheless, Joyce's amoralism comes at the price of an internal incoherence in her attitudes.

It is worth emphasizing that Commitment-to-Incoherence allows us to make room for the existence of amoralists while accounting for the motivating power of moral judgements only by exploiting the distinguishing features of the inferential expressivist account of moral vocabulary. We can make room for the existence of amoralists since the expression of belief that lying is wrong is not identified with or reduced to disapproval of lying. At the same time, we can account for the motivating power of moral judgements since we are not severing the connection between expression of belief that lying is wrong and disapproval of lying. Indeed, there is an inferential connection between the two in virtue of the way the meaning of *wrong* is inferentially explained.

The following is an oft-told story about the meta-ethical landscape. Moral cognitivism takes moral sentences to express beliefs but, unlike traditional expressivism, has difficulty in explaining the motivating power of moral judgements. Traditional ethical expressivism takes moral sentences to express non-cognitive attitudes, but, unlike cognitivism, falls prey to the Frege–Geach Problem. Hybrid approaches attempt to have 'the best of both worlds' by taking moral sentences to express both beliefs and non-cognitive attitudes. However, such approaches, in the end, fall prey to the standard problems for cognitivism or expressivism, and fail to properly explain the connection between the two alleged components of moral judgements. Inferential expressivism too involves both beliefs and non-cognitive attitudes, but without taking moral sentences to express both types of attitudes at once. Inferential expressivism is a form of cognitivism in that it takes typical moral sentences to express beliefs, and only beliefs. It is nonetheless a form of expressivism in that it takes the meaning of moral sentences to be inferentially explained in terms of non-cognitive attitudes. Like traditional cognitivism, inferential expressivism can account for the logical properties and epistemic value of moral discourse. Like traditional expressivism, it explains the motivating power of moral judgements. It may be the best of both worlds.

6
Attitudes

Inferential expressivism solves the conditional version and the negation version of the Frege–Geach Problem plaguing traditional expressivism. Indeed, inferential expressivism solves the Frege–Geach Problem as arising from embeddings under any of the core logical constants as well as epistemic modals.

But ethical expressivism's problems with embeddings do not end here. One can embed moral vocabulary in all sorts of linguistic contexts. Given the central role of attitudes in the expressivist project, one particularly important case appears to be that of embeddings under propositional attitude ascriptions, such as when one says that someone hopes that lying is wrong. The problem of giving an account of the meaning of moral vocabulary in propositional attitude ascriptions is one aspect of the *Many Attitudes Problem* (Schroeder 2008a, 2013; Shiller 2017). Another aspect of the problem is that of explaining what it is to *have* a non-doxastic attitude with moral content. Finally, in a multilateral setting, the Many Attitudes Problem also challenges one to explain what it is to *express* a non-doxastic attitude with moral content. In this chapter, we present a solution to the Many Attitudes Problem in all its aspects. Our solution includes a recipe for assigning meanings to attitude verbs, which illustrates another area of application of the inferential expressivist approach to semantics. The discussion also allows us to clarify what can be said from an inferential expressivist viewpoint about what it is to have attitudes and the nature of attitudes themselves.

6.1 Many attitudes, few contents

One can take all sorts of attitudes towards lying being wrong. One can, for instance, believe, desire, regret, hope, or fear that lying is wrong. What is it to take these attitudes towards lying being wrong? And what is the meaning of ascriptions of these attitudes towards lying being wrong? In the case of belief, the traditional expressivist has a natural story to tell: to believe that lying is

Reasoning with Attitude: Foundations and Applications of Inferential Expressivism. Luca Incurvati and Julian J. Schlöder, Oxford University Press. © Oxford University Press 2023.
DOI: 10.1093/oso/9780197620984.003.0006

wrong is to disapprove of lying, and *Ann believes that lying is wrong* is true just in case she disapproves of lying. However, it is not clear how this story is to be extended to the other attitudes one can take towards lying being wrong, such as desire: to desire that lying is wrong cannot be to desire to disapprove of lying, but what other options does the expressivist have? This is the Many Attitudes Problem as we have introduced it in Chapter 1.

It is helpful to first consider why the referentialist does not seem to have a problem here. According to the referentialist, the meaning of *Lying is wrong* can be given by the set of possible worlds in which the denotation of *lying* is in the extension of *wrong*. To explain the meaning of *Ann desires that lying is wrong* one then appeals to the possible worlds ranked highest according to Ann's desires: *Ann desires that lying is wrong* is true just in case *Lying is wrong* is true at all these worlds. By varying this recipe, one can generate truth conditions for all attitude ascriptions.

The referentialist recipe does not explain what desire is. Rather, the referentialist explains what it is for Ann to desire something in terms of worlds ranked according to what Ann desires. In this explanation, *desire* occurs in the *explanans*, so it cannot be the *explanandum*. Similarly, the referentialist can explain what it is for Ann to suppose something by appealing to the set of worlds compatible with Ann's suppositions, what it is for Ann to fear something by appealing to the set of worlds containing something Ann fears, or what it is for Ann to hope something by ranking worlds according to what Ann hopes. None of these explanations tell us what it is to suppose, fear, or hope. But the referentialist succeeds in providing explanations of what it is for something to be the content of some propositional attitude.

This is then the yardstick against which the expressivist's dealings with propositional attitudes should be measured. A solution to the Many Attitudes Problem does not require one to explain what a particular attitude *is*, but to explain what it is for something to be the content of that attitude. But this is already a thorny task.

The referentialist explanation relies on a notion of propositional content that is independent of any attitude one may take towards it. The meaning of a declarative sentence is taken to be a proposition with particular truth conditions. It is then not too difficult to tell a story about what it is to have an attitude whose content is such a proposition. On the face of it, this strategy is not available to the expressivist. Traditional expressivists explain the meaning of declarative sentences in terms of the attitudes expressed by uttering a sentence. It is unclear whether one can extract from this an attitude-independent notion of the content of a sentence.

Some sophisticated versions of expressivism, notably Gibbard's (2003), might seem to provide an immediate solution to the Many Attitudes Problem. Gibbard, recall, associates every declarative sentence with a set of fact-prac worlds representing the judgement expressed by that sentence. Just as the referentialist gives an account of propositional attitudes as particular relations to sets of worlds, could the Gibbardian not give an account of those attitudes as particular relations to sets of fact-prac worlds?

Such an account, Beddor (2020: 2793ff.) observes, would fail to be explanatory. Gibbard gives an account of what it is to have a descriptive or normative *belief*: it is to be in the particular evidential and planning state that is represented by a set of fact-prac worlds. Thus, a set of fact-prac worlds represents a belief. It is therefore unclear what explanation the Gibbardian expressivist could provide for a set of fact-prac worlds being the content of, say, the attitude of hope. Indeed, it is unclear whether such an explanation could even make sense. To be sure, the Gibbardian expressivist could insist that one hopes that lying is wrong if the set of fact-prac worlds representing the belief that lying is wrong is among one's highest *hope*-ranked sets of fact-prac worlds. But while this would allow one to assign a semantic value to sentences such as *Ann hopes that lying is wrong*, the Gibbardian expressivist would still owe an interpretation—an answer to the Interpretation Question—of what it is for a planning state to be the content of a hoping attitude.

By contrast, referentialists use sets of worlds to represent not beliefs but propositions, which can then be the *contents* of beliefs. It is reasonably clear, moreover, what it means to rank a world over another: that one prefers one world to be actual over the other. The problem is that, although Gibbard's fact-prac worlds are formally akin to the referentialist's worlds, they are to be interpreted very differently. Sets of fact-prac worlds represent attitudes, whereas sets of worlds represent attitude-independent contents. Mimicking the referentialist strategy may be a formal possibility, but it clashes with the intended interpretation of sets of fact-prac worlds. Gibbard, of course, is aware of the problem. Thus, instead of explaining propositional attitudes as relations to particular contents, he plans to explain propositional attitudes as relations to particular *beliefs*.

> What, then, of contentful states of mind other than belief, such as doubt and hope. These I treat in relation to belief states. Where a belief is possible, there could be a corresponding doubt or hope. Instead of a function that takes, say, pieces of content to states of hope . . we have a function that takes states of belief to states of hope. (Gibbard 2012: 290)

While the referentialist associates every declarative sentence with a content and provides a recipe for explaining attitudes in terms of truth functions operating on these contents, Gibbard associates every declarative sentence with a belief and suggests that a recipe can be given for explaining attitudes in terms of functions operating on these beliefs. But, unfortunately, Gibbard has not provided such a recipe beyond its outlines.

Not all possible ways of implementing Gibbard's plan are equally promising. Beddor (2020) suggests solving the Many Attitudes Problem by providing *reductive* accounts of the attitudes in terms of the notion of a normative belief. But as we mentioned in Chapter 1, expressivists would be in a dialectically uncomfortable position if they were forced to commit to reductionism about the attitudes (see Baker 2022: 3–4). One of the core motivations for expressivism, we take it, is that it allows us to give due to the richness of our inner lives—indeed, the *many attitudes* we can have—by acknowledging it in our semantic theorizing. Reductionism takes away from this and, we suspect, disastrously so.

Baker (2022) points out that a reductive approach such as Beddor's is unnecessary. Indeed, it is not even necessary to provide an account of what the attitudes *are*. The referentialist has only provided an account of what it is for an attitude to have a particular content and the expressivist need do nothing more. Baker suggests, and we concur, that the semantic aspect of the Many Attitudes Problem is simply a version of the Frege–Geach Problem, namely the version in which vocabulary receiving expressivist treatment is embedded under propositional attitude verbs. Solutions to the Frege–Geach Problem, Baker continues, take a general shape. If belief that lying is wrong is functionally equivalent to disapproval of lying, then belief that lying is not wrong is functionally equivalent to tolerating lying. What it is for lying to be the content of a tolerating attitude is then explained as tolerance of lying 'bearing appropriate functional or rational relations to disapproval of lying' (Baker 2022: 6).

This style of explanation—appealing to appropriate relations between attitudes—is not just applicable to the negation version of the Frege–Geach

Problem, but to any version whatsoever. So, Baker goes on to argue, we can use the same strategy to explain what it is for anything to be the content of any attitude. In particular, we can use it to explain what it is to hope that lying is wrong, or what it is for *lying is wrong* to be the content of any other attitude. Baker's proposal is as follows.

> For any propositional attitude other than belief, $\mathcal{A}$, the content of $\mathcal{A}$ is p if and only if $\mathcal{A}$ stands in the right kind of functional relations to the belief that p. (Baker 2022: 7)

Like the referentialist, Baker offers an explanation which features the attitude $\mathcal{A}$ in both the *explanans* and the *explanandum*. So this is an explanation not of what $\mathcal{A}$ is, but of what it is for p to be the content of $\mathcal{A}$. Of course, the explanation also features p in both the *explanans* and the *explanandum* but, as Gibbard observed, the expressivist is by this stage assumed to already have an explanation of what it is to have a normative belief that p. Like Gibbard, therefore, Baker suggests explaining what it is to have an attitude with a certain content in terms of a belief with the same content. But, also like Gibbard, Baker does not say exactly how this explanation is going to proceed. He does offer some useful clarifications. For instance, he notes (p. 8) that the functional relations are not necessarily to the belief that p, but 'to other attitudes which are semantically related to p', of which belief is merely a special case. But he does not say what these functional relations *are* for an attitude such as hope.

6.2 Expressing many attitudes

Inferential expressivism puts some flesh on the bones of Baker's suggestion. From an inferential expressivist standpoint, the functional relations between attitudes—or, in the case of inferential expressivism, between attitude expressions—are *inferential* relations. Thus, we have explained the meaning of *Lying is wrong* under the expression of belief in terms of its inferential relation to the expression of disapproval of lying. But what about the meaning of *Lying is wrong* under attitude expressions other than belief? In our multilateral framework, we are already dealing with such expressions of attitudes, namely the expressions of disbelief, refraining from believing, and refraining from disbelieving. The reason why we did not need to specify the meaning of *Lying is wrong* under these attitude expressions is that this is already taken care of by the coordination principles. The coordination principles are a concrete and precise implementation of Gibbard's suggestion that 'instead of a function that takes, say, pieces of content to states of hope... we have a function that takes states of belief to states of hope' as applied to the case of expressions of disbelief, refraining from believing, and refraining from disbelieving.

But what about expressions of attitudes such as hope itself, desire, fear, and regret? It seems plausible that one can directly express hope that it is raining by answering *Hopefully* in response to *Is it raining?*, much in the way in which one can express belief that it is raining by answering *Yes* in response to the same question. Indeed, *hopefully* passes the usual tests for determining whether an expression is a force indicator. Similarly, it seems plausible that one can express desire by means of optatives such as *Oh, if only it were raining*. Once again, such uses of *if only* seem to pass the usual test for determining whether a locution indicates force. For instance, *if if only it were raining, then ...* is gibberish. Finally, it seems plausible that answers such as *Oh no* and *Ugh*, when given in the appropriate tone, to the question *What if it rains?* express fear that it is raining, and that answers such as *Regretfully, yes* to the question *Did you go to the cinema?* express regret that one went to the cinema.

Once we have primitive attitude expressions of hope, desire, fear, and regret, we must coordinate them with the other speech acts. Thus, for example, we can lay down a coordination principle stating that it is inconsistent to express fear and hope towards the same sentence. And we can lay down another coordination principle ensuring that regret and belief stand in an entailment relation, so that one cannot express regret without being committed to expressing belief towards the same sentence. The interaction of desire and belief expressions is a more complex example. Someone who expressed desire that their sink be fixed and who expressed belief that if they call a plumber, their sink will be fixed, is committed, all things being equal, to expressing desire towards calling a plumber. This appears to be an instance of the general principle that from the belief expression towards a conditional and the expression of desire towards its consequent one can infer a commitment to expressing desire towards its antecedent.

As in the previous chapter, we avoid having an explanatory problem by only appealing to attitudes that can be directly expressed. One can identify the proper coordination principles by attending to the data about language use and other expressive behaviour. We cannot undertake a full survey here, but such behaviour has been studied in the rich literature on what has been called *folk psychology* (see, e.g., Sellars 1956; Fodor 1987; Churchland 1989a). We do not take a stand here on whether folk psychology is a correct theory of what goes on in people's minds or indeed on whether folk psychology is best seen as a theory at all. We take it to be largely uncontroversial that folk psychology, however it is best understood, provides an approximately reliable way to explain and predict people's behaviour. This suffices for folk psychology to be a viable guide to finding how attitude expressions interact.

However, the principles of folk psychology are recognized to be *defeasible* (see, e.g., Davidson 1974). Thus, if folk psychology is a guide to finding the right coordination principles, we need to make room for defeasible rules within the multilateral framework.[1] This can be done by distinguishing between the types of commitment codified by inference rules. By expressing an attitude, a speaker explicitly commits to that attitude. Strict, non-defeasible inference rules then unpack the implicit commitments that strictly follow from this. Thus, when the speaker is presented with these commitments, they must explicitly undertake them or retract one of their earlier commitments. By contrast, defeasible inference rules unpack implicit commitments that follow from a speaker's explicit commitments *ceteris paribus*. Thus, when the speaker is presented with these commitments, they must undertake them, or retract one of their earlier commitments, *or* strongly assert a defeater—a reason why *ceteris* are not *paribus* (see, e.g., Pollock 1987). Hence, we can include defeasible coordination principles in our commitment-preserving proof theory, staying mindful of their difference with monotonic principles.

We have explained what can be said, from an inferential expressivist viewpoint, about the expressions of attitudes other than belief towards *Lying is wrong*. This, however, does not yet deal with all aspects of the Many Attitudes Problem. For we still need to explain the meaning of moral vocabulary in attitude *ascriptions*. In other words, we have so far given an account of the meaning of *Lying is wrong* under expressions of hope, but we have not given an account of the meaning of *John hopes that lying is wrong* under expressions

[1] The use of defeasible rules in a broadly inferentialist setting was advocated by Sellars (1956), who employs *ceteris paribus* clauses when formulating the rules governing the behaviour of perceptual vocabulary.

of attitudes, including belief. In the next section, we present a recipe for providing inferential expressivist semantics for attitude ascriptions.

6.3 Ascribing many attitudes

From an inferential expressivist point of view, to develop a semantics for attitude ascriptions—that is, sentences such as *John believes it is raining*, *Mary hopes lying is wrong*, or *Mira fears that lying is wrong and often unpunished*—we must provide introduction and elimination rules for attitude ascriptions under attitude expressions.

It will again be useful to take a pragmatist perspective. We use attitude expressions to establish *mutual* attitudes, such as mutual beliefs or mutual moral attitudes. This is useful for social coordination and joint action. For instance, when we plan to meet up, we can establish a mutual belief about where and when, so we can expect of each other that we will arrive at the time and place we agreed upon. Attitude ascriptions are likewise useful for social coordination, since they allow us to coordinate on what to expect of *individuals*. In uttering *Franz wants to meet on Thursday*, we are not attempting to establish a mutual belief about the time of the meeting, but are saying something about what we can expect from Franz. Indeed, we are saying what attitude we would expect Franz to express, were he present. In particular, we can expect Franz to be willing to make it a mutual belief that the meeting is on Thursday. This allows us to take into account Franz's attitude in our deliberations, despite his absence.

We take the functional role of attitude ascriptions to allow us to coordinate on what attitudes we can expect individuals to express under suitable circumstances. By the Pragmatist Razor, the semantic explanation of attitude ascriptions should only appeal to features that are required to account for their functional role. Hence, to comply with the Razor, we formulate the rules for the introduction and elimination of attitude ascriptions by suitably relating those ascriptions to their expression. Thus, the introduction rule for belief ascription states that when someone is observed to express belief towards A, we may ascribe the belief that A to them, presuming the expression was sincere. The elimination rule for belief ascription states that when someone is ascribed the belief that A, one may draw the same conclusions about them as if one had observed them expressing belief towards A and presumed this expression to be sincere. The same rules can be given, *mutatis mutandis*, for other attitude ascriptions.

Thus, the introduction rules for attitude ascriptions are *language-entry* rules. They allow us to infer that someone has an attitude from the observation of certain worldly circumstances, namely their having performed an attitude expression. Conversely, the elimination rules for attitude ascriptions are *language-exit* rules. They allow us to derive practical consequences from someone having an attitude, for instance about how they are expected to behave. From someone sincerely expressing belief towards *I will be in Amsterdam next Tuesday*, we learn something about how they are expected to behave—we learn that we can expect them to be in Amsterdam next Tuesday—which may bear on our own planning. From someone being ascribed the belief that they will be in Amsterdam next Tuesday, we learn the same information.

Rules linking language to the observation of worldly circumstances and the expectation of action are not news to inferentialism. As we saw in Chapter 2, some inferentialists have suggested to use language-entry and language-exit rules to give the meaning of colour terms. According to their proposed introduction rule for *red*, for instance, one can strongly assert that an object is red having observed that it is red. And according to their proposed elimination rule, having learned that an object is red, one can draw the same practical consequences one could draw if one had observed it to be red.

Of course, we often ascribe attitudes to people who have not expressed those attitudes and whom we expect to behave in ways suitably connected with the attitude, even though we do not necessarily expect them to directly express the attitude. This can be explained on our account. When we observe someone behaving in a particular way, we may infer that they would express an attitude in the appropriate circumstances, and use this fact to introduce the corresponding attitude ascription. Conversely, if we have ascribed an attitude to someone, we can infer that they would express it in the right circumstances and, on the basis of this, infer further predictions about their behaviour. Thus, we can account for the meaning of attitude ascriptions in terms of attitude expressions even though we routinely ascribe attitudes on the basis of testimony or indirect evidence, such as behaviour revealing that someone who has not expressed a certain belief does have that belief.

There is nothing special about the case of attitude ascriptions here. Consider again the case of colour terms. We routinely predicate redness of things on the basis of testimony or indirect evidence. But this does not undermine the inferentialist account of colour terms according to which the meaning of *red* is partly given in terms of being presented with something red. For one can nonetheless predicate redness of an object on the basis of testimony or indirect evidence by inferring from testimony or indirect evidence that the object is one that would be perceived as red in the appropriate circumstances. In saying this, we are not subscribing to this specific account of the meaning of colour terms. The example serves to illustrate the point that our account of attitude ascriptions is compatible with the observation that we are willing to ascribe attitudes even to agents whom we have not observed to express the attitude.[2]

What about *self*-ascriptions of attitudes such as *I believe that it is raining*? Self-ascriptions are sometimes called *avowals* and taken to express the attitude which one is self-ascribing. For instance, on Dorit Bar-On's (2004, 2015) account, in uttering *I believe that it is raining* one may express the belief that it is raining as well as the belief that one believes that it is raining. On our approach, by contrast, self-ascriptions are no different from other-ascriptions in the attitude they express: in uttering *I believe that it is raining* one expresses the belief that one believes that it is raining and nothing else, just as in uttering *John believes that it is raining* one expresses the belief that John believes that it is raining. Nonetheless, the expression of belief that one believes that it is raining is inferentially related to the expression of belief that it is raining. In particular, in self-ascribing the belief that it is raining one makes it public that one would sincerely express the belief that it is raining under the right circumstances. This does not mean that the inference from $+A$ to $+\mathbf{B}_S A$—where $\mathbf{B}_S$ is to be read as *S believes that A*—is valid. The premiss of the introduction rule for $+\mathbf{B}_S A$ is an observation about behaviour, not the expression of belief that A, which is what is denoted by $+A$. If the premiss of the introduction rule for $+\mathbf{B}_S A$ were $+A$, the rule would be a language-internal rule, not a language-entry rule. Nevertheless, someone who expresses an attitude is expected to self-ascribe the same attitude and vice versa. For when one expresses some attitude, one is in particular displaying the behaviour that would typically sanction the premiss of the introduction rule for the self-ascription of that attitude.

[2] One may also wonder whether our account of attitude ascriptions is compatible with ascribing attitudes to non-human animals. For example, one might ascribe to one's cat the belief that there is food in the pantry or the desire to be fed. One option to deal with the issue is to take certain verbalizations of a non-human animal to be attitude expressions and apply our semantics directly. Another option is to extend the behaviour licensing attitude ascriptions to encompass what Davis (2003) calls evidential expression (see fn. 7 in Chapter 1), that is behaviour that is indicative of holding a particular attitude.

We can clarify the distinction between the expression and the self-ascription of an attitude on our account by considering again the case of Moorean sentences. In Chapter 4, we argued that the reason why it is odd to utter *It is raining and I do not believe that it is* is that the speaker undertakes an immediate commitment to believing that it is raining while announcing that they do not satisfy this commitment. When the speaker immediately commits to believing that it is raining, the listener presumes that they want to be recognized as holding this belief, but this presumption is immediately violated by the speaker's declaration that they do not believe that it is raining. The assertion of *It is raining and I do not believe that it is* is therefore pragmatically incoherent, although not inconsistent.

We can now expand on this explanation of the characteristic oddness of Moorean sentences. In particular, we can see that the listener is indeed entitled to presume that the speaker wants to be recognized as holding the belief that it is raining. For in uttering *It is raining and I do not believe that it is*, the speaker is immediately committing to displaying behaviour—namely, the overt expression of belief that it is raining—that suffices to infer that they believe that it is raining, by the introduction rule for belief ascriptions. Thus, we can rephrase our explanation of the characteristic oddness of Moorean sentences by saying that in uttering *It is raining and I do not believe that it is*, the speaker immediately commits to both displaying behaviour suitable for inferring the strong assertion of a belief ascription to themselves and to strongly rejecting this ascription. This, to stress, is pragmatically incoherent but not inconsistent. In particular, in uttering *It is raining and I do not believe that it is* one is not committed to expressing incompatible attitudes, since to display behaviour associated with the belief that one believes that it is raining is not to express this belief.

By contrast, it is inconsistent to assert an outright contradiction, such as *It is raining and it is not raining*, since in so doing one expresses the belief that it is raining and it is not raining, which immediately commits one to holding incompatible attitudes. The inconsistency is due to the attitude expressed, not what can be inferred from the fact that the speaker has expressed that attitude. The difference matters in suppositional contexts. Uttering *Suppose it is raining and it is not raining* is still inconsistent, since in so doing one still commits to holding incompatible attitudes, namely making as if one believes and disbelieves that it is raining. Uttering *Suppose it is raining and I do not believe that it is* is however neither inconsistent nor incoherent. We said in Chapter 4 that this is because there is no presumption in this case that the speaker wants to be recognized as someone who believes that it is raining. We can now clarify why this presumption is not present: the behaviour associated with expressing making as if one believes that it is raining is different from the behaviour associated with believing that it is raining. One is making as if one has a belief, but one has not displayed the behaviour associated with expressing a belief. Thus, one cannot apply the introduction rule for belief ascription.

We have now given an answer, suitable to inferential expressivism and broadly in line with the plan outlined by Baker (2022), to the question of what the *meaning* of attitude ascriptions such as *John fears that lying is wrong* is. As we have presented it so far, our account takes the meaning of attitude ascriptions to be given in terms of the corresponding attitude expressions. But, one might object, there could be attitude ascriptions for which there is no corresponding attitude expression. The English language contains many attitude verbs, such as *desire*, *want*, *wish*, *imagine*, *contend*, *suspect*, and *expect*. Our linguistic practices might assign to all of them a distinct meaning. Nonetheless, there need not be a primitive attitude expression corresponding to all of these attitude verbs. We already encountered a putative case of this kind in the previous chapter, when we discussed tolerance. Certainly, *John tolerates lying* is a meaningful sentence, but there may be no direct way for John to express tolerance of lying. On the account of moral vocabulary we defended, John can nonetheless express his belief that lying is not wrong, which is suitably related to the expression of disapproval of lying. We can use this fact to provide introduction and elimination rules for ascriptions of tolerance: the meaning of *John tolerates lying* is constrained by the fact that if John expresses belief towards *Lying is not wrong*, it is *ceteris paribus* correct to ascribe tolerance of lying to John. Thereby ascriptions of tolerance, despite this attitude not being directly expressible, are related via the semantics of *wrong*, to disapproval. Thus, in general, the introduction and elimination rules for the ascription of any attitudes that cannot be directly expressed may be given in terms of the expression of other attitudes.

6.4 Having many attitudes

We have presented an inferential expressivist semantics for attitude ascriptions. According to this semantics, the meaning of *Bahram hopes that lying is wrong* is given in terms of Bahram's expression of hope towards lying being wrong. Moreover, we have explained what it is for Bahram to express hope that lying is wrong. But there is still one aspect of the Many Attitudes Problem that we have not addressed: what is it for Bahram to *have* the attitude of hope that lying is wrong?

Recall that the referentialist provides the following answer to the question of what it is for Bahram to have the attitude of hoping that it is raining: Bahram hopes that it is raining just in case *Bahram hopes that it is raining* is true. And, in turn, *Bahram hopes that it is raining* is true just in case *It is raining* is true at all possible worlds ranked highest according to Bahram's hopes.

As a form of sophisticated expressivism, inferential expressivism subscribes to minimalist accounts of notions such as truth, property, and proposition. Could the inferential expressivist also endorse a minimalist account of having an attitude with a certain content? According to the minimalism about the attitudes we endorsed in the previous chapter, someone who sincerely expresses an attitude towards some sentence has the same attitude towards the content expressed by that sentence. Minimalism about the attitudes underwrites the introduction rule for attitude ascriptions which we laid down in the previous section, stating that if someone is observed to express a certain attitude towards *A*, we may infer that they have that attitude towards the content expressed by *A*, presuming the expression was sincere. Minimalism about the attitudes provides a sufficient condition for someone to have an attitude with a particular content. Having also laid down an elimination rule for attitude ascriptions, we are now in a position to provide necessary and sufficient conditions for someone to have an attitude with a particular content. In particular, we can take the meaning of *S has the attitude of belief towards A* to be given by rules allowing us to pass from the strong assertion of *S believes that A* to the strong assertion of *S has the attitude of belief towards A and vice versa*, and similarly for strong rejection. The same can be said, *mutatis mutandis*, about attitude verbs other than belief.

This provides a minimalist account of what it is to have an attitude with a certain content. By inspecting the rules for attitude ascriptions, we can clarify what this kind of minimalism amounts to. According to the necessary and sufficient conditions we have provided, for someone to have an attitude with a certain content is for them to be able to sincerely express that attitude. This can be seen from the fact that they actually express the attitude when they can be presumed to be sincere or that they would express the attitude sincerely in appropriate circumstances. Thus, for Bahram to have the attitude of hope that it is raining is for Bahram to be able to sincerely express hope that it is raining. There is nothing special about the content towards which Bahram is expressing hope here, of course, so the account also encompasses attitudes held towards moral contents. Hence, for Bahram to hope that lying is wrong is for Bahram to be able to sincerely express hope that lying is wrong.

Nothing more, we contend, needs to be said to put inferential expressivism on equal footing with referentialism when it comes to the aspect of the Many Attitudes Problem concerning what it is to have an attitude with a particular content. According to referentialism, *lying is wrong* is the content of a *hoping* attitude held by Bahram if the ordering of worlds determined by Bahram's mental state is one in which only worlds in which *Lying is wrong* is true are ranked the highest. According to our account, *lying is wrong* is the content of a *hoping* attitude if Bahram's mental state is such that he can sincerely express hope towards *Lying is wrong*. Both referentialists and inferential expressivists can provide an answer to the question of what it is to have an attitude with a particular content.

Nonetheless, the similarities between the referentialist answer and the inferential expressivist answer should not be exaggerated. For the referentialist, one has an attitude with a particular content if certain worldly circumstances obtain: the circumstances that are modelled by the fact that one stands in a certain relation with possible worlds or orders them in a certain way. These circumstances are typically taken to be certain substantial facts about the mind, such as, for instance, the obtaining of certain configuration in a Fodorian Language of Thought (Fodor 1975, 1987), corresponding to the arrangement of possible worlds which, according to the referentialist, models the meaning of the relevant attitude ascription. Now, the circumstances modelled by relations to possible-world arrangements may or may not obtain. Indeed, it is compatible with the referentialist picture that those circumstances never obtain, so that our mentalistic talk is typically false. In other words, the referentialist story is compatible with an error theory about our folk-psychological talk. This error theory can take the form of a fictionalist account of the mind, according to which mentalistic terms, while part of our everyday practice of making sense of ourselves and others, strictly speaking fail to refer (Lycan 2022) or refer only within a fiction (Wallace 2022). Or it can take the form of an eliminativist account such as the one given by Patricia and Paul Churchland (Churchland 1981, 1989b), who claim that folk psychology might be radically mistaken and many of the attitudes postulated by folk psychology do not exist (see also Ramsey et al. 1990).

Regardless of whether it is coupled with an error-theoretic account or with a realist account, it remains the case that the referentialist story about the meaning of attitude ascriptions and what it is to have an attitude with a certain content appeals to the worldly features that are modelled by relations to possible-world arrangements. Even if these worldly features, as a matter of fact, never obtain, it is nonetheless a mistake, from our pragmatist point of view, to appeal to them in explaining the meaning of attitude ascriptions and the resulting account of what it is to have an attitude with a particular content. For these features are thoroughly unnecessary to explain the *raison d'être* of our attitude talk, as several referentialists such as Fodor and Bill Lycan themselves are willing to recognize. The functional role of our attitude talk lies in our need to ascribe commitments to expressing attitudes to ourselves and others for the purposes of coordination. By the Pragmatist Razor, our explanation of the meaning of attitude ascriptions and the resulting account of what it is to have an attitude with a particular content should not appeal to features that are not needed to account for this functional role.

We can contrast the referentialist predicament with the combination of our account of the meaning of attitude ascriptions and a suitable minimalism about having attitudes. Minimalism about having attitudes commits the inferential expressivist to the existence of attitudes in a minimal sense of *attitude*, at least to the extent that, in some circumstances, people display the kind of behaviour that licenses the ascription of attitudes to them and that there is a presumption that they are being sincere. Minimalism about having attitudes is indeed minimal, but it does rule out certain versions of error-theoretic positions in the philosophy of mind, to the extent that the presumption that people are sincere is, in at least some circumstances, correct.

Fictionalists such as Lycan hold that, very strictly speaking, our attitude talk is systematically false because 'no one is ever literally in a mental state' (Lycan 2022: 72). But from the perspective of a minimalist approach to having attitudes, there is nothing more for the world to do for someone to have an attitude with a certain content than for them to be able to sincerely express that attitude in the right circumstances—which can be the case, literally and strictly speaking. For the same reason, minimalism about having attitudes appears to rule out the Churchlands' (Churchland 1981) eliminativism.[3]

The case of eliminativism about the mind also allows us to clarify an important aspect of minimalism about having attitudes. Besides claiming that folk psychology is radically mistaken, the Churchlands also claim that scientific progress might result in mentalist talk being phased out from natural language as we learn to explain ourselves and others in the scientifically proper way. If folk psychology is overturned by anti-mentalistic science, Paul Churchland contends, then:

> it is not inconceivable that some segment of the population, or all of it, should become intimately familiar with the [new] vocabulary . . . and displace the use of [folk psychology] altogether, even in the marketplace.
> (Churchland 1981: 86)

We are sceptical about the possibility of mentalistic talk being phased out of natural language, no matter how radical the conclusions of a future neuroscience. Our mentalistic talk has evolved so as to be particularly suited to fulfil the functional role of ascribing commitments to attitude expressions to ourselves and others. And indeed, if our account is on the right track, the meaning of attitude ascriptions is given directly in terms of attitude expressions, thus satisfying the Pragmatist Razor in the strictest possible way. But in any case, being a minimalist about having attitudes does not entail that the existence of attitudes is dependent on attitude talk being used by some speakers, any more than minimalism about truth or properties entails that the existence of truths or properties is dependent on truth or property talk being used by some speakers.

To the extent that speakers could, in at least some circumstances, sincerely express attitudes in suitable circumstances, minimalism about the attitudes commits inferential expressivists to the existence of attitudes in a minimal sense of *attitude*. However, this does not mean that inferential expressivists are committed to holding that being able to sincerely express attitudes in

[3] Having said this, if error theorists about the mind only intend to rule out the existence of attitudes in a more substantial, non-minimal sense, and such a non-minimal sense of *attitude* can be made out, then minimalism about having attitudes is compatible with their view. Nonetheless, the combination of inferential expressivism with minimalism about having attitudes would still remain incompatible with error-theoretic approaches to the mind. For inferential expressivism about the attitudes, *qua* form of expressivism, rejects the referentialist underpinnings of the error theory. See the discussion of the Problem of Creeping Minimalism in Chapter 1.

suitable circumstances is all there is to having attitudes. As we did in the ethical case, we can distinguish here between a quasi-realist approach and a realist one. According to a quasi-realist account of having attitudes, all there is to having attitudes is captured by the minimalist account of having attitudes. Thus, quasi-realists add to the minimalist claim about what it takes for someone to have an attitude with a particular content the negative metaphysical claim that there is nothing more to having attitudes.

However, quasi-realism about having attitudes is not forced upon inferential expressivists simply on account of them endorsing the minimalist account of having attitudes. Indeed, we submit, inferential expressivism is compatible with realist accounts of what it is to have an attitude with a certain content. Consider, for instance, an account *à la* Fodor (1975, 1987), according to which to have an attitude with a particular content is to satisfy a certain configuration in the Language of Thought. Inferential expressivism is compatible with such a thoroughly realist account, as long as anyone who satisfies the configuration corresponding to a certain content could, under suitable circumstances, sincerely express an attitude with that content and vice versa. For this to be plausible, it must be the case that just as folk psychology is a reliable guide to determining the right principles coordinating attitude expressions, it is also a reliable guide to determining the configurations in the Language of Thought that speakers are in. But this is indeed what Fodor claims.

Fodor's realism is, as Daniel Dennett (1991: 30) puts it, 'industrial-strength'. Thus, several philosophers have been attracted to less demanding—'mild', to again use Dennett's term—versions of realism about what it is to have attitudes. For instance, according to interpretivists (Dennett 1987, 1991; Williams 2020), for someone to have an attitude with a certain content is for them to be best interpreted as having that attitude. Interpretivists are not committed to and indeed typically reject the Fodorian idea of a Language of Thought. Nonetheless, they do insist that attitudes are real. For instance, on Dennett's (1991) view, the best interpretation of an agent is the one that is best predictive of their behaviour, and the predictive success of an interpretation is objective and based on the existence of real patterns in the activities of intelligent creatures, patterns which exist independently of what anyone says. Inferential expressivism is compatible with such a mild version of realism about having attitudes too, as long as anyone who is best interpreted as having an attitude with a certain content could, under suitable circumstances, express an attitude with that content and vice versa. For this to be plausible, it must be the case that folk psychology provides a reliable guide to what the best interpretation of an agent's behaviour is, which is precisely the purpose of folk psychology. Thus, the case for the compatibility of inferential expressivism and interpretivist versions of realism is even stronger than the case for the compatibility of inferential expressivism with Fodorian realism.

6.5 Semantics in its proper place

Minimalist accounts of some concept are often coupled with a negative thesis: that there is nothing more to being something that falls under that concept than what it is entailed by minimalism. We have endorsed a form of minimalism about having an attitude with a certain content, but have argued that a negative thesis about the attitudes is not thereby forced upon us. Inferential expressivism is compatible with there being nothing to the attitudes beyond what is entailed by a minimalist account of what it is to have them. However, inferential expressivism is also compatible with various brands of realism about the attitudes. Even when coupled with a form of minimalism about having an attitude with a certain content, inferential expressivism remains neutral on the question of realism about the attitudes beyond a minimal sense of *attitude*.

This, we contend, is how it should be: semantic theorizing ought to remain neutral on such matters. Nonetheless, one may have extra-semantic reasons to endorse further positive or negative claims about the nature of the attitudes and what it is to have them. Inferential expressivism does not stand in the way, as long as these further claims are extensionally compatible with the necessary and sufficient conditions for having an attitude that follow from the inferential expressivist semantics of attitude ascriptions and minimalism about having attitudes. To be sure, as we argued in Chapter 3, inferential expressivism is naturally understood as making not only semantic claims, but also meta-semantic ones. Thus, one might well wonder where these extra-semantic commitments are coming from. This leads to the question of realism from an inferential expressivist perspective, which we already touched upon in Chapter 1, and to which we will return in the next chapter.

It might appear as if the possibility of endorsing further extra-semantic claims makes our semantics of attitude ascriptions hostage to empirical refutation. Suppose, for instance, that one goes beyond minimalism about having attitudes and endorses a form of realism according to which attitudes are natural kinds and it is the task of some science to determine their precise properties and extent. Such a science might then provide us with the means to determine when and whether someone is able to sincerely express an attitude. But this opens up the possibility of finding out that someone, without being incoherent, can sincerely express attitude $\mathcal{A}_1$ but not attitude $\mathcal{A}_2$ despite the fact that expressing $\mathcal{A}_1$ immediately commits one to also expressing $\mathcal{A}_2$ according to some coordination principle. In such an event, we are mistaken about what the correct coordination principles are, but not necessarily about how the meaning of mentalistic talk is given. We claimed that folk psychology can serve as a guide to finding the correct coordination principles. Should some putative future science refute one of the principles of folk psychology on the basis of which the inferential expressivist has formulated some coordination principles, then the coordination principle must be revised or replaced. But the inferential expressivist strategy for assigning meaning to mentalistic vocabulary is not thereby refuted.

In order to refute the inferential expressivist account of mentalistic vocabulary, the putative science would have to show that there is no way of formulating the coordination principles so that the relations among attitudes are correctly described. But this would mean that the putative science in question would not be a science of the attitudes at all. We argued in the previous section that minimalism about having attitudes is not refuted by the putative neuroscientific discovery that there are no attitudes. In a similar fashion, the inferential expressivist account of mentalistic vocabulary is not refuted by the putative scientific discovery that there are no systematic relations between attitudes. Such discoveries are about *something else*. To give an account of the meaning of attitude ascriptions, we adopted a pragmatist perspective and took as our starting point the function of mentalistic vocabulary, including its use in ascribing properties that aid social coordination. Mentalistic vocabulary can fulfil this function only if there are systematic relations between the attitudes that can be captured by means of coordination principles. The alleged discovery that there are no such relations would only show that the relevant science is not a science of what attitude talk is about.

Pragmatist themes have already made their way into the philosophy of mind under the label of *normative functionalism* (see Beisecker 2012 and Maher 2012 for expositions of the lineage). Normative functionalists such as Brandom (1994) hold that for someone to have an attitude with a certain content is for them to be responsive to certain norms, in particular those governing our linguistic practice. Now, on our account to have an attitude with a certain content is to be able to sincerely express it, and attitude expressions are governed by coordination principles. Since the coordination principles codify normative relations holding between attitude expressions—they tell us what attitude expressions a speaker is *committed to* in virtue of being committed to other attitude expressions—it follows that our account of what it is to have an attitude implies that in order to have an attitude with a certain content, one must occupy a certain normative position. This is in line with normative functionalism.

Nonetheless, our account does not *entail* normative functionalism, since it is not committed to the further claim, to which normative functionalists subscribe, that all there is to have an attitude is to occupy a certain normative position. Unlike normative functionalism, our account does not rule out the possibility that occupying a normative position is in fact a mere consequence of what it is, in fact, to have an attitude. As we have been emphasizing, it is compatible with our account, for instance, that what it is to have an attitude with a certain content is to be in a certain configuration of the Language of Thought, which can be understood in purely non-normative terms.

Conversely, normative functionalism does not entail our account of what it is to have an attitude with a certain content either. We have taken language use and other expressive behaviour studied in folk psychology to provide a guide to determining the coordination principles codifying the normative relations between attitude expressions. Normative functionalists, however, need not accept this role for folk psychology. Thus, the combination of normative functionalism and inferential expressivism is a distinctive position on which to have an attitude is to be responsive, in a certain way, to norms discoverable by attending to language use and other expressive behaviour.

We have provided a solution to the Many Attitudes Problem in all of its aspects. We have explained what it is to express an attitude with a certain content by appealing to the coordination principles. We have explained the meaning of attitude ascriptions in terms of attitude expression. And we have explained what it is to have an attitude with a certain content by appealing to our account of attitude ascriptions and a suitable form of minimalism about having attitudes. As we have been stressing throughout, sophisticated expressivists are in fact wont to subscribe to various forms of minimalism. The most well-known form of minimalism is perhaps the one about truth. In the next chapter, we will present our take on this form of minimalism. This will also allow us to illustrate how the strategy we used in Chapter 4 to tackle the triviality results for epistemic modals can be employed to provide a uniform and comprehensive approach to the semantic paradoxes.

7
Truth

In this chapter, we present an inferential expressivist account of the truth predicate. Inferential expressivism about truth explains the meaning of the truth predicate not by appealing to its purported referent, but in terms of its inferential connections to the speech acts of strong assertion and strong rejection. Once again, our starting point is a form of traditional expressivism about truth. Unsurprisingly, traditional expressivism about truth succumbs to the Frege–Geach Problem. More surprisingly, the problem affects typical formulations of the deflationary theory of truth. For deflationists, as we shall see, can be seen as subscribing to a form of expressivism about truth on the basis of their functionalist approach to truth and subsequent application of the Pragmatist Razor. Inferential expressivism allows one to retain the functionalist insights of deflationary accounts of truth while avoiding the Frege–Geach Problem. Moreover, applying the multilateral methodology to the case of the truth predicate reveals that the truth rules preserve commitment but not evidence. This leads to a new diagnosis and solution of the Liar Paradox, which can be naturally extended to a host of revenge paradoxes. The upshot is a theory of truth which bears deep similarities with supervaluational approaches to truth (Kripke 1976; Cantini 1990). We conclude by returning to the question of whether and to what extent inferential expressivism is compatible with a realist view of the area of discourse to which it is applied.

Before we start, a preliminary remark is in order. Truth can be predicated of many things. Suppose, for instance, that Thomas declares *It is raining outside*. We look out of the window and can see that it is, in fact, raining outside. We can describe the situation by saying that Thomas's *claim* is true or that the *sentence* he uttered is true. Philosophers often describe the situation by saying that the *proposition* expressed by the sentence uttered by Thomas is true. In what follows, we take sentences to be the primary bearers of truth. The question of what the primary truth bearers are is a vexed one (Künne 2003: 249ff.) and one we need not take a stand on here. Our discussion can be recast by taking the primary truth bearers to be, for instance,

Reasoning with Attitude: Foundations and Applications of Inferential Expressivism. Luca Incurvati and Julian J. Schlöder, Oxford University Press. © Oxford University Press 2023.
DOI: 10.1093/oso/9780197620984.003.0007

propositions or claims, given a suitably minimalist understanding of these notions.

7.1 Traditional expressivism about truth

As pointed out in Chapter 1, Ayer was a traditional expressivist about ethics. At the semantic level, traditional ethical expressivism holds that moral predicates do not have a referential semantic value but indicate the expression of an attitude. While Ayer's endorsement of an expressivist analysis of moral predicates is well known, what is perhaps less well known is that he also endorsed an expressivist analysis of the truth predicate.

> Reverting to the analysis of truth, we find that in all sentences of the form 'p is true,' the phrase 'is true' is logically superfluous. When, for example, one says that the proposition 'Queen Anne is dead' is true, all that one is saying is that Queen Anne is dead. And similarly, when one says that the proposition 'Oxford is the capital of England' is false, all that one is saying is that Oxford is not the capital of England. Thus, to say that a proposition is true is just to assert it, and to say that it is false is just to assert its contradictory. And this indicates that the terms 'true' and 'false' connote nothing, but function in the sentence simply as marks of assertion and denial. And in that case there can be no sense in asking us to analyse the concept of 'truth.' (Ayer 1936: 88)

Ayer is here subscribing to a form of traditional alethic expressivism. In saying that 'the terms "true" and "false" connote nothing,' Ayer is explicitly endorsing the No Referential Semantics thesis of traditional expressivism as applied to the truth and falsity predicates: these predicates do not have a referential semantic value. In saying that these terms 'function in the sentence simply as marks of assertion and denial,' Ayer is explicitly endorsing the semantic thesis of traditional expressivism as applied to the truth and falsity predicates: these predicates indicate the expression of an attitude.

We argued in Chapter 1 that traditional expressivism is best understood as a combination of semantic and meta-semantic ingredients. Ayer's focus is on the semantic aspects of the truth predicate. P. F. Strawson's (1949) focus is on the meta-semantic aspects. He describes several non-descriptive uses of the truth predicate, such as expressing commendation or endorsement of what is said. By what can be regarded as an application of the Pragmatist Razor, Strawson (1949: 83) concludes that 'to say that an assertion is true is not to make any further assertion at all; it is to make the same assertion'. The truth predicate has no referential semantic value, but indicates an expression of commendation or endorsement.

However, the truth predicate displays the embedding behaviour of ordinary predicates. In particular, it can freely embed in conditional antecedents. But this means that traditional alethic expressivism falls prey to the Frege–Geach Problem. For example, and *contra* traditional alethic expressivism, in uttering *If it is true that it is raining, we should take an umbrella* one is not thereby expressing the belief that is raining, or endorsing or commending the statement that it is raining.[1]

Strawson's account of the function of the truth predicate makes it appear as though its presence in the language is superfluous. Indeed, according to Strawson, when we say *That's true*, in response to an assertion, we could replace what we said with *I agree*, *I endorse this*, or analogous locutions. 'What commends the word as, e.g., a confirmatory device is its economy' (Strawson 1949: 95). According to deflationists, the presence of the truth predicate in our language is not simply a matter of economy: it allows us to perform linguistic moves we could not otherwise make. First, the truth predicate may be used for *indirect* endorsements. We may want, for instance, to voice our agreement with Thomas even if we have forgotten what he said exactly. With a truth predicate, we can do this by referring to Thomas's claim as *what Thomas said* and predicating truth of it. Second, the truth predicate may be used for *compendious* endorsements. We may want, for instance, to endorse every instance of the Law of Excluded Middle. With a truth predicate, we can do this by universally quantifying over all sentences of the form *A or not A* and predicating truth of them.[2]

[1] That Strawson's account of truth faces the Frege–Geach Problem was in effect already noted by Searle (1962). Searle argues that Strawson's analysis of *true* is guilty of what he later (Searle 1969: 136–141) calls the *Speech Act Fallacy*, namely the fallacy of inferring properties of the meaning of an expression from its use to perform certain speech acts. Searle points out that *true* can be used in embedded contexts in which it cannot be taken to indicate commendation or agreement. As our endorsement of the Pragmatist Razor should make clear, however, we disagree with Searle that it is a fallacy to infer properties of the meaning of expressions from their use to perform certain speech acts.

[2] The terminology of 'indirect' and 'compendious' endorsement is Wright's (1992). The idea that the truth predicate is a device for indirect and compendious endorsement goes back at least as far as Quine 1970 and possibly Ramsey 1927.

Thus, like traditional expressivists, deflationists take the truth predicate to be a device of endorsement. Unlike traditional expressivists such as Strawson, they take the presence of this device in our languages not to be simply a matter of economy. For the truth predicate allows us to make endorsements we could not otherwise make, namely indirect and compendious endorsements. Indeed, according to deflationists, the truth predicate has come to be part of our languages precisely in response to the need to make such endorsements. The expressive *function* of the truth predicate is its *raison d'être*. Other mechanisms, such as substitutional quantification, could have realized this function, but our languages have evolved so that it is realized by the truth predicate (Horwich 1998: 124ff.).

By focusing on indirect and compendious endorsements, deflationists can give a better account than Strawson of the function of the truth predicate. By an application of the Pragmatist Razor (see, e.g., Horwich 1998), deflationists conclude that the meaning of the truth predicate should not go beyond what is needed to explain its function as a device of indirect and compendious endorsement. However, if, at the semantic level, the truth predicate still indicates an expression of belief or endorsement, the account is not going to fare any better than traditional expressivism with regards to the Frege–Geach Problem. In uttering *If what Thomas said is true, we should take an umbrella*, one is not expressing belief towards or endorsing what Thomas said (see Soames 1999: 237; Picollo and Schindler 2018: 329). In the next section, we show that the resources of inferential expressivism allow us to retain deflationism's functionalist insights while explaining the meaning of the truth predicate in all contexts.

7.2 The meaning of the truth predicate

Deflationists tell a functionalist story about the truth predicate. The truth predicate serves to fulfil an expressive need. As we saw in Chapter 3, Price (1990) tells a similarly functionalist story about negation. Negation too serves to fulfil an expressive need, namely to register perceived incompatibilities. By an application of the Pragmatist Razor, Price concludes that negation indicates strong rejection and expresses disbelief. This traditional expressivist account of negation, however, faces the Frege–Geach Problem. We argued in the same chapter that one can vindicate Price's functionalist insights without falling prey to the Frege–Geach Problem by adopting inferential expressivism about negation. Although negation does not indicate rejection, as traditional expressivism contends, its meaning is inferentially explained in terms of rejection, a primitive operation that serves to register incompatibilities.

The same strategy can be used to vindicate deflationists' functionalist insights without succumbing to the Frege–Geach Problem. Although the truth predicate does not express belief or endorsement, its meaning is inferentially explained in terms of strong assertion, which expresses belief. In particular, according to inferential expressivism about truth, the meaning of the truth predicate is given by the following *strongly asserted truth rules*, which allow us to move from the strong assertion of *A* to the strong assertion of *A is true* and vice versa.

$$(+Tr\text{I.})\ \frac{+A}{+Tr\ulcorner A\urcorner} \qquad (+Tr\text{E.})\ \frac{+Tr\ulcorner A\urcorner}{+A}$$

Inferential expressivism about truth addresses the Frege–Geach Problem for truth in a by now familiar way. In uttering *If what Thomas said is true, we should take an umbrella* one is indeed not expressing belief towards what Thomas said. Nonetheless, the utterance does commit one to expressing belief towards *We should take an umbrella* should one express belief towards what Thomas said. If one expresses such a belief towards what Thomas said simply by uttering *What Thomas said is true*, this commitment follows by *modus ponens*. If, by contrast, one endorses what Thomas said by explicitly repeating his utterance, this commitment follows by an application of the strongly asserted truth rules and a subsequent application of *modus ponens*. Similarly to the case of negation, we can accommodate the Frege–Geach point that in certain contexts the truth predicate does not express belief. Nonetheless, the meaning of the truth predicate is explained in terms of strong assertion, which expresses belief. This suffices to vindicate deflationists' functionalist insight that the truth predicate is a device to perform certain kinds of endorsement.

However, as it stands, the deflationist story about the function of the truth predicate is incomplete. For as Kevin Scharp (2013: 63) noted, the truth predicate is not only needed for indirect and compendious endorsement, but also for indirect and compendious opposition. We can give this observation its due by appropriately relating truth to strong rejection, the speech act expressing disbelief. Thus, we take the meaning of the truth predicate to be given not only by the strongly asserted truth rules, but also by the following *strongly rejected truth rules*, which allow one to move from the strong rejection of A to the strong rejection of *A is true* and vice versa.

$$(-Tr\text{I.})\ \frac{-A}{-Tr\ulcorner A\urcorner} \qquad (-Tr\text{E.})\ \frac{-Tr\ulcorner A\urcorner}{-A}$$

The *truth rules*—that is, the strongly asserted truth rules and the strongly rejected truth rules—jointly entail that truth is bivalent, that is that $+Tr\ulcorner\neg A\urcorner$ is derivable from $+\neg Tr\ulcorner A\urcorner$, as witnessed by the following derivation.

$$\cfrac{\cfrac{\cfrac{\cfrac{+\neg Tr\ulcorner A\urcorner}{-Tr\ulcorner A\urcorner}\,(+\neg\text{E.})}{-A}\,(-Tr\text{E.})}{+\neg A}\,(+\neg\text{I.})}{+Tr\ulcorner\neg A\urcorner}\,(+Tr\text{I.})$$

Scharp's example of opposition expressed by using the truth predicate is that one can oppose the continuum hypothesis by uttering *The continuum hypothesis is not true*. As seen in the first three steps of the above derivation, it is indeed the case that if one asserts *The continuum hypothesis is not true* it follows that one strongly rejects the continuum hypothesis. In a multilateral setting, one can achieve the same result by strongly rejecting *The continuum hypothesis is true*.

Indeed, a multilateral setting makes it clear that Scharp's observation extends further. We can use the truth predicate to perform compendious and indirect endorsements that are weak, for instance by uttering *Perhaps what Thomas said is true*. And we can similarly use the truth predicate to perform compendious and indirect oppositions that are weak. It is not necessary to add further rules to account for these uses of the truth predicate, since the following rules are derivable from the truth rules using Smileian *reductio*.

$$(\ominus Tr\text{I.})\ \frac{\ominus A}{\ominus Tr\ulcorner A\urcorner} \qquad (\ominus Tr\text{E.})\ \frac{\ominus Tr\ulcorner A\urcorner}{\ominus A}$$

$$(\oplus Tr\text{I.})\ \frac{\oplus A}{\oplus Tr\ulcorner A\urcorner} \qquad (\oplus Tr\text{E.})\ \frac{+Tr\ulcorner A\urcorner}{\oplus A}$$

By taking the meaning of the truth predicate to be partly given by the strongly asserted truth rules, inferential expressivism can take up traditional expressivist and deflationist insights about the meaning and function of truth talk whilst avoiding the Frege–Geach Problem. By taking the meaning of the truth predicate to be also given by the strongly rejected truth rules, it can furthermore incorporate additional insights about the expressive function of truth talk such as Scharp's. However, accepting the unrestricted truth rules lands us in dangerous territory. Paradox looms.

7.3 Evidence and the truth rules

To explore the consequences of accepting the truth rules, it will be helpful to set epistemic modals aside. We return to the complications brought about by epistemic modals further below. So suppose we extend BML, the basic multilateral logic from Chapter 3, with the truth rules. By employing a Liar sentence l materially equivalent to its negation—that is, a sentence l such that $+l \subset\supset \neg Tr\ulcorner l\urcorner$—we can derive a contradiction. We first use the strongly asserted truth rules to show that the Liar is interderivable with its own negation. In particular, using ($+Tr$I.) we can show that $+l \vdash^{\mathsf{BML}} +\neg l$.

$$\dfrac{\dfrac{+l}{+Tr\ulcorner l\urcorner}\,(+Tr\mathrm{I.}) \qquad +l \subset\supset \neg Tr\ulcorner l\urcorner}{+\neg l}\,(\text{Contraposition})$$

And using ($+Tr$E.) we can show that $+\neg l \vdash^{\mathsf{BML}} +l$.

$$\dfrac{\dfrac{+\neg l \qquad +l \subset\supset \neg Tr\ulcorner l\urcorner}{+Tr\ulcorner l\urcorner}\,(\text{Contraposition})}{+l}\,(+Tr\mathrm{E.})$$

Then, since $+l \vdash^{\mathsf{BML}} +\neg l$ and negation registers incompatibility, it follows that $+l \vdash^{\mathsf{BML}} \bot$. By Smileian *reductio**, it follows that $\vdash^{\mathsf{BML}} -l$. This entails $\vdash^{\mathsf{BML}} +\neg l$ by the bilateral rules for negation. But since $+\neg l \vdash^{\mathsf{BML}} +l$, we can conclude that $\vdash^{\mathsf{BML}} \bot$. The extension of BML with the truth rules is inconsistent.

The standard reaction is to blame the strongly asserted truth rules, which are used to establish the interderivability of the Liar and its negation. The strongly asserted truth rules, however, are central to our inferential expressivist explanation of the meaning of the truth predicate and to our solution to the Frege–Geach Problem. Hence, these rules should not be lightly given up. The standard reaction is however correct that the interderivability of the Liar and its negation deserves closer scrutiny.

If $+\neg l$ is derivable from $+l$, then so is $+l \wedge \neg l$. Similarly, if $+l$ is derivable from $+\neg l$, then so is $+l \wedge \neg l$. In the presence of the Law of Excluded Middle, which is a theorem of BML, we can therefore derive $+l \wedge \neg l$ from no assumptions. This means that if (i) as a matter of the meaning of the truth predicate, the derivations from $+l$ to $+\neg l$ and from $+\neg l$ to $+l$ preserve evidence and (ii) we have, as a matter of logic, evidence for *l or not l*, then (iii) we also have, as a matter of logic and the meaning of the truth predicate, evidence for *l and not l*. Defenders of a paraconsistent approach to the semantic paradoxes might accept (Beall 2009) or indeed welcome (Priest 1979) the conclusion that we have, as a matter of logic and the meaning of the truth predicate, evidence for *l and not l*. Defenders of a paracomplete approach (see, e.g., Field 2008) might instead reject the assumption that we have, as a matter of logic, evidence for *l or not l*.

But there is a third option, namely that of rejecting the assumption that the derivations witnessing the interderivability of the Liar and its negation, as a matter of the meaning of the truth predicate, preserve evidence. For within our framework, inference need not preserve evidence, but only commitment. Our framework therefore opens up the possibility of holding on to the interderivability of the Liar and its negation whilst retaining the appealing idea that, as a matter of logic, we have evidence for every instance of the Law of Excluded Middle and for no counterexample to the Law of Non-Contradiction. We can take the strongly asserted truth rules to be valid in that they preserve commitment, but insist that, like the rules of $\Diamond$-Elimination (see Chapter 4), they do not preserve evidence. As a result, the derivations witnessing the interderivability of the Liar and its negation, albeit valid, do not preserve evidence either. This allows us to give a new diagnosis of the Liar paradox: what goes wrong in its derivation is that it applies Smileian *reductio** to inferences that only preserve commitment, whereas Smileian *reductio** can only be legitimately applied to inferences that preserve evidence. Since the derived rule of disjunction elimination inherits the restrictions on Smileian *reductio**, it also follows that the putative derivation of $+l \wedge \neg l$ from $+l \vee \neg l$ preserves neither evidence nor commitment and is therefore invalid.

We have focused on a reconstruction of the Liar Paradox using the strongly asserted truth rules, but one can also derive a version of the paradox using the strongly rejected truth rules. Our diagnosis is analogous: what goes wrong in this version of the Liar Paradox is that it involves applications of the strongly rejected truth rules, which are not evidence-preserving, within Smileian *reductio**.

Thus, considerations relating to the Liar Paradox tell in favour of taking the truth rules to preserve commitment but not evidence. But there are other reasons, quite independent of the semantic paradoxes, to hold that the truth rules fail to preserve evidence. These reasons have to do with the function of the truth predicate as a device for indirect and compendious endorsement. A thought experiment presented by Stewart Shapiro (2003) to refute a certain version of deflationism provides a case in point. In the thought experiment, a disciple sits at the feet of a guru while a logician stands nearby. The disciple is extremely faithful, as made clear by his utterance of *Everything the guru says is true*. The guru and the logician speak both the languages of arithmetic and set theory; the disciple only speaks the language of arithmetic and is incapable of grasping the language of set theory. Every now and then, the guru asserts a sentence, which may be arithmetical, set-theoretical, or a combination of the two. Suppose that the sentences the guru utters include the usual bridge principles between arithmetic and set theory. The faithful disciple therefore endorses *The bridge principles are true*. At some point, the guru asserts a set-theoretic sentence *A* that has no translation, in any reasonable sense, in the language of arithmetic. The disciple, faithful as ever, endorses *A is true*. The logician walks over to the disciple and tells him that *A*, together with the bridge principles, entails an arithmetical sentence *B* which the guru has so far neither asserted nor denied. The disciple, trusting the logician's skills, comes to believe that *B* is true.

According to Shapiro, this scenario spells trouble for deflationists, such as Hartry Field (1994), who restrict the range of application of the truth predicate to sentences the speaker understands. The problem is that it appears to be the case that we do make meaningful attributions of truth to sentences in languages we do not understand. For instance, as Shapiro notes, someone who does not understand a word of ancient Greek can announce that much of what Plato wrote is true. Field (1994: §8) discusses several options to handle cases of this sort. What these options have in common is that the range of application of the truth predicate is extended to sentences which have reasonable translations in the speaker's language. However, Shapiro (2003: 121) points out, this does not help to explain the disciple's reasoning in the thought experiment. The experiment is designed so that the set-theoretic sentence asserted by the guru has no reasonable translation in any language the disciple speaks. Hence, the sentence is outside the scope of the range of application of the disciple's truth predicate, even if this range is extended in the way envisaged by Field.

Shapiro's thought experiment shows that the restrictions imposed by Field on the range of application of the truth predicate, even when relaxed in the way adumbrated, prevent some legitimate uses of the truth predicate as a device of indirect and compendious endorsement. To account for these uses, we take the truth predicate to apply to all sentences whatsoever. Thus, we do not restrict the range of the truth predicate to sentences that the speaker can understand or that have reasonable translations in the speaker's language. However, this entails that, while preserving commitment, the truth rules do not preserve evidence. For a speaker who does not, and cannot, understand a sentence *A* may nevertheless understand *A is true*. In this case, they can be justified in asserting *A is true*, since, like the disciple in Shapiro's thought experiment, they may have evidence for *A is true*. However, they cannot be justified in asserting *A*, since they do not understand *A* and hence cannot even grasp what would constitute evidence for *A*. It follows that in the inference from *A is true* to *A*, evidence is not preserved, hence the rule (+*Tr*E.) is not evidence-preserving.

Shapiro's thought experiment makes it clear that someone who does not understand *A* but believes that *A* is true can nevertheless come to believe logical consequences of *A*. That the truth rules preserve commitment allows us to account for this fact. Consider again the disciple from Shapiro's thought experiment. He expresses belief towards, and is therefore explicitly committed to, *Everything the guru said is true*. Since the bridge principles and *A* are among the things the guru says, the disciple is also committed to expressing belief towards *The bridge principles are true* and *A is true*. By the truth rules, he is therefore committed to expressing belief towards the bridge principles and *A*. And since the bridge principles and *A* jointly entail *B*, the disciple is also committed to expressing belief towards *B*. But not only is he committed to expressing belief towards *B*, he can recognize that this is the case. For he is aware that in expressing belief towards *Everything the guru says is true*, he has undertaken a commitment towards what the guru says—whatever that is. Moreover, he is aware of the fact that the bridge principles and *A* are among the things which the guru says and that they jointly entail *B*, since this is what the logician points out to him. On the basis of his grasp of the fact that the truth rules preserve commitment and his overall grasp of the dynamics of commitment, the disciple realizes that he is committed to expressing belief towards *B*. Aware of this commitment and unwilling to abandon his sincere faith in the pronouncements of the guru, he comes to believe that *B*.

Thus, by taking legitimate inference to preserve commitment but not necessarily evidence, our account can explain how someone can have evidence for *A is true* without having evidence for *A* but can nonetheless come to believe facts that follow from *A*: it is because the truth rules fail to preserve evidence but are nonetheless valid in that they preserve commitment. Thought experiments analogous to Shapiro's can be constructed involving the strongly rejected truth rules. Thus, there are reasons independent of the semantic paradoxes to hold that the truth rules do not preserve evidence if we allow, as we should, the truth predicate to be applied to sentences the speaker does not understand. It is worth stressing that, like the property of truth preservation in classical logic, evidence preservation and commitment preservation are, strictly speaking, properties of rules, not of particular inferences. Thus, the fact that one may have evidence for the premiss of a truth rule without having evidence for its conclusion suffices to establish that the truth rules are not evidence-preserving. That there are instances of the truth rules, such as the inference from *'5+7=12' is true* to *5+7=12*, in which one has evidence for the conclusion whenever one has evidence for the premiss is neither here nor there.

Having said this, it is of course possible to consider whether the truth rules preserve evidence when restricted to applications of the truth predicate to sentences the speaker understands. Shapiro's thought experiment shows that such a restriction of the range of application of the truth predicate would be inadvisable. But in any case, it is possible to construct cases in which one may have evidence for the premiss of a truth rule without having evidence for its conclusion even when such a restriction is in place, by exploiting the fact that one may have evidence for, say, *What the guru said is true* without knowing what exactly the guru said but being in principle in a position to understand what he did say. At this point, one might try to further restrict the truth rules to cases of pure disquotation. This would be even less advisable for deflationists, since it would amount to completely giving up on their usual story about the function of the truth predicate as a device for indirect and compendious endorsement. What we have suggested is that the semantic paradoxes show that the truth rules fail to preserve evidence even when restricted to cases of pure disquotation.

7.4 The truth predicate in multilateral logic

Our diagnosis of the Liar Paradox is that its derivation uses the asserted truth rules as if they preserved evidence, but these rules only preserve commitment. This diagnosis leads to a cure. Since the truth rules preserve commitment, we may add them to BML. But since they do not preserve evidence, we must exclude their application from Smileian *reductio**, which blocks the derivation of the Liar Paradox.

Formally, we extend the language of BML with a truth predicate and add the truth rules, both strongly asserted and strongly rejected, to the system. We then disallow application of the truth rules within Smileian *reductio**, much in the way in which we disallowed application of ◇-Elimination rules within Smileian *reductio** when moving from BML to EML. The resulting version of Smileian *reductio** is as follows.

$$(\mathrm{SR}_1{}^*)\ \frac{\begin{array}{c}[+A]\\ \vdots{*}\\ \bot\end{array}}{-A}\ \text{if the inference to } \bot \text{ uses no truth rules and no premisses signed with } \ominus$$

$$(\mathrm{SR}_2{}^*)\ \frac{\begin{array}{c}[-A]\\ \vdots{*}\\ \bot\end{array}}{+A}\ \text{if the inference to } \bot \text{ uses no truth rules and no premisses signed with } \ominus$$

Call BML_{Tr} the system that results from extending BML with the truth rules but excluding their application from Smileian *reductio**. The treatment of the Liar Paradox in BML_{Tr} is as follows. From the fact that $+l \vdash^{\mathsf{BML}_{Tr}} +\neg l$ it follows that $+l \vdash^{\mathsf{BML}_{Tr}} \bot$ and so, by Smileian *reductio*, that $\vdash^{\mathsf{BML}_{Tr}} \ominus l$. Thus the Liar sentence ought to be rejected. Similarly, it follows from the fact that $+\neg l \vdash^{\mathsf{BML}_{Tr}} +l$ that $+\neg l \vdash^{\mathsf{BML}_{Tr}} \bot$ and so that $\vdash^{\mathsf{BML}_{Tr}} \ominus\neg l$. Thus the Liar's negation ought to be rejected too. But $\ominus l$ and $\ominus\neg l$ are jointly consistent. And since the derivations of $\ominus l$ and $\ominus\neg l$ apply a rule that is not evidence-preserving, we cannot infer $+\neg l$ or $+l$ by Smileian *reductio**. Indeed, there can be no derivation of a Liar Paradox in BML_{Tr}, as can be shown by providing a model for the theory by extending the techniques used in the appendix to Chapter 3 to provide a model theory for BML (see Incurvati and Schlöder Forthcoming).

One distinctive feature of our approach is that it reconciles the truth rules with the classical laws of negation. Given a map σ from propositional atoms to the sentences of BML_{Tr} and a propositional logic formula A, let $\sigma[A]$ denote the BML_{Tr} sentence obtained by uniformly replacing every atom p in A with $\sigma(p)$. The following theorem is then an immediate consequence of the result that the valid arguments involving only asserted premisses and conclusions in BML are exactly the classically valid arguments (see Chapter 3).

Theorem 7.4.1. *Let* Γ *be a set of formulae in* BML_{Tr} *and* A *be a formula in* BML_{Tr}. $\Gamma \models^{CPL} A$ *if and only if* $\{+\sigma[B] \mid B \in \Gamma\} \vdash^{\mathsf{BML}_{Tr}} +\sigma[A]$.

The theorem tells us that the BML_{Tr} logic of strong assertion validates all substitution-instances of classically valid arguments. In particular, BML_{Tr} proves all instances of the Laws of Excluded Middle and Non-Contradiction. However, it does not validate all the classical *meta-rules*. For example *reductio*, proof by cases and conditional proof for the material conditional are not generally valid, since they inherit the restrictions of Smileian *reductio**. The failure of the meta-rules was, of course, already a feature of BML, but now this failure extends to cases in which applications of the truth rules are involved. To stress, the proper versions of the meta-rules in BML_{Tr} ban any application of the truth rules, including seemingly innocent applications such as the one allowing us to move from *'5+7=12' is true* to *5+7=12*. Doesn't this irredeemably restrict the range of possible uses of the truth predicate?

The failure of the meta-rules entails that truth is not fully *transparent*: one cannot intersubstitute $Tr\ulcorner A\urcorner$ and A in all contexts. Overall, we take this to be a good thing. Denying full transparency plays a crucial role in avoiding Liar paradoxes as well as certain revenge paradoxes, as we will see later in the chapter. The concept of truth is however employed in many areas of inquiry. It has been argued that deflationism cannot properly account for all these applications (Boyd 1983). Deflationists have responded to this at length (Williams 1988; Horwich 1998). Examination of these responses shows that they do not require full transparency: it suffices that A and $Tr\ulcorner A\urcorner$ can be inferred from one another.

Nonetheless, a problem would appear to remain. There are seemingly legitimate pieces of ordinary reasoning, which, on the face of it, would seem to violate the restrictions on the meta-rules. For instance, from *s or t* and *not t is true* it ought to follow that *s*. A natural formalization of this inference is as follows, using a derivable rule of Explosion.

$$\#\ \dfrac{+s \vee t \qquad [+s]^1 \qquad \dfrac{\dfrac{[+t]^2 \qquad \dfrac{\dfrac{+Tr\ulcorner\neg t\urcorner}{+\neg t}\,(+Tr\text{E.})}{-t}\,(+\neg\text{E.})}{\bot}\,(\text{Strong Rejection})}{+s}\,(\text{Explosion})}{+s}\,(+\vee\text{E.})^{1,2}$$

However, the application of disjunction elimination in the last step is disallowed by the restrictions on (+∨E.). Nevertheless, the inference from $+s \vee t$ and $+Tr\ulcorner\neg t\urcorner$ to $+s$ is valid in BML_{Tr}, as witnessed by the following derivation.[3]

$$\dfrac{\dfrac{+Tr\ulcorner\neg t\urcorner}{+\neg t}\,(+Tr\text{E.}) \qquad \dfrac{\dfrac{+s \vee t \qquad [+s]^1 \qquad \dfrac{\dfrac{[+t]^2 \qquad \dfrac{[+\neg t]^3}{-t}\,(+\neg\text{E.})}{\bot}\,(\text{Strong Rejection})}{+s}\,(\text{Explosion})}{+s}\,(+\vee\text{E.})^{1,2}}{+\neg t \supset s}\,(+\supset\text{I.})^3}{+s}\,(+\supset\text{E.})$$

The method exemplified by this derivation consists in assuming the desired conclusion of a truth rule (in this case, $+\neg t$) in a restricted context and discharging it by applying the truth rule in the global proof context. The method generalizes: applications of truth rules can be 'moved outside' a restricted proof context in many cases. Specifically, if $+A$ does not depend on dischargeable assumptions, then any argument that would be valid except for the application of a truth rule to $+A$ can be rewritten into a valid argument with the same conclusion. The restrictions on the meta-rules are therefore much less restrictive than they appear to be at first sight.

[3] The derivation involves successive applications of the introduction and elimination rules for ⊃ and hence contains a local peak. The peak can be levelled by unpacking the derivations of the rules for ⊃ and applying the usual reduction procedures.

7.5 Truth and supervaluation

The meta-rules that characteristically fail in supervaluationist logic are exactly those that fail in the multilateral logics we have been developing in this book. Indeed, as we noted in Chapter 4, the similarities between supervaluationist logic and multilateral logic run deep, since EML axiomatizes the consequence relation of the supervaluationist logic of vagueness. The case of truth shows that the similarities between supervaluationist logic and multilateral logic run deeper still.

In 'An outline of a theory of truth' (1976), Saul Kripke presents a recursive procedure for assigning an extension to the truth predicate so that the truth rules hold. At each stage, the recursive procedure assigns sentences either to the extension or the anti-extension of the truth predicate. Sentences that end up being in neither the extension nor the anti-extension are not assigned a determinate truth value. The Liar sentence is never assigned to either the extension or the anti-extension of the truth predicate. Truth-value gaps are handled according to the Kleene scheme. It follows that *l is true or l is not true* does not have a determinate truth value and so Kripke's theory does not validate the Law of Excluded Middle.

In the same paper, Kripke already suggested the possibility of handling truth-value gaps by means of different schemes. One of these is Bas van Fraassen's (1971) supervaluational scheme. The application of the supervaluational approach to the case of vagueness is familiar. We assign extensions to predicates relative to precisifications—ways of making the predicates precise. An object may be in the extension of a predicate according to all, some or no precisification. So, for instance, Harry may be in the extension of *bald* on all precisifications, in which case *Harry is bald* is true; Harry may be in the extension of *bald* on no precisification, in which case *Harry is bald* is false; or Harry may be in the extension of *bald* in some but not all precisifications, in which case *Harry is bald* is neither true nor false. But even if *Harry is bald* is neither true nor false, *Harry is bald or Harry is not bald* is true on all precisifications, so the Law of Excluded Middle is validated. The application of the supervaluational approach to the case of truth is similar, except that one supervaluates over assignments of sentences to extensions and anti-extension of the truth predicate. While it remains the case that the Liar sentence is never assigned to either the extension or the anti-extension of the truth predicate, *l is true or l is not true* has a determinate truth value and indeed the modification of Kripke's theory using van Fraassen's scheme validates the Law of Excluded Middle.

Now, the Kripke construction and its modification based on van Fraassen's scheme are typically used to compute an extension of the truth predicate in True Arithmetic, the theory consisting of all sentences in the language of arithmetic which are true in the standard model. This means that in order to determine the exact relation between our theory of truth and the supervaluational approach to truth based on van Fraassen's scheme, we need to suitably extend BML_{Tr}. First, since we are dealing with the language of arithmetic, we need to consider a first-order version of BML_{Tr}. This can be straightforwardly obtained by expanding the language with the universal quantifier in the obvious way and adding the standard natural deduction rules for the universal quantifier prefixed with the sign for strong assertion.

$$(+\forall\text{I.})\ \frac{+A[y/x]}{+\forall x.A}\ \begin{array}{l}\text{if } y \text{ does not occur free in premisses}\\ \text{or undischarged assumptions used}\\ \text{to derive } A[y/x]\end{array} \qquad (+\forall\text{E.})\ \frac{+\forall x.A}{+A[y/x]}$$

Second, since the construction based on van Fraassen's scheme is used to compute an extension of the truth predicate in True Arithmetic, we need to add the *ω-Rule*, which allows one to infer a universally quantified sentence from all its instances for the natural numbers. As usual, boldface numerals are canonical names for the numbers.

$$(\omega\text{-Rule})\ \frac{+A(\mathbf{0}) \qquad +A(\mathbf{1}) \qquad +A(\mathbf{2}) \qquad \ldots}{+\forall n\, A}$$

One can then prove (see Incurvati and Schlöder Forthcoming) that the sentences that the resulting theory proves from True Arithmetic are exactly those that are determined to be arithmetical truths by the Kripke construction based on van Fraassen's scheme. Thus, a natural extension of BML_{Tr} axiomatizes the supervaluational approach to truth for True Arithmetic.

Variations on the supervaluational scheme employed in the recursive procedure to define the extension of the truth predicate are possible. One such variation was considered by Andrea Cantini (1990), another one already by Kripke (1976: 711–712). It is possible to further extend BML_{Tr} to obtain axiomatizations of the resulting supervaluational approaches to truth as well (see Incurvati and Schlöder Forthcoming).

An appealing picture of the relationship between multilateral logic and supervaluationist model theories emerges. The model theories for multilateral logics are typically given in terms of conditions on a Kripke model or similar mathematical structure. Force markers are interpreted using *global* conditions, concerning what happens at all points of the model. Embeddable predicates or operators are interpreted using *local* conditions, concerning what happens at a point of the model. The relationship between global and local conditions is governed by the inference rules. Embeddable predicates or operators bring global conditions down to the local level. In some cases, such as negation, this requires no restriction on the application of the relevant rules in hypothetical contexts. In other cases, some restrictions are required, which results in the relationship between local and global conditions being governed by supervaluational schemes. In particular, when strong assertion is brought down to the local level by means of an operator, we obtain the *must* operator, whose logic is the same as that of the *definitely* operator as standardly defined in the supervaluationist logic of vagueness; when strong assertion, together with strong rejection, is brought down to the local level by means of a predicate, we obtain a truth predicate definable by a recursive procedure based on some supervaluational scheme.

7.6 Classical recapture and revenge

Any solution of the Liar paradox that admits all instances of the asserted truth rules must depart from classical logic in some way. However, this departure from classical logic prompts the obvious challenge of accounting for ordinary pieces of reasoning, for instance within mathematics, that appear to be impeccable but make use, on the face of it, of the classical principles rejected by non-classical approaches to the paradoxes. Proponents of these approaches have often suggested that this challenge can be met by a procedure which has come to be known as *classical recapture*, the idea that the classical principles are acceptable under certain conditions. Since these conditions are satisfied by domains of discourse such as mathematics, classical reasoning is in order in these domains. However, Julien Murzi and Lorenzo Rossi (2019) have provided a recipe for turning the classical recapture strategy into a revenge paradox: the classical recapture strategy, they argue, leads to disaster. Now the theory of truth we have presented departs from classical logic in that it rejects some classically valid meta-rules, so the recipe can be put to work here. We show that the revenge paradox is avoided because the rules governing the paradoxicality predicate used in the revenge sentence do not preserve evidence.

Let us first describe Murzi and Rossi's recipe, using the paracomplete approach to the semantic paradoxes as a running example. The recipe uses two ingredients. The first ingredient is the *recapture principles*, which give the conditions under which classical reasoning is in order. Paracompletists avoid the Liar Paradox by rejecting the Law of Excluded Middle. However, they claim to be able to recapture classical logic in domains of inquiry in which all instances of the Law of Excluded Middle hold. Accordingly, the recapture principles for a paraconsistent theory specify that, in the presence of the Law of Excluded Middle, one can reason classically. In particular, Murzi and Rossi lay down recapture principles that sanction unrestricted use of *reductio* and conditional proof for sentences that obey the Law of Excluded Middle.

The second ingredient of the recipe is a *paradoxicality predicate*, that is a predicate *Par* that says of a sentence that it leads to contradiction if one reasons classically with that sentence. In the case of paracomplete approaches, $Par\ulcorner A\urcorner$ means that if $A \vee \neg A$ is the case, then absurdity follows. The Liar sentence is paradoxical and so we have, as expected, $Par\ulcorner l\urcorner$. However, Murzi and Rossi point out, by considering a sentence p such that $p \subset\supset (Tr\ulcorner p\urcorner \supset Par\ulcorner p\urcorner)$ we can obtain a contradiction.

How does our theory fare against this argument? Our logic validates all classically valid arguments, but is revisionary to the extent that it rejects certain classically valid meta-rules. Now, to recapture classical reasoning involving these meta-rules and account for seemingly impeccable pieces of ordinary reasoning that would seem to fail on our approach, we have not explicitly laid down recapture principles. Instead, in Section 7.4, we described a proof-theoretic procedure which shows that the restrictions we have placed on the meta-rules are much less restrictive than they appear at first sight. However, the reason why we did not have to explicitly lay down recapture principles is that our theory already validates something akin to them: if a sentence A is asserted and satisfies $+A \subset\supset Tr\ulcorner A\urcorner$, then we can reason classically with that sentence. For each application of the truth rules in suppositional contexts can then be replaced with applications of *modus ponens* for the material conditional. This suggests using $+A \subset\supset Tr\ulcorner A\urcorner$ to characterize a paradoxicality predicate and define $Par\ulcorner A\urcorner$ to mean that from the assumption that A is materially equivalent to $Tr\ulcorner A\urcorner$ absurdity follows. Formally, and letting $\vdots^*$ denote evidence-preserving inference, we can give the meaning of *Par* via the following *paradoxicality rules*.

$$
(+Par\text{I.})\ \dfrac{\begin{matrix}[+A \subset\!\supset Tr\ulcorner A\urcorner]\\ \vdots^{*}\\ \bot\end{matrix}}{+Par\ulcorner A\urcorner}
\qquad\qquad
(+Par\text{E.})\ \dfrac{+Par\ulcorner A\urcorner \qquad +A \subset\!\supset Tr\ulcorner A\urcorner}{\bot}
$$

Note that since *Par*⌜*A*⌝ says that reasoning with *A* classically leads to absurdity, the subderivation in the introduction rule for *Par* must be restricted to evidence-preserving inferences, since this is the fragment of our proof theory that accords fully with classical canons of reasoning.

Following Murzi and Rossi's recipe, we can now derive a contradiction by considering a sentence *p* such that $p \subset\!\supset (Tr\ulcorner p\urcorner \supset Par\ulcorner p\urcorner)$.[4] The derivation proceeds in three steps. To ease readability, we do not label applications of classically valid transformations in the formal derivations. In the first step, we show that $+p$ is derivable from $+p \subset\!\supset Tr\ulcorner p\urcorner$.

$$
\dfrac{\dfrac{\dfrac{\dfrac{+p \subset\!\supset Tr\ulcorner p\urcorner \qquad +p \subset\!\supset (Tr\ulcorner p\urcorner \supset Par\ulcorner p\urcorner)}{+Tr\ulcorner p\urcorner \subset\!\supset (Tr\ulcorner p\urcorner \supset Par\ulcorner p\urcorner)}}{+Tr\ulcorner p\urcorner \supset (Tr\ulcorner p\urcorner \supset Par\ulcorner p\urcorner)}}{+Tr\ulcorner p\urcorner \supset Par\ulcorner p\urcorner} \qquad +p \subset\!\supset (Tr\ulcorner p\urcorner \supset Par\ulcorner p\urcorner)}{+p}
$$

In the second step, we use this result to derive $+Par\ulcorner p\urcorner$ from no assumptions.

$$
\dfrac{\dfrac{[+p \subset\!\supset Tr\ulcorner p\urcorner]^{1} \qquad \dfrac{\dfrac{[+p \subset\!\supset Tr\ulcorner p\urcorner]^{1}}{+p} \qquad +p \subset\!\supset (Tr\ulcorner p\urcorner \supset Par\ulcorner p\urcorner) \quad [+p \subset\!\supset Tr\ulcorner p\urcorner]^{1}}{+Par\ulcorner p\urcorner}}{\bot}\,(+Par\text{E.})}{+Par\ulcorner p\urcorner}\,(+Par\text{I.})^{1}
$$

In the third step, we prove that an absurdity can be derived from $+Par\ulcorner p\urcorner$.

$$
\dfrac{+Par\ulcorner p\urcorner \qquad \dfrac{\dfrac{\dfrac{\dfrac{\dfrac{+Par\ulcorner p\urcorner}{+Tr\ulcorner p\urcorner \supset Par\ulcorner p\urcorner} \quad +p \subset\!\supset (Tr\ulcorner p\urcorner \supset Par\ulcorner p\urcorner)}{+p}}{+Tr\ulcorner p\urcorner}\,(+Tr\text{I.})}{+p \wedge Tr\ulcorner p\urcorner}}{+p \subset\!\supset Tr\ulcorner p\urcorner}}{\bot}\,(+Par\text{E.})
$$

[4] This derivation is due to Lucas Rosenblatt, who substantially improved on our original version of the classical recapture paradox within our setting.

It is the second step of the derivation that is suspicious: it involves an application of the elimination rule for *Par* within a subderivation ending in an application of the introduction rule for *Par* itself. However, the subderivation of the introduction rule for *Par* is restricted to evidence-preserving inferences. We observed that extending Basic Multilateral Logic may result in new inferences that do not preserve evidence. Adding the truth predicate was one such case: the truth rules do not preserve evidence, so their application should be excluded from proof contexts restricted to evidence-preserving inferences. But the paradoxicality rules do not preserve evidence either, so when we add them to BML, their application too should be excluded from such contexts. The reasons we gave for thinking that the truth rules fail to preserve evidence are really reasons for thinking that any rules that allow us to disquote fail to preserve evidence. In the case at hand, this can be seen by noting that under the coordination principles, the elimination rule for *Par* is equivalent to a rule allowing us to move from $+Par\ulcorner A\urcorner$ and $+Tr\ulcorner A\urcorner$, where A is quoted, to $\ominus A$, where it is not. More vividly, we could present an analogue of Shapiro's thought experiment in which the disciple believes that the guru, besides the bridge principles, utters only paradoxical sentences instead of only true sentences and use the modified thought experiment to argue that the paradoxicality rules preserve commitment but not evidence.

Thus when we add the paradoxicality rules, we should take heed to exclude them from restricted proof contexts, including the proof context of the introduction rule for *Par* itself. But once we do this, the sentence p does not appear to generate a paradox. Are the restrictions on *Par* introduction too restrictive?

Murzi and Rossi argue for the unrestricted rule of *Par* introduction in a paracomplete setting, but their example can be easily recast in our framework. So consider Lois, a logic student who is learning how to reason in BML_{Tr}. Reasoning in this theory, Lois derives a contradiction from the assumption that $+l \subset\supset Tr\ulcorner l\urcorner$ holds. She concludes that l is paradoxical, that is $Par\ulcorner l\urcorner$. Murzi and Rossi claim that Lois must be using the unrestricted rule of *Par* introduction. However, this does not follow. For a contradiction can be derived in BML_{Tr} from the assumption that $+l \subset\supset Tr\ulcorner l\urcorner$ without using the truth rules, let alone the paradoxicality rules. Thus, the restricted version of *Par* introduction which disallows applications of the truth rules and of the paradoxicality rules suffices to validate Lois's reasoning. Our rules of *Par* introduction and elimination sanction the Liar sentence l as paradoxical, as they should. Thus, *contra* Murzi and Rossi, the inferential expressivist about truth *can* 'express one of the basic lessons of the semantic paradoxes (namely, that certain sentences trivialize one's theory if reasoned with classically, while others don't)' (Murzi and Rossi 2019: 171).

However, the third step of the derivation of a contradiction using the unrestricted paradoxicality rules only makes use of the rule of *Par* elimination. Hence, the restrictions on *Par* introduction do not affect the validity of this step. Thus, the supposition that $Par\ulcorner p\urcorner$ is asserted leads to a contradiction: $\ominus Par\ulcorner p\urcorner$ is a theorem of the calculus of BML_{Tr} extended with the paradoxicality rules. Thus, BML_{Tr} extended with the paradoxicality rules cannot prove $+Par\ulcorner p\urcorner$, on pain of absurdity. Murzi and Rossi might therefore insist that the restrictions on *Par* introduction prevent us from assigning to the paradoxicality predicate its intended meaning: reasoning with *p* classically (i.e., treating it as materially equivalent to its own truth) leads to absurdity, so we *should* be able to conclude $+Par\ulcorner p\urcorner$.

But to demand that *p* falls in the extension of *Par* is to demand too much. We introduced *Par* as the paradoxicality predicate for the theory BML_{Tr} and it is satisfied by all the sentences in the language of BML_{Tr} that entail a contradiction when reasoned with classically. Extending BML_{Tr} with the paradoxicality rules yields a new theory in which there are additional paradoxical sentences such as *p*. We can introduce a new predicate Par' for this new theory, satisfied by all of its sentences that entail a contradiction when reasoned with classically. In the extension of BML_{Tr} with both *Par* and Par' there will be another new paradoxical sentence, for which one can introduce a new predicate, and so on.

In saying that some sentence is paradoxical, we are saying something about that sentence *with respect to a proof theory*, namely that under the assumption that the sentence can be reasoned with classically, an absurdity can be derived from the sentence in the proof theory. When the proof theory is extended, a new predicate is needed to capture paradoxicality with respect to the extended proof theory. As Rosenblatt (2023: 9) notes, 'the notion of paradoxicality at stake is that of *paradoxicality-in-S*, and not paradoxicality simpliciter'. The predicate *Par* expresses paradoxicality relative to a system and is therefore more akin to a predicate expressing provability relative to a system than to an unrelativized truth predicate.

So, while it will be possible to find a revenge paradox for the theory containing the truth rules and the paradoxicality rules, there is good reason to believe that such a paradox will again be treatable by our strategy, as can any further analogous revenge paradoxes. This—to treat revenge paradoxes with the same cure 'all the way up' the chain of revenges—is, we submit, the best one can hope for. Now, the classical recapture paradox is a revenge paradox that targets the non-classical features of our theory of truth. In the next section, we deal with another class of revenge paradoxes for our theory, which target its use of rejection.

7.7 Rejectability and revenge

We argued that the Liar sentence and its negation ought to be weakly rejected. In fact, this is a theorem of BML_{Tr}. This suggestion appears to give rise to a revenge paradox. What about the Rejecter sentence which says of itself that it is weakly rejectable (Wright 2012; Bacon 2015)?

Andrew Bacon (2015: 340) formulates a Rejecter Paradox by using a predicate for strong assertibility: the Rejecter is the sentence *a* materially equivalent to *a is not strongly assertible*. We have taken the meaning of the truth predicate to be given by the truth rules, which specify that truth is disquotational under both strong assertion and strong rejection. To specify the meaning of the strong assertibility predicate *StA*, we instead lay down the *strong assertibility rules*, which only specify that strong assertibility is disquotational under strong assertion. That is, the meaning of the strong assertibility predicate is given by rules which tell us that from *A* we can infer *A is strongly assertible* and vice versa.

$$(+StAI.)\ \frac{+A}{+StA\ \ulcorner A \urcorner} \qquad (+StAE.)\ \frac{+StA\ \ulcorner A \urcorner}{+A}$$

Strong assertibility differs from truth in that it is not bivalent: from a sentence being not strongly assertible it does not follow that its negation is strongly assertible. This is as it should be, since it might be, for instance, that both the sentence and its negation have the same unmet presupposition, in which case neither sentence is strongly assertible. From a deflationist perspective, the difference between strong assertibility and truth is that although both can be used to express indirect or compendious endorsement, only the truth predicate can be used to express indirect or compendious opposition, thanks to the strongly rejected truth rules.

Since strong assertion and weak rejection stand in a relation akin to contradictoriness, we can then define weak rejectability as the negation of strong assertibility, that is we take *WRej* to be shorthand for $\neg StA$. The Rejecter sentence is then the sentence a such that $+a \subset\supset WRej\ulcorner a \urcorner$. In the extension of WBL_{Tr} with the strong assertibility rules, one can then show that $+a$ entails $+\neg a$ and vice versa. Hence from the assumption that $+a$ one can derive a contradiction and therefore by Smileian *reductio** that $-a$. From this it follows that $+\neg a$, which entails a contradiction.

This reasoning is analogous to the one used to derive a contradiction from the Liar sentence by using the strongly asserted truth rules in the absence of the appropriate restrictions on Smileian *reductio**. And our diagnosis is the same: the paradox rests on an application of Smileian *reductio** in a derivation that does not preserve evidence. The proofs that $+a$ entails $+\neg a$ and vice versa crucially involve the strong assertibility rules, but the same arguments that showed that the truth rules do not preserve evidence show that the strong assertibility rules do not preserve evidence either. Thus, if we extend WBL_{Tr} with the strong assertibility rules, we must exclude them from Smileian *reductio**. It is possible to show that once we do so, no contradiction follows from the Rejecter sentence a. Like the Liar, both the Rejecter and its negation are to be weakly rejected, but $\ominus a$ and $\ominus\neg a$ are jointly consistent. Our strategy of banning the application in Smileian *reductio** of rules that fail to preserve evidence extends to the Rejecter Paradox.

Following Bacon, we have defined weak rejectability as the negation of strong assertibility. It may appear, however, that our logical framework provides an alternative route to defining weak rejectability. BML includes a weak rejection sign, so one might attempt to characterize weak rejectability directly in terms of it. The obvious way to do so would be to lay down the following rules.

$$(+WRej\text{I.})\ \frac{\ominus A}{+WRej\ulcorner A \urcorner} \qquad (+WRej\text{E.})\ \frac{+WRej\ulcorner A \urcorner}{\ominus A}$$

Murzi and Carrara (2015) propose rules for weak rejectability similar to these and use them to give a version of the Rejecter Paradox which differs from Bacon's. We cannot deal with this version of the paradox in the way we dealt with Bacon's, since the addition of the proposed rules for the weak rejectability predicate to BML_{Tr} leads to inconsistency even if we ban their application within Smileian *reductio**.

The problem for the proposed rules for the rejectability predicate is instead that the introduction rule does not preserve commitment. For we have taken the range of application of the truth predicate to include sentences one does not and cannot understand, and the same must hold for the rejectability predicate. And when A is a sentence one does not and cannot understand, it is perfectly coherent to weakly reject A as well as *A is weakly rejectable*. For instance, in the absence of the guru, the disciple must weakly reject any set-theoretic sentence as well as any sentence saying that a set-theoretic sentence is correctly rejectable. With regards to set-theoretic sentences, the disciple appears to be in a radical case of a *quandary*, a situation 'where we do not know, do not know how we might come to know, and can produce no reason for thinking that there is any way of coming to know what to say or think' (Wright 2001: 71).

We submit, therefore, that weak rejectability ought to be characterized so that it is coextensive with the negation of strong assertibility. We can obtain this result simply by defining weak rejectability as the negation of strong assertibility, as we have shown. But we can also obtain this result by defining weak rejectability directly in terms of weak rejection once we help ourselves to the force marker for weak assertion.

$$(\oplus WRej\text{I.})\ \frac{\ominus A}{\oplus WRej\ulcorner A\urcorner} \qquad (\oplus WRej\text{E.})\ \frac{\oplus WRej\ulcorner A\urcorner}{\ominus A}$$

If *WRej* is defined as $\neg StA$, these rules are derivable from the strong assertibility rules. It follows that these rules must preserve commitment. Nonetheless, by now familiar arguments show that the proper rules for the weak rejectability predicate fail to preserve evidence and must therefore be excluded from Smileian *reductio**. This suffices to avoid the Rejecter Paradox.

Nonetheless, the Rejecter Paradox brings to the fore the fact that certain seemingly valid rules turn out not to be such once we admit sentences that we do not and cannot understand. This aspect is important to address another potential source of paradoxes for our theory of truth, based on its interaction with our theory of epistemic modality.

7.8 Epistemic Liars

We have presented our account of truth using BML as a base theory. The resulting formal theory of truth, BML_{Tr}, is consistent, since it has models obtainable by extending the models for BML. However, if we develop our account of truth using EML as our base theory instead, paradox looms once more: we can use the expressive resources of EML, and in particular the $\Diamond$ and the $\Box$, to formulate *Epistemic Liars*. The Weak Epistemic Liar says of itself that it might be false. That is, it is the sentence $l^{\Diamond}$ such that $+l^{\Diamond} \subset\supset \Diamond\neg Tr\ulcorner l^{\Diamond}\urcorner$. The Strong Epistemic Liar says of itself that it must be false. That is, it is the sentence $l^{\Box}$ such that $+l^{\Box} \subset\supset \Box\neg Tr\ulcorner l^{\Box}\urcorner$. As we show in the Appendix, the assumption that either $l^{\Diamond}$ or $l^{\Box}$ exist leads to contradiction in the system obtained by extending EML with the truth rules but excluding them from Smileian *reductio**.

Both the derivation of contradiction using $l^{\Diamond}$ and the one using $l^{\Box}$ make use of the derivable rule of weakly asserted $\Box$ elimination, which allows us to infer $+A$ from $\oplus\Box A$. The only modal rule employed in the derivation of the weakly asserted $\Box$ elimination rule is the rule of strongly asserted $\Diamond$ introduction, which allow us to infer $+\Diamond A$ from $\oplus A$. These two rules share the following feature: they allow us to pass directly from a weak speech act to a strong one, in this case from a weak to a strong assertion.

However, this feature of the rules is suspicious once we admit sentences in the language that we cannot understand and give rise to quandaries. As we saw in the previous section, the right response to a sentence *A* we cannot understand appears to be to suspend all judgement concerning that sentence, which means that one should weakly reject, and indeed weakly assert, *A* as well as that *A* is weakly rejectable and that *A* is weakly assertible. The same goes for sentences stating that *A* is true, that *A* might be true, or that *A* must be true. But this means that the strongly asserted $\Diamond$ introduction rule is not unrestrictedly valid: it allows us, in effect, to move from the weak assertion of the sentence to the strong assertion of its modalized version.

This by no means entails a wholesale rejection of the rule of strongly asserted $\Diamond$ introduction. In Chapter 4 we defended the rule on the grounds that the weak assertion of *A* and the strong assertion of *It might be that A* appear to be inferentially equivalent: it seems possible to draw the same inferences from these two utterances. The case of Epistemic Liars reveals that this argument must be restricted to sentences that we understand and that have no part that we do not understand, and indeed the examples we provided in Chapter 4 to support the inferential equivalence of the weak assertion of *A* and the strong assertion of *It might be that A* used sentences of this kind.

The foregoing considerations do entail, however, that when we extend EML with a truth predicate, whose range may include sentences that we cannot understand and give rise to quandaries, we must restrict the application of the strongly asserted $\Diamond$ introduction to contexts that do not involve semantic predicates such as *is true*. This means, in particular, that the application of the derived rule of weakly asserted $\Box$ elimination in the derivations of the Epistemic Liar Paradoxes is invalid.

Our solution to the Epistemic Liar Paradoxes accords with the suggestion made in the literature that modal axioms or rules should be restricted in a context in which the language includes a truth predicate. The Weak Epistemic Liar Paradox is, in effect, a version of *Montague's Paradox* (Montague 1963) in our setting. Montague's Paradox is obtained in contexts in which necessity is treated as a predicate by considering a sentence n that says of itself that it is not necessary, that is a sentence n such that $n \subset\supset \neg\Box n$. By identifying $\Box$ with $\neg\Diamond\neg$, we obtain the Weak Epistemic Liar $l^{\Diamond}$. A common diagnosis of Montague's Paradox is that the standard axioms of modal logic are acceptable for modal *operators*, but not for modal *predicates*. Hannes Leitgeb suggests that

> this is the very reason why the systems for modal operators are not affected by paradoxes: since their languages are much more restricted syntactically than languages with modal predicates are, the former can only yield instantiations of [the rules and axioms of modal logic] which are equally restricted. (Leitgeb 2008: 76)

When possibility and necessity are treated as predicates, however, the syntactic options widen to allow, for example, diagonals, and the standard axioms must be restricted.

> [I]n the syntactically more liberal case of languages with modal predicates, we might search for plausible sets of restricted instances of [the rules and axioms of modal logic]. (Leitgeb 2008: 76)

Now, EML sanctions the rules and axioms of the modal logic **S5**. By extending EML with a truth predicate, however, we increase the expressive power afforded by the modal *operator* $\Diamond$: by combining this operator with a truth predicate and using expressions of the form $\Diamond Tr$, we obtain the expressive power of a modal *predicate* for epistemic possibility. But in the presence of a modal predicate, so the diagnosis of Montague's Paradox goes, the received rules of modal logic must be restricted.

The proof-theoretic approach allows us to locate the required restriction precisely. As noted, proof analysis of the Weak and Strong Epistemic Liar Paradoxes reveals that the crucial step in their derivation is an application of the derived rule allowing one to infer $+A$ from $\oplus\Box A$. By inspecting the derivation of this rule, we find that the only modal principle required is the rule allowing us to infer $+\Diamond A$ from $\oplus A$. We have argued on independent grounds that this rule is illicit in contexts involving a truth predicate, which are exactly those in which the expressive power of modal predicates can be operative.[5] Thus, we can develop our theory of truth using EML as the base theory by exercising the requisite care.

7.9 The question of realism

Inferential expressivism about truth holds that the meaning of the truth predicate is given by the truth rules. But what about the *property* of being true?

In a sense, there is a straightforward answer to this question. There are obvious similarities between the semantic thesis of inferential expressivism about truth and the minimalist view that the meaning of the truth predicate is exhausted by all instances of the T-Schema. Now, as noted in Chapter 1, minimalism has a tendency to creep in. Thus, besides subscribing to some form or another of minimalism about truth, sophisticated expressivists tend to also subscribe to minimalist approaches to properties and propositions. It is natural for inferential expressivists to follow suit and adopt minimalist approaches to our talk of properties and propositions, suitably cast within a multilateral framework. In particular, we can take the meaning of the predicate *has the property of being F* to be given by rules allowing us to move from strong assertion or strong rejection of *a is F* to strong assertion or strong rejection of *a has the property of being F* and vice versa. Using these rules, from the assertion of A we can infer the assertion of *A is true*, from which

[5] We argue elsewhere (Incurvati and Schlöder 2022b) that the rule must be completely jettisoned when giving an account of a logic of the *definitely* operator which is not subject to the paradoxes of higher-order vagueness.

in turn we can infer the assertion of *A has the property of being true*. Thus, like other sophisticated expressivists, the inferential expressivist subscribes to the existence of the property of being true, in an appropriately minimal sense of *property*.

But what about a more substantive sense of *property*? The semantic claim of inferential expressivism about truth can be motivated, using the Pragmatist Razor, on the basis of the deflationist insight that the function of the truth predicate is that of performing indirect and compendious endorsements. However, many deflationists do not rest content with this semantic claim. Nor do they rest content with the positive claim that there is a property of truth in the minimal sense of *property*. They are also wont to endorse the negative metaphysical claim that there is nothing more to the property of truth than what is guaranteed by the meaning of the truth predicate as they conceive of it. Thus, for instance, minimalists about truth (Horwich 1998) add that truth is not a 'substantive property', perhaps because it has no explanatory role, perhaps because it is a mere logical property (Field 1999: 534), or perhaps because it is a property without a nature or essence (Horsten 2009: 556).

But the negative metaphysical claim is not forced upon us, and, we contend, is one that the inferential expressivist is not committed to. At the same time, inferential expressivism is not committed to *substantivalism about truth* (Sher 2016), the idea that there exists a property of truth beyond the minimal sense of *property*. Inferential expressivism about truth is neutral about the existence of a substantive property of truth.

The dialectical situation is familiar. Given their acceptance of minimalism about *property*, sophisticated ethical expressivists subscribe to the positive claim that there exist moral properties, in a minimal sense of *property*. Now many ethical expressivists are also quasi-realists: they endorse the negative metaphysical claim that there are no moral properties beyond the minimalist's sense. But this, we argued in Chapter 1, is not part and parcel of the expressivist approach to the meaning and function of moral talk: ethical expressivists can just as well be realists and endorse the positive metaphysical claim that there are substantial moral properties. Ethical expressivism is, strictly speaking, neutral with respect to ontological questions about the existence of moral properties beyond what follows from minimalism.

We encountered another example in the previous chapter. Inferential expressivists about mentalistic talk may endorse a form of minimalism about having attitudes. Given this minimalism, they subscribe to the existence of attitudes in a minimal sense of *attitude*. However, we argued, inferential expressivism is neutral on ontological questions about the existence of attitudes beyond what follows from minimalism.

Thus, inferential expressivism is compatible with both a realist and a quasi-realist understanding of the relevant domain, be it morality, truth, or the mind. But the question we posed in Chapter 1 for non-referentialism remains: is this a distinction with a difference? That is, can a sensible distinction be drawn between a realist inferential expressivist and a quasi-realist inferential expressivist?

Consider the case of truth. The quasi-realist, known in the domain of truth as a minimalist, argues that there is no substantive property of truth. The realist, known in the domain of truth as the substantivalist, argues that the truth predicate latches onto a substantive property. The issue, however, is what it means to say that the truth predicate latches onto a substantive property. Substantivalists typically say that truths correspond to the facts. But talk of *facts* and *correspondence* is easily deflated. We seem to have made no progress in our attempt to draw a distinction.

Similarly, consider the case of morality. The moral quasi-realist argues that there are no substantive moral properties. The moral realist, for her part, holds that moral predicates latch onto substantive properties. The issue, however, is what it means to say that moral predicates latch onto substantive properties rather than simply minimal ones. Indeed, talk of latching on and talk of substantive properties is easily deflated. This, of course, is just the Problem of Creeping Minimalism. The quasi-realist and the realist appear to endorse exactly the same collection of ontological claims. But then, our question rears its head again: is there conceptual space to be a realist inferential expressivist and also space to be a quasi-realist inferential expressivist?

We argued in Chapter 1 that Baker is correct that there is an important distinction to be drawn between those that take the meaning of certain expressions to be explainable solely in terms of their referents and those that take the meaning of those expressions to be explainable without appealing to their referents. This distinction, we argued, is best understood as the distinction between referentialists and non-referentialists. Clearly, however, the distinction between referentialism and non-referentialism will not help to distinguish between an inferential expressivist who is a realist and one who is a quasi-realist, since both characters are non-referentialists: inferential expressivism is committed to the idea that the meaning of a linguistic expression can be explained without appealing to its referent.

If the distinction between inferential expressivists who are realists and ones who are quasi-realists cannot be drawn in terms of their semantic commitments, it is natural to look at their *meta-semantic* commitments. Dreier's original explanation explanation already attempted to distinguish anti-realists from realists in terms of their meta-semantic commitments: unlike the realist, the anti-realist holds that what makes it the case that an expression has the meaning that it does can be explained without appealing to its referent. Thus, in effect, Dreier claims that what distinguishes anti-realists from realists is their answer to the Meaning Determination Question. To be sure, a sensible distinction between those anti-realists who are expressivists, such as quasi-realists, and those anti-realists who are not expressivists, such as error theorists, must also be drawn. But that distinction can be drawn, following Baker, by using the fact that expressivism rejects the error theory's commitment to referentialism. Hence, one may try to use Dreier's meta-semantic explanation to distinguish quasi-realist expressivism from realist expressivism.

However, Dreier's original explanation explanation cannot be used to distinguish between realist inferential expressivists and quasi-realist inferential expressivists either. For inferential expressivism also endorses the meta-semantic claim that the meanings of expressions are determined by inferential relations between attitude expressions. Thus, inferential expressivists give an answer to the Meaning Determination Question that would make them, according to Dreier's original explanation explanation, quasi-realists.

Nonetheless, inferential expressivists may disagree over other meta-semantic claims. In particular, we suggest, it is natural to make sense of the disagreement between realist inferential expressivists and quasi-realist inferential expressivists over whether certain expressions latch onto objects or properties in terms of their meta-semantic commitments about the *etiology* of the meanings of those expressions. Inferential expressivists who are realists and those that are quasi-realists disagree over whether the explanation of the origin of the meanings of terms receiving inferential expressivist treatment must appeal to objects and properties. Thus, for example, an inferential expressivist who is a substantivalist about truth might insist that the natural history of our endorsing practices involves an appeal to the property of truth (see, e.g., Devitt 1997: 325–330). And an inferential expressivist who is a realist about morality might insist that our practices of approving and disapproving (or indeed the very attitudes themselves) evolved in response to the moral properties—that, for instance, there was evolutionary pressure to approve of, hence be motivated to pursue, what is right and to disapprove of, hence be motivated to avoid, what is wrong.

Thus, we can make sense of the realist's claim that our talk about a certain domain latches onto certain objects or properties as the claim that the etiology of that talk involves those objects or properties. This is compatible with subscribing to the semantic and meta-semantic claims associated with inferential expressivism. Quasi-realist inferential expressivists, for their part, will insist that the etiology of our talk about a certain domain need not involve appeal to the objects and properties characteristically associated with that domain. For instance, a quasi-realist inferential expressivist might tell a game-theoretic story about the practices surrounding moral terms being the contingent outcome of a coordination game with multiple stable equilibria, without invoking moral properties. The battleground where realist inferential expressivists and quasi-realist inferential expressivists meet lies within etiology.

Now, there are etiological questions to be asked and possibly they are to be answered. However, answers to the etiological questions do not directly bear on how we use moral language, are motivated by our moral judgements, or know moral claims. Thus, several central themes in meta-ethics do not appear to hinge on the etiological questions and are therefore not informed by the realism/quasi-realism distinction within inferential expressivism. Moreover, there might be no good way of answering whether our practices involving moral attitudes have evolved in response to substantive properties, or are merely outcomes of a coordination game that could have had other outcomes—or whether being a contingent outcome of a coordination game *is* a substantive property. Considerations of this sort might lead one not simply to push back the question of realism into etiology, but to *dissolve* it into etiology. The result would be a view that is neither realist nor quasi-realist, but rather quietist in nature.

Thus, as we argued in Chapter 1, Baker (2021) is right that to distinguish expressivism from the error theory or, more generally, from views that may be considered realist to the extent that they endorse referentialism, one should look at the semantic level. But Dreier (2004) was nevertheless right that important distinctions are to be made at the meta-semantic level.

One avenue for characterizing realism at the meta-semantic level is by focusing, as Dreier did, on the Meaning Determination Question within meta-semantics. We have suggested that another avenue, which makes conceptual room for realist and quasi-realist forms of inferential expressivism, consists in focusing on the etiology of meaning. Realist inferential expressivism is characterized by the claim that some referential talk must be invoked into the etiology of meanings, which captures the idea that those meanings latch onto the relevant objects and properties. Quasi-realist inferential expressivism is characterized by the claim that the correct etiology of meanings does not involve referential talk, which captures the idea that meanings do not latch onto objects and properties. Quietism is to decline to engage in the etiological debate. Naturally, one can be a realist in one domain and a quasi-realist or quietist in another.

In this chapter, we have presented an inferential expressivist account of the meaning of the truth predicate. The account naturally leads to a formal theory of truth which has, we have argued, several appealing features. Notably, the theory avoids the Liar Paradox and its treatment of this paradox can be naturally extended to several revenge paradoxes. The general strategy to avoid the semantic paradoxes is the same as the one that allowed our theory of epistemic modality to avoid modal collapse, namely placing principled and motivated restrictions on certain meta-rules of classical logic, including the conditional proof rule. In the next chapter, we will see that this strategy also proves fruitful when developing an inferential expressivist account of conditionals.

7.10 Appendix

We present the proofs of the Epistemic Liar Paradoxes. We begin with the Weak Epistemic Liar, which uses the sentence $l^{\Diamond}$ such that $+l^{\Diamond} \subset\supset \Diamond\neg Tr\ulcorner l^{\Diamond}\urcorner$. We first show that $\vdash^{\mathsf{EML}} -l^{\Diamond}$.

$$
\dfrac{\dfrac{\dfrac{\dfrac{\dfrac{[+l^{\Diamond}]^1 \qquad +l^{\Diamond} \subset\supset \Diamond\neg Tr\ulcorner l^{\Diamond}\urcorner}{+\Diamond\neg Tr\ulcorner l^{\Diamond}\urcorner}\,(+\supset\text{E.})}{\oplus\neg Tr\ulcorner l^{\Diamond}\urcorner}\,(+\Diamond\text{E.})}{\ominus Tr\ulcorner l^{\Diamond}\urcorner}\,(\oplus\neg\text{E.}) \qquad \dfrac{[+l^{\Diamond}]^1}{+Tr\ulcorner l^{\Diamond}\urcorner}\,(+Tr\text{I.})}{\bot}\,(\text{Weak Rejection})}{\ominus l^{\Diamond}}\,(\text{SR}_1)^1
$$

We then use this result to derive a contradiction in EML from the assumption that $l^{\Diamond}$ exists.

$$\dfrac{\dfrac{\dfrac{\dfrac{\dfrac{\ominus l^{\Diamond} \qquad \dfrac{+l^{\Diamond} \subset\!\supset \Diamond\neg \mathit{Tr}\ulcorner l^{\Diamond}\urcorner \qquad \dfrac{[-\neg\Diamond\neg \mathit{Tr}\ulcorner l^{\Diamond}\urcorner]^1}{+\Diamond\neg \mathit{Tr}\ulcorner l^{\Diamond}\urcorner]^1}\,(+\neg\mathrm{I.})}{+l^{\Diamond}}\,(+\supset\mathrm{E.})}{\bot}\,\text{(Weak Rejection)}}{\oplus\Box \mathit{Tr}\ulcorner l^{\Diamond}\urcorner}\,(\mathrm{SR}_4)^1}{+\mathit{Tr}\ulcorner l^{\Diamond}\urcorner}\,(\oplus\Box\mathrm{E.})}{+l^{\Diamond}}\,(+\mathit{Tr}\mathrm{E.}) \qquad \ominus l^{\Diamond}}{\bot}\,\text{(Weak Rejection)}$$

The Strong Epistemic Liar Paradox, which makes use of the sentence $l^{\Box}$ such that $+l^{\Box} \subset\!\supset \Box\neg \mathit{Tr}\ulcorner l^{\Diamond}\urcorner$, proceeds in a similar fashion. We first show that $\vdash^{\mathsf{EML}} \oplus l^{\Box}$.

$$\dfrac{\dfrac{[-l^{\Box}]^1 \qquad \dfrac{\dfrac{\dfrac{\dfrac{[-l^{\Box}]^1}{-\mathit{Tr}\ulcorner l^{\Box}\urcorner}\,(-\mathit{Tr}\mathrm{I.})}{+\neg \mathit{Tr}\ulcorner l^{\Box}\urcorner}\,(+\neg\mathrm{I.})}{+\Box\neg \mathit{Tr}\ulcorner l^{\Box}\urcorner}\,(+\Box\mathrm{I.}) \qquad +l^{\Box} \subset\!\supset \Box\neg \mathit{Tr}\ulcorner l^{\Box}\urcorner}{+l^{\Box}}\,(+\supset\mathrm{E.})}{\bot}\,\text{(Strong Rejection)}}{\oplus l^{\Box}}\,(\mathrm{SR}_4)^1$$

We then use this result to derive a contradiction in EML from the assumption that $l^{\Box}$ exists.

$$\dfrac{\dfrac{\dfrac{\dfrac{\dfrac{\oplus l^{\Box} \qquad \dfrac{\dfrac{\dfrac{+l^{\Box} \subset\!\supset \Box\neg \mathit{Tr}\ulcorner l^{\Box}\urcorner}{+\neg l^{\Box} \subset\!\supset \neg\Box\neg \mathit{Tr}\ulcorner l^{\Box}\urcorner}\,\text{(Contraposition)} \qquad \dfrac{[-\Box\neg \mathit{Tr}\ulcorner l^{\Box}\urcorner]^1}{+\neg\Box\neg \mathit{Tr}\ulcorner l^{\Box}\urcorner}\,(+\neg\mathrm{I.})}{+\neg l^{\Box}}\,(+\supset\mathrm{E.})}{-l^{\Box}}\,(+\neg\mathrm{E.})}{\bot}\,\text{(Weak Assertion)}}{\oplus\Box\neg \mathit{Tr}\ulcorner l^{\Box}\urcorner}\,(\mathrm{SR}_4)^1}{+\neg \mathit{Tr}\ulcorner l^{\Box}\urcorner}\,(\oplus\Box\mathrm{E.})}{-\mathit{Tr}\ulcorner l^{\Box}\urcorner}\,(+\neg\mathrm{E.})}{-l^{\Box}}\,(-\mathit{Tr}\mathrm{E.}) \qquad \oplus l^{\Box}}{\bot}\,\text{(Weak Assertion)}$$

8
Conditionals

We have so far dealt with speech acts, such as assertion, whose force applies to a single sentence. A natural generalization of the multilateral framework makes use of *binary* speech acts, such as *conditional assertion*, whose force relates two sentences. In this chapter, we use the idea of a binary speech act to provide an inferential expressivist treatment of indicative and subjunctive conditionals. Employing a by now familiar strategy, we explain the meaning of embeddable conditional operators in terms of their inferential relation to unembeddable conditional speech acts. We show how the resulting account of the indicative conditional meets the challenge posed by the Gibbard Collapse Argument and, when combined with the theory of truth from the previous chapter, solves Curry's Paradox.

8.1 Binary speech acts

As observed in Chapter 4, linguistic items which appear to modify the force of a speech act, rather than its content, can occur in conditional consequents. This includes force modifiers such as *perhaps* and force indicators such as answers to self-posed questions.

(1) a. If it is going to rain, perhaps we should stay in.
    b. If it is going to rain, should we stay in? Yes!

A natural way to account for these examples is to treat them not as *speech acts with conditional content* but as *conditional speech acts*: the first sentence would serve to perform a conditional weak assertion, and the latter to perform a conditional strong assertion—or, for brevity, a conditional assertion. Quine articulates the view that *If A, then B* ought to be analysed as a conditional assertion.

Reasoning with Attitude: Foundations and Applications of Inferential Expressivism. Luca Incurvati and Julian J. Schlöder, Oxford University Press. © Oxford University Press 2023.
DOI: 10.1093/oso/9780197620984.003.0008

> An affirmation of the form "if p then q" is commonly felt less as an affirmation of a conditional than as a conditional affirmation of the consequent. If, after we have made such an affirmation, the antecedent turns out true, then we consider ourselves committed to the consequent, and are ready to acknowledge error if it proves false. (Quine 1950: 12)

Applied to one of our examples, this means that someone uttering *If it is going to rain, then should we stay in? Yes!* is not categorically asserting the conditional content *if it is going to rain, we should stay in*. Rather, they are asserting the categorical content *we should stay in* conditionally on *it is going to rain*. If it is not going to rain, the speaker has not asserted anything at all. The latter part of the proposal has however raised some eyebrows. For Quine (perhaps too hastily) continues as follows.

> If on the other hand the antecedent turns out to have been false, our conditional affirmation is as if it had never been made. (Quine 1950: 44)

This is, in Dorothy Edgington's (1995: 19) words, 'absurd'. But it is 'not absurd if we delete the word "conditional" from it' (p. 19). If the antecedent turns out to be false, no assertion has been made, but a *conditional* assertion has nevertheless been made, or so Edgington contends. The distinction matters since, even if the antecedent turns out to be false, a conditional assertion can feature in certain inferences. For example, 'someone who believes the speaker, and knows the consequent to be false, may infer the falsity of the antecedent' (Dummett 1991b: 115).

Others disagree. Simon Goldstein (2019) formalizes a theory of conditional assertion that takes Quine literally. On Goldstein's account, a conditional assertion is 'screened off' (p. 300) by the negation of its antecedent. Goldstein implements this idea within a dynamic semantic framework in which meanings are given by how epistemic states are updated. The content of a sentence is an *update potential*, that is an instruction for updating one's epistemic state by ruling out alternatives. Goldstein defines a conditional assertion operator ⇒ so that updating a context with $A \Rightarrow B$ and $\neg A$ is the same as updating it with just $\neg A$. It follows that when a conditional assertion is followed by the negation of its antecedent, it 'might as well never have happened' (p. 300), at least as far as the state of the context is concerned. Goldstein is of course aware of the type of criticism raised by Edgington and

Dummett. He cites an example by Lycan (2006), which we can phrase as a conditional assertion as follows.

(2) If Congress passes a health-care bill, will I sign it? Yes!

Uttered by the president, (2) may be said to affect the context even if Congress ultimately does not pass a health-care bill. For example, its utterance might influence future votes or tell us something about the dispositions of the speaker. These context changes effected by (2) are not screened off by learning or asserting the negation of its antecedent. Goldstein responds that such context updates are inferred from *the utterance* of (2), but not from the *speech act* of conditional assertion that is performed by (2). For we may assume that the speaker

> takes themselves to satisfy the normative requirements on uttering [(2) which] requires the speaker to have certain intentions with respect to health-care repeal in general. Even if the possession of such intentions is not part of the actual meaning of the conditional above, such information can still be gleaned from an utterance of it. (Goldstein 2019: 299, fn. 16)

This is not a convincing response to the challenge posed by cases such as (2). Goldstein can tell a story involving the speaker's own intentions only because the consequent of (2) features an assertion related to the expression of these intentions. But the problem remains when we consider cases which do not have this feature.

(3) If Congress passes a health-care bill, the president will sign it.

From this utterance, we learn something about *the president* even if Congress does not pass a health-care bill. From the speaker taking themselves to satisfy the normative requirements for uttering (3) we can infer something about *their beliefs* about the president. But this falls short of learning something about the president. Thus, (3) makes a contribution to the context not merely in virtue of the attitudes of the speaker, but in virtue of its content, even if the antecedent is false. It is not as if its utterance never happened when Congress fails to pass a health-care bill.

The lesson, we contend, is that *conditional assertion*—or, strictly speaking, *conditional strong assertion*—is a *sui generis* speech act. It affects the context in a distinctive way, even if the antecedent is false. It therefore cannot be analysed by simply giving an account of assertion and a story about how speech acts can be performed conditionally. Approaching a formalization will make the picture clearer. Compare the following two possible ways of formalizing conditional assertion.

(4) $A \Rightarrow +B$.

(5) $\Rightarrow_+ (A, B)$.

In the first formalization, which uses $\Rightarrow$ to denote some operation to conditionalize speech acts, we have the assertion of B conditional on some sentence A. In the second formalization, by contrast, A and B occur as the contents to which a single force marker $\Rightarrow_+$ for conditional assertion is applied. It is the second formalization that gives the correct picture of conditional assertion. We say that $\Rightarrow_+$ is a *binary force marker* since it is used to indicate a *single* speech act with two content sentences.[1] At least for the case of conditional questions, the idea was already anticipated by Dummett.

> [T]he whole context 'If A, then (?...)' should be taken as constituting a single force-operator, namely one signalizing the asking of a question conditionally upon its being the case that A; or, better, the context 'If..., then (?...)' should be regarded as a force-indicator with two argument-places. (Dummett 1973a: 339)

But what is denoted by $\Rightarrow_+ (A, B)$ if not an assertion of B, performed conditionally on A? To answer this question, we distinguish, as usual, between two dimensions of speech acts, the attitude they express and their essential effect on the conversation. Let us start from the attitude expressed by conditional assertion. In making a conditional assertion, one expresses an attitude of conditional belief, rather than a belief with conditional content.[2] We can

[1] Once we countenance force markers of different arities, we can also give a clearer explanation of $\bot$ as a *nullary force marker*. It indicates a force—the force of announcing *Contradiction!*—but has no content. Thus, we can distinguish the act of announcing contradiction (a nullary speech act with no content) from the act of asserting *a* contradiction (a unary speech act with a particular, contradictory content).

[2] See Ferrero 2009 and Lennertz 2021 for further discussion of conditional attitudes as distinct from attitudes with conditional content.

shed light on conditional belief by starting from the Quinean picture. Quine is essentially correct that if the speaker asserts *B* conditionally on *A* and *A* is the case, then they are committed to *B*. Only a minor correction is required. It is not that the speaker is committed to *B* if *A* is the case, but only if they are committed to *A*. For if the speaker is in a position to reject *A*, then they are not committed to *B*. By making the commitment to *attitudes* towards sentences explicit, this delivers the principle that if someone conditionally asserts *B* on *A* and they are committed to expressing belief towards *A*, then we may infer that they are also committed to expressing belief towards *B*. This principle is formally encapsulated by the following *modus ponens*-esque inference rule.

$$(\Rightarrow_{+}\mathrm{MP})\ \frac{+A \qquad \Rightarrow_{+}(A, B)}{+B}$$

Similar considerations apply with regards to the essential effect of conditional assertion on the conversation. In particular, the essential effect of a conditional assertion of *B* on *A* is a proposal to change the context so that *B* becomes common ground upon *A* becoming common ground.[3]

We now have all ingredients needed to apply the multilateral methodology to extend the multilateral framework with a binary force marker $\Rightarrow_{+}$ for conditional strong assertion. We have a linguistic realization of conditional strong assertion by means of utterances such as *If it is going to rain, then should we stay in? Yes!*. And we have a coordination principle specifying how this speech act interacts with others, namely ($\Rightarrow_{+}$MP). Once embedded within the multilateral framework, the coordination principle ($\Rightarrow_{+}$MP) allows us to explain Dummett's *modus tollens*-esque inferences. For the coordination principle immediately yields the following derived rule.

$$\frac{+\neg B \qquad \Rightarrow_{+}(A, B)}{+\neg A}$$

The derivation is a straightforward application of Smileian *reductio**.

[3] This can be contrasted with the essential effect of asserting a conditional *If A, then B*, which consists in a proposal to add this conditional to the common ground. This also commits the speaker to proposing or accepting a proposal that the context is changed so that *B* becomes common ground upon *A* becoming common ground, but this is not an *essential* effect, but a consequence of the essential effect of assertion and the meaning of the embeddable conditional.

$$\dfrac{\dfrac{+\neg B}{-B}\,(+\neg\text{E.}) \qquad \dfrac{\Rightarrow_{+}(A,B) \qquad [+A]^{1}}{+B}\,(\Rightarrow_{+}\text{MP})}{\dfrac{\dfrac{\bot}{-A}\,(\text{SR}_1{}^{*})^{1}}{+\neg A}\,(+\neg\text{I.})}\,(\text{Strong Rejection})$$

We can describe further conditional speech acts. In particular, we will make use of binary force markers for conditional strong rejection $\Rightarrow_{-}$, conditional weak assertion $\Rightarrow_{\oplus}$ and conditional weak rejection $\Rightarrow_{\ominus}$. Their interactions with other speech acts are also governed by versions of *modus ponens.*

$$(\Rightarrow_{-}\text{MP})\ \dfrac{+A \qquad \Rightarrow_{-}(A,B)}{-B} \qquad (\Rightarrow_{\oplus}\text{MP})\ \dfrac{+A \qquad \Rightarrow_{\oplus}(A,B)}{\oplus B}$$

$$(\Rightarrow_{\ominus}\text{MP})\ \dfrac{+A \qquad \Rightarrow_{\ominus}(A,B)}{\ominus B}$$

Moreover, conditional speech acts can be performed using the indicative and the subjunctive mood. We may utter *If it is going to rain, should we stay in? Yes!* and thereby perform an indicative conditional assertion. But we may also utter *If it were going to rain, would we stay in? Yes!* and thereby perform a subjunctive conditional assertion. Thus in addition to the indicative conditional force markers ($\Rightarrow_{+}, \Rightarrow_{-}, \Rightarrow_{\oplus}, \Rightarrow_{\ominus}$) we must consider their subjunctive analogues ($\Box\!\!\Rightarrow_{+}, \Box\!\!\Rightarrow_{-}, \Box\!\!\Rightarrow_{\oplus}, \Box\!\!\Rightarrow_{\ominus}$), whose behaviour is similarly governed by a version of *modus ponens.*

Besides being governed by a *modus ponens*-esque coordination principle, the behaviour of each binary force marker is also governed by a conditional proof-esque rule, specifying when a speaker is committed to the attitude expressed by a conditional speech act. We return to these additional coordination principles and the differences between subjunctive and indicative conditionals below. First, we must address a more pressing issue. Giving a formal account of conditional speech acts does not suffice to explain the meaning of conditionals in general. This is because conditionals can embed, leading to a familiar problem.

8.2 Frege–Geach for conditionals

Like *might* and *wrong*, conditionals can embed under negations and in conditional antecedents. So a version of the Frege–Geach argument is relevant here (see Dummett 1973a: 346–354 for an early discussion of this problem as it relates to conditionals).

(6) If the light goes on if you press the switch, the electrician has been.[4]

(7) It is not the case that if you press the switch, the light will go on.

If one wishes to analyse assertions of conditionals as conditional assertions, one must explain the meaning of (6) and (7). Someone asserting (6) need not believe that the light will go on conditionally on the switch being pressed, and similarly for (7). Thus, neither embedded conditional can be analysed as a conditional assertion. This is the Frege–Geach Problem for conditional assertions.

However, unlike other versions of the Frege–Geach Problem we have discussed so far, there is some space to resist the embedding data. Conditionals in conditional antecedents and conditionals under negation may not sound entirely natural. Gibbard (1980: 235) claims that many conditionals embedded in conditional antecedents are hard to understand. 'Suppose I tell you, of a conference you don't know much about, *If Kripke was there if Strawson was, then Anscombe was there*. Do you know what you have been told?' Dummett (1991b: 171) notes that it is neither 'our normal practice to apply negation to an entire conditional statement' nor to use 'conditional sentences ...in which the antecedent is itself a conditional'. Hence, he contends, 'we cannot grasp [their] content'. Edgington (1995: 284) agrees and concludes that '[c]onditionals do not go into truth-functional contexts, or into each other, easily'. All acknowledge, however, that some felicitous embeddings of conditionals are possible, which remains in need of explanation.

Embeddings under negation and in conditional antecedents are the historically most prominent versions of the Frege–Geach Problem (see Chapter 1 for the conditional version and Chapter 5 for the negation version). For current purposes, however, we can obtain a less controversial instance of embeddability of *if* by embedding a conditional in the *consequent* of another.

[4] (6) is a minor variation of an example by Edgington (1995: 283).

(8) If the electrician has been, then if you press the switch, the light will go on.

If we were to insist that *if* is a force indicator for conditional assertion, the interpretation of (8) would be a conditional, conditional assertion. At first, this may not appear to be particularly problematic. We have argued that a conditional assertion is not an assertion performed conditionally, but a *sui generis* speech act. So in principle, it could be possible to conditionally perform a conditional assertion as well. However, this would again be a *sui generis* speech act. By iterating this procedure, we would then obtain a *sui generis* speech act of conditional^n assertion for arbitrary *n*. This is implausible. Although not exactly analogous to the Frege–Geach argument, examples such as (8) do seem to support the claim that *if* does not only function as a force indicator, but also as an embeddable operator.

There is still room for manoeuvre, however. One can read a conditional with a conditional consequent as a simple conditional assertion with a conjunctive antecedent, combining the antecedent of the embedded and the antecedent of the embedding conditional. Indeed, (8) might appear to have the same meaning as the following.

(9) If the electrician has been and you press the switch, then the light will go on.

The claim that (8) is *equivalent* to (9) is known as the *Import-Export Principle*, which will become relevant below. To avoid the embedding problem, however, something stronger than equivalence is required. For to avoid having to tell a story about embedded conditionals, it must be that *if* just *does not* embed. If an embedded *if* is merely equivalent to an unembedded *if*, one still needs to assign a meaning to the embedded *if* to explain the equivalence. If one does not assign any meaning to embedded *if*, then (8) is meaningless instead of equivalent to the meaningful (9). To explain *away* the embedding problem posed by (8), one needs to defend the claim that (8) is analysed as having the same logical form as (9) and that in this form there is no embedded conditional. Likely, this would take the form of a syntactic explanation as to why (8) is parsed as (9). To our knowledge, no such explanation has been attempted. A similar but more promising strategy can be applied to cases such as (7), where *if* appears to be embedded under negation. According to Kratzer's (1981) influential account of the indicative conditional, the surface syntax of a negated conditional is misleading. Properly parsed, the story goes, the negation is scoped inside the conditional's consequent: the proper logical form of (7) is the same as the one suggested by the grammatical form of the following conditional.

(10) If you press the switch, it is not the case that the light will go on.

Again, it does not suffice to claim that the contents of (7) and (10) are equivalent, since in that case one still needs to assign a content to the embedded conditional in (7). And, again, claims of equivalence are connected with a famous logical principle, in this case the *Law of Conditional Excluded Middle*, which states that for any *A* and *B*, either *If A, then B* or *If A, then not B*. By disjunctive syllogism it follows from this principle that the negation of a conditional entails the conditional with its consequent negated (and the converse is uncontroversial). The debate surrounding the Law of Conditional Excluded Middle notwithstanding, Kratzer's analysis involves the right kind of syntactic explanation to explain away conditionals under negation. Her claim is not that negated conditionals are equivalent to conditionals with negative content, but rather that there are syntactic reasons as to why a negation that on the surface seems to scope over a conditional must in fact be analysed as scoping inside the consequent.

This strategy can be naturally extended to account for conditionals under other scope-taking operators, such as epistemic modals. The suggestion would be that *It might be that if A, then B* is interpreted as *If A, then it might be that B* (see Ciardelli 2021 for a recent development of this strategy). But suppose that conditionals in conditional antecedents are set aside, that a syntactic strategy is successful in explaining conditionals under negation and modals, and that a similar strategy is used to account for conditionals in conditional consequents. There remain nonetheless linguistic data that cannot be explained if one takes *if* to always denote a force indicator, since conditionals can embed under quantifiers.

(11) There is a man here who, if you ask him, will help you.

(12) Every woman here, if you ask her, will help you.

The quantifier cannot be moved into the consequent of these conditionals since the scope of the quantifier includes their antecedents.[5]

A variant of this argument is due to Max Kölbel (2000: 100–101). He observes that even if conditionals embedded under negation, disjunction, or other conditionals are explainable or suspect, conditionals embedded under conjunction are beyond reproach.

(13) If I give him food, he will wag his tail and if I don't give him food he will bark.

One might be tempted to say that this is not an assertion of two embedded conditionals, but two assertions expressed by a single conjunctive sentence. But this is hopeless. Asserting a conjunction is distinct from making two assertions. In particular, as Kölbel notes, this can be seen by embedding (13) under a quantifier.

(14) There is a dog who, if I give him food, will wag his tail and if I don't give him food, will bark.

There is no good way to read an utterance of (14) as the performance of two distinct assertions.[6] From the functionalist point of view that we have adopted, furthermore, the data presented by Kölbel suggest that the *point* of having an embeddable conjunction operator is to allow embeddings of the kind he considers. One cannot embed two assertions, but the meaning-conferring rules for conjunction ensure that conjunction is suitably inferentially related to making two assertions, which allows us to utter sentences such as (14). So embeddings under conjunction are genuine, with all the implications for the Frege–Geach Problem that this brings.[7]

[5] There is perhaps a reading of *A man here will help you if you ask him* as equivalent to (11) where the quantifier is scoped inside the conditional consequent and *him* is resolved by anaphoric binding. This does not work for (12). The only available reading of *Every woman here will help you if you ask her* takes the quantifier as scoping over the entire conditional.

[6] Lennertz (2021) argues in favour of *quantificational attitudes* that are not reducible to simple attitudes held towards quantificational contents. One might therefore attempt to analyse (11) and (12) as expressing quantificational, conditional attitudes. But Kölbel shows that this is hopeless since example (14) would have to be analysed as expressing a quantificational, conjunctive, conditional attitude. Thus, such a strategy brings us back to having logical vocabulary generating logically complex attitudes, so the strategy would be confronted with the explanatory challenge raised by the Frege–Geach Problem (see Chapter 5).

[7] This does not call for a binary 'conjunctive speech act', since the meaning of conjunction is perfectly well explained in terms of strong assertion.

Thus, the Frege–Geach argument has teeth here after all. Conditionals embed at least under conjunction and quantification, while conditional assertions cannot embed in this way. Having established that conditionals can embed, we will admit the whole range of possible embeddings. That is, we will assume that conditionals also embed under negation, modals, and in other conditionals. Even sceptics about embedded conditionals such as Dummett and Edgington admit that cases such as (6) or (7) have some fringe uses. But fringe or not, it seems clear that the Frege–Geach argument puts pressure on the idea of analysing conditional utterances *only* as conditional speech acts.

At this point, one might conclude that the embedding data refute the idea of conditional assertions altogether. But this does not follow. There is a difference between embeddable uses of *if* and the linguistic realization of conditional assertion, *If A, then is it B? Yes!*. Trying to embed a conditional assertion results in gibberish, as witnessed by the following examples for the cases of quantification and negation.

(15) # There is a man here who, if you ask him, will he help you? Yes!

(16) # It is not the case that, if you press the switch, will the light go on? Yes!

So the fact that *if* embeds does not prima facie threaten the analysis of *If A, then is it B? Yes!* as serving to perform a conditional assertion.[8] It does however show that we need to account for embeddable uses of *if*. Our solution, as can be expected at this point, is to take such uses of *if* not to serve to perform conditional speech acts. Nonetheless, the meaning of embeddable *if* can be inferentially explained in terms of conditional speech acts.[9] The following rules allow us to pass from indicative conditional strong assertion to the strong assertion of an indicative conditional and vice versa.

[8] One may worry whether this argument overgenerates. Can we also show that there are binary force indicators for conjunctive and disjunctive speech acts by finding a realization in terms of self-posed questions and observing that embedding these results in gibberish? We cannot, since this plan fails in its first step. Trying to disjoin two assertions already results in gibberish (*# Is it raining? Yes! Or is it hailing? Yes!*) and having an assertion depend disjunctively on a sentence in the same way in which in a conditional assertion an assertion depends conditionally on a sentence results in gibberish too (*# It is raining or is it hailing? Yes!*). Analogous considerations apply to the case of conjunction.

[9] Why, then, are conditional assertions and assertions of conditionals both realized using *if*? It is plausible enough that very similar or even on the surface identical expressions are used both for a speech act and for the corresponding embeddable operator. Thus, it may be a simple matter of economy that we only find a single expression for conditionals in spite of its two uses being different. Possibly, *if* in English functions as an underspecified expression of conditionality *tout court*.

$$(+\rightarrow\text{I.})\ \frac{\Rightarrow_+ (A,B)}{+A \rightarrow B} \qquad (+\rightarrow\text{E.})\ \frac{+A \rightarrow B}{\Rightarrow_+ (A,B)}$$

And the following rules allow us to pass from indicative conditional strong rejection to the strong rejection of an indicative conditional and vice versa.

$$(-\rightarrow\text{I.})\ \frac{\Rightarrow_- (A,B)}{-A \rightarrow B} \qquad (-\rightarrow\text{E.})\ \frac{-A \rightarrow B}{\Rightarrow_- (A,B)}$$

Analogous rules govern the inferential relation between indicative conditional weak assertion and the weak assertion of an indicative conditional and between indicative conditional weak rejection and the weak rejection of an indicative conditional. Furthermore, we have analogous rules governing the inferential relation between the subjunctive versions of these conditional speech acts and embedded subjunctive conditionals.

Before moving on, let us highlight some salient and immediate advantages of defining the meaning of embedded conditionals in this fashion. Edgington (1995: 283) suggested that when interpreting a case such as (6), where a conditional occurs as a conditional antecedent, one selects a non-conditional sentence *D* that is the 'basis' for the embedded conditional and reads *If, if A, B, then C* as *If D, then C*. For (6)—where *if A, B* is *if you press the switch, the light goes on*—Edgington suggests that *D* is *the power is on*. But this is not very promising as a general strategy. One may utter (6) in contexts where it is not at stake whether there is power, but whether the switch is broken. So the selection of an appropriate *D* is ad hoc and Edgington's suggestion leaves the meaning of an embedded conditional underdetermined. It is also unclear whether a basis *D* that is not itself a conditional can be found in all cases. To be sure, this may just be grist to Edgington's mill. Her claim is that the interpretation of embedded conditionals is heuristic so she may not be worried about the selection of *D* being ad hoc and sometimes even impossible. But given the embedding data we surveyed here, we find this unsatisfying.

In any case, the Frege–Geach Problem shows that Edgington was right that embedded conditionals cannot be analysed as conditional assertions. She was also right that one must identify a sentence *D* equivalent to the conditional assertion *if A, B* to interpret embedded uses. But it is not necessary to demand that *D* is not itself a conditional, as long as it is not a conditional *speech act*. On our account, the 'basis' for the embedded *if A, B* is the conditional $A \rightarrow B$, whose assertion is equivalent, but not identical, to the conditional assertion $\Rightarrow_+ (A, B)$. This, we submit, is the only principled and uniform manner to select a sentence D for embedded use whose assertion is equivalent to $\Rightarrow_+ (A, B)$.

Another immediate advantage of defining the meaning of embedded conditionals in terms of the corresponding conditional speech acts concerns Edgington's (1995: 283) defence of the claim that conditionals do not embed under negation. Edgington defends this claim by noting that negating a conditional sounds the same as negating its consequent. As already mentioned, we agree that negating a conditional and negating its consequent give rise to *equivalent* conditionals. But we contend that this does not entail that the negated conditional can be disregarded. Rather, the equivalence between the negated conditional and the conditional with the negated consequent is something to be explained. In endorsing and aiming to explain the equivalence, we also depart from Kratzer, who analyses away negated conditionals. Within our framework, we can explain the equivalence in a straightforward and immediate way. The following derivation shows that asserting the negation of a conditional is equivalent to conditionally asserting the conditional with its consequent negated.

$$
\dfrac{\dfrac{\dfrac{\dfrac{\dfrac{\dfrac{+\neg(A \rightarrow B)}{-(A \rightarrow B)}\,(+\neg\text{E.})}{\Rightarrow_- (A, B)}\,(-\rightarrow\text{E.}) \qquad [+A]^1}{-B}\,(\Rightarrow_-\text{MP})}{+\neg B}\,(+\neg\text{I.})}{\Rightarrow_+ (A, \neg B)}\,(\Rightarrow_+\text{CP})^1}{+(A \rightarrow \neg B)}\,(+\rightarrow\text{I.})
\qquad
\dfrac{\dfrac{\dfrac{\dfrac{\dfrac{+(A \rightarrow \neg B) \quad [+A]^1}{+\neg B}\,(\Rightarrow_+\text{MP})}{-B}\,(+\neg\text{E.})}{\Rightarrow_- (A, B)}\,(\Rightarrow_-\text{CP})^1}{-A \rightarrow B}\,(-\rightarrow\text{I.})}{+\neg(A \rightarrow B)}\,(+\neg\text{I.})
$$

These derivations make use of the principles ($\Rightarrow_+$CP) and ($\Rightarrow_-$CP), which have the same shape as the conditional proof rule for the material conditional but govern the behaviour of the conditional speech-act force indicators. We now turn to the discussion and precise formulation of principles of this kind for both indicative and subjunctive conditional speech acts.

8.3 Counterfactuals and counterepistemics

In the presence of the inferential equivalence of $+A \rightarrow B$ and $\Rightarrow_+ (A, B)$, the inference rule ($\Rightarrow_+$MP) tells us immediately how to eliminate $\rightarrow$ under $+$.

But how are we to introduce $\rightarrow$ under $+$? An attractive option is to take indicative conditional assertions to be governed by *Conditional Proof*.

$$\frac{\begin{array}{c}[+A]\\ \vdots \\ +B\end{array}}{\Rightarrow_{+}(A,B)}$$

The adoption of a conditional proof rule can be motivated by appealing to a particular interpretation of the Ramsey test, so called because it was first enunciated by Ramsey in 'General propositions and causality', written in 1929 and published posthumously. Ramsey writes:

> If two people are arguing 'If *p* will *q*?' and are both in doubt as to *p*, they are adding *p* hypothetically to their stock of knowledge and arguing on that basis about *q*. (Ramsey 1931a: 247)

In presenting his test, Ramsey makes use of a conditional performance of a speech act, namely a conditional question. The purpose of the test is to determine whether to reply *Yes!* to the conditional question, so whether or not to perform a conditional assertion. The test can be used to establish whether an agent is committed to expressing belief towards *B* conditionally on *A*. In particular, we hypothetically add expressing belief towards *A* to the agent's commitments and check whether we can then show that the agent is committed to expressing belief towards *B*. If so, we may consider the agent committed to the conditional belief expressed by the conditional assertion *If A, then B? Yes!*.

But how can we introduce a subjunctive conditional assertion, $\square\!\!\Rightarrow_{+}$? If we take both subjunctive conditional assertions and indicative conditional assertions to obey unrestricted versions of the conditional proof rule, it is not clear how we can distinguish between indicative and subjunctive conditionals. To address this problem, we must attend to the difference between indicative and subjunctive conditionals. This will reveal that a suitable version of the conditional proof rule for subjunctive conditional assertion should disallow the use of arbitrary side premisses.

Thus, like Ginger Schultheis (Forthcoming: 2), we will develop an account on which 'all of the semantic differences between indicative conditionals and counterfactual conditionals boil down to differences in what [background assumptions are] held fixed'. This claim will be true in a very literal sense in our semantics, since the only difference between the inference rules governing indicatives and those governing subjunctives will be in the side premisses allowed by their respective versions of conditional proof. Schultheis—following what we take to be a near consensus on the matter—goes on to make the further claims that when we evaluate 'an indicative conditional, we hold fixed all of our knowledge' (p. 2) and that this is how indicatives differ from counterfactual conditionals, which are evaluated by only holding 'fixed a contextually-determined subset of our knowledge' (p. 2). We will take issue with these further claims, since there are also some background assumptions one cannot use when evaluating indicative conditionals.

To appreciate why it is a mistake to allow arbitrary side premisses when introducing subjunctive conditionals, consider the following famous examples by Ernest Adams (1970).

(17) a. If Oswald did not shoot Kennedy, someone else did.
b. If Oswald had not shot Kennedy, someone else would have.

One would typically accept (17a), but reject (17b). When we try to introduce (17a), we are considering the hypothetical scenario in which Oswald did not shoot Kennedy. Since we know that Kennedy was in fact shot, we conclude that, in this scenario, someone else did. Thus, by a version of the Ramsey test or the conditional proof rule, we can introduce the indicative conditional (17a). Hence, we accept it. When we try to introduce (17b), we are also considering a hypothetical scenario in which Oswald did not shoot Kennedy, but set aside our knowledge that he was in fact shot. Since we have no reason to assume that in the hypothetical scenario anyone shot Kennedy, we cannot introduce the subjunctive conditional (17b). Hence we reject it.

From an inferentialist perspective, the difference between indicative and subjunctive conditionals exhibited by Adams's examples is a difference in the availability of background assumptions in the hypothetical scenario in which one assumes the antecedent and attempts to reason towards the conclusion. When trying to introduce subjunctive conditionals, one cannot make use of certain background assumptions, such as the assumption that Kennedy was shot, that are available when evaluating indicative conditionals.

Nonetheless, as anticipated, it is also a mistake to allow arbitrary side premisses when introducing indicative conditionals. Consider the following indicative conditional.

(18) If it is not raining, it might be hailing.

This conditional can be uttered in a context in which the speaker is unsure whether it is raining, hailing, or something else is going on, for instance because they can hear pitter-patter on the roof but have no other information about the situation outside.

How can we introduce an indicative conditional such as (18)? The intuitive reasoning would proceed along the following lines. It is a background assumption that there is pitter-patter on the roof, which entails, by general knowledge about what produces pitter-patter sounds, that it might be raining and that it might be hailing. We may then consider the hypothetical scenario in which it is not raining and conclude from the background assumptions that it might be hailing. Hence, we can introduce the indicative conditional (18) by a version of the Ramsey test. However, in the same context and in the same way, we may also conclude from the background assumptions that it might be raining. But it would be a mistake to introduce the indicative conditional *If it is not raining, then it might be raining* by considering the hypothetical scenario in which it is not raining and using the background assumption that it might be raining.

Thus, both types of conditional, indicative and subjunctive, only permit a restricted set of background assumptions in their respective versions of the Ramsey test. Subjunctives are *counterfactual* in that we must disregard some factual knowledge when evaluating them. Indicatives are *counterepistemic* in that we must disregard some epistemic modal knowledge when evaluating them. This gives rise to the following principles for inferring indicatives and subjunctives.

Indicative Conditional Proof. If from *A* and epistemically compatible assumptions one can infer *B*, infer *If A, then B*.

Subjunctive Conditional Proof. If from *A* and factually compatible assumptions one can infer *B*, infer *If it were the case that A, then it would be the case that B*.

These rules permit the use of compatible background assumptions, since clearly not all facts must be disregarded when trying to introduce a subjunctive conditional. For instance, suppose one is outside and it is sunny. One may import the fact that one is outside into the hypothetical context in which it is raining to conclude that one is going to get wet. By Subjunctive Conditional Proof, one may thereby introduce the subjunctive conditional *If it were raining, I would get wet*. But in the same scenario one may not import the fact that it is sunny to conclude *If it were raining, it would be sunny*. The reason why one may import the fact that one is outside but not the fact that it is sunny is that the former is factually compatible with *It is raining* whereas the latter is not. The same goes for indicative conditionals and epistemic compatibility. Not all epistemic background assumptions are prohibited, only incompatible ones.

But what makes an assumption factually or epistemically compatible with some hypothetical scenario? We can characterize this in terms of the force marker $\mathbb{S}$ for supposition we introduced in Chapter 4. There we took $\mathbb{S}$ to be governed by the following coordination principle.

$$(\mathbb{S}\text{-Inference})\ \frac{\mathbb{S}A \qquad \begin{matrix}[+A]\\ \vdots \\ +B/\bot\end{matrix}}{\mathbb{S}B/\bot}$$ where the derivation of $+B/\bot$ may only use premisses of the form $+C$ where $\mathbb{S}C$ is derivable in the proof context of $\mathbb{S}A$

We can now see that this is too coarse. For we must distinguish between two distinct modes of supposition, one counterfactual and one counterepistemic. In particular, the following sentences realize different speech acts of supposition, expressing distinct attitudes (see also Eva et al. 2022).

(19) a. Suppose Oswald did not shoot Kennedy.
b. Suppose Oswald hadn't shot Kennedy.

Such suppositional constructions elicit the same differences in judgements as the subjunctive and indicative conditionals in (17). We therefore distinguish between subjunctive supposition, denoted by the force marker $\mathbb{S}^{\Box}$, and indicative supposition, denoted by the force marker $\mathbb{S}^{I}$. These distinct speech acts are governed by different coordination principles. The first set of principles specifies how the two distinct types of supposition interact with the corresponding conditional assertions.

$$(\mathbb{S}^I/\Rightarrow_+)\ \frac{\mathbb{S}^I A \qquad \Rightarrow_+(A,B)}{\mathbb{S}^I B} \qquad (\mathbb{S}^\Box/\Box\!\!\Rightarrow_+)\ \frac{\mathbb{S}^\Box A \qquad \Box\!\!\Rightarrow_+(A,B)}{\mathbb{S}^\Box B}$$

The coordination principle governing the interaction between indicative supposition and indicative conditional assertion states that indicatively supposing *A* and indicatively asserting *B* conditionally on *A* commits one to indicatively supposing *B*. The coordination principle governing the interaction between subjunctive supposition and subjunctive conditional assertion is analogous. These coordination principles can be motivated on the basis of the account we gave in Chapter 4 of the essential effect of supposition on the conversation. In particular, we argued that the essential effect of supposition is a proposal to temporarily update the common ground. Moreover, a proposal to updating the common ground with some sentence *A* is the essential effect of an assertion of *A*. So if one proposes to perform a temporary update of the common ground with *A* and has already committed to asserting *B* conditionally on *A*, then this temporary update to the common ground will also include *B*. That this is so is guaranteed by the coordination principles governing the interaction between supposition and conditional assertion.

We have explained how the two types of supposition interact with the corresponding types of conditional assertion. We must now specify how the two types of supposition interact with categorical assertion. For subjunctive supposition, we can adopt the $\mathbb{S}$-Inference principle unchanged.

$$(\mathbb{S}^\Box\text{-Inference})\ \frac{\mathbb{S}^\Box A \qquad \begin{array}{c}[+A]\\ \vdots \\ +B/\bot\end{array}}{\mathbb{S}^\Box B/\bot}$$

where the derivation of $+B/\bot$ may only use premisses of the form $+C$ where $\mathbb{S}^\Box C$ is derivable in the proof context of $\mathbb{S}^\Box A$

This principle states that, when we reason under the subjunctive supposition that *A*, we may include as background assumption anything else we have subjunctively supposed. By combining this principle with the coordination principle governing the interaction between subjunctive supposition and subjunctive conditional assertion, it follows that we may include as background assumption any $+C$ such that the subjunctive conditional $\Box\!\!\Rightarrow_+(A, C)$ has already been accepted.

The principle governing the interaction between indicative supposition and strong assertion is subject to similar, but more permissive conditions. All factual premisses may also be included as background assumptions, where, for current purposes, we can take a premiss to be factual just in case it is a Boolean combination of atomic sentences.

$$(\mathbb{S}^I\text{-Inference})\ \frac{\mathbb{S}^I A \qquad \begin{array}{c}[+A]\\ \vdots\\ +B/\bot\end{array}}{\mathbb{S}B/\bot}$$

where the derivation of $+B/\bot$ may only use premisses of the form $+C$ where (i) $\mathbb{S}^I C$ is derivable in the proof context of $\mathbb{S}^I A$ **or** (ii) $+C$ is derivable in the proof context of $\mathbb{S}^I A$ and C is a Boolean combination of atoms

Having laid down the coordination principles governing the interaction between supposition and the speech acts of assertion and conditional assertion, we are now in a position to formally state principles for inferring conditional speech acts. These principles are, again, coordination principles, since they feature no operator but specify how conditional speech acts interact with strong assertion and supposition. For brevity's sake, we only present the rules for $\Rightarrow_+$, $\Rightarrow_-$, and $\Box\!\!\Rightarrow_+$, since all other cases are analogous.

$$(\Rightarrow_+\text{CP})\ \frac{\begin{array}{c}[+A]\\ \vdots\\ +B\end{array}}{\Rightarrow_+ (A, B)}$$ if for any premiss $+C$ used to derive $+B$, we have $\mathbb{S}^I A \vdash \mathbb{S}^I C$

$$(\Rightarrow_-\text{CP})\ \frac{\begin{array}{c}[+A]\\ \vdots\\ -B\end{array}}{\Rightarrow_- (A, B)}$$ if for any premiss $+C$ used to derive $-B$, we have $\mathbb{S}^I A \vdash \mathbb{S}^I C$

$$(\Box\!\!\Rightarrow_+\text{CP})\ \frac{\begin{array}{c}[+A]\\ \vdots\\ +B\end{array}}{\Box\!\!\Rightarrow_+ (A, B)}$$ if for any premiss $+C$ used to derive $+B$, we have $\mathbb{S}^\Box A \vdash \mathbb{S}^\Box C$

There may now appear to be a close similarity between the suppositional account of conditionals (see, e.g., Mackie 1973; Edgington 1995; Carter 2021) and our inferential expressivist account. Notably, the following rule is equivalent to $(\Rightarrow_+\text{CP})$.

$$
\text{(Suppositional CP)}\ \frac{\begin{array}{c}[\mathbb{S}^I A]\\ \vdots \\ \mathbb{S}^I B\end{array}}{\Rightarrow_+ (A, B)}
$$

But writing the Conditional Proof rules in terms of inferences between suppositions only works for the conditional assertion markers $\Rightarrow_+$ and $\Box\!\!\Rightarrow_+$. It is to provide conditional proof rules for arbitrary conditional speech acts that we have first characterized compatibility in terms of supposition and then provided conditional proof rules using the resulting notion of compatibility.

In any case, it is not clear whether Suppositional CP is a proof-theoretic version of the suppositional account as has been intended in the literature. Sam Carter (2021) develops the only fully formalized proposal on the table and defines a conditional *If A, then B* to mean that supposing *A* entails *B*, not that supposing *A* entails supposing *B*, as Suppositional CP would have it. Moreover, Carter does not allow conditionals to be embedded in any context except conditional consequents. So the similarities between our account and the suppositional account, while striking, should not be exaggerated.

But why should compatibility be characterized in the manner specified by the restrictions in the Conditional Proof rules? We have motivated the Conditional Proof rules by taking them to be a proof-theoretic implementation of the Ramsey test. In the Ramsey test, one begins by hypothetically assuming the antecedent. Given our characterization of supposition, this is the same as *temporarily asserting* the antecedent, which results in a proposal to temporarily add the antecedent to the common ground. The speech act of supposition enables us to reason about what the consequences would be of adding a sentence to the common ground by allowing us to temporarily add a sentence to the common ground. The condition that $\mathbb{S}^I A \vdash \mathbb{S}^I C$ ensures that in the common ground resulting from the temporary update of the current common ground with *A*, the background assumption that *C* is available, and similarly for subjunctives.

One might have expected compatibility to be characterized by means of a condition such as $\mathbb{S}^I(A \wedge B) \nvdash \bot$. Such a characterization of compatibility would leave our semantics for conditionals underdetermined, since non-derivability is not generally decidable. But such a characterization of compatibility is also not necessary for current purposes. What is at stake when using one of the Conditional Proof rules is whether a background assumption that is already accepted in the global context can be imported in the hypothetical context. Thus, we do not require a characterization of compatibility that tells us when two assumptions are compatible *tout court*, but only a characterization that tells us which of our already accepted claims can be imported in a hypothetical context.

Still, there might appear to be an obvious circularity in the coordination principles we have laid down. The coordination principles for supposition involve the conditional speech acts, and the coordination principles for the conditional speech acts involve supposition. This is not so much a circularity as it is a mutual dependence. And if supposition and conditionals are indeed as closely connected as it appears, such a dependence is just what one would expect.

The coordination principles governing supposition and the conditional speech acts might also appear to be too weak. Nothing in the rules for $\mathbb{S}^{\Box}$ or for $\Box\!\!\Rightarrow_{+}$ can tell us that *I am outside* is compatible with *It is raining*, but *It is sunny* is not. But it would be too much to demand that the coordination principles establish such facts. Facts about what is compatible with what form part of our general world knowledge. The coordination principles for supposition and the conditional speech acts, by contrast, state how we use this knowledge in hypothetical reasoning. The coordination principles for supposition and conditional speech acts cannot explain why, for example, when we suppose that a match is wet, we conclude that it will not light. There may be nothing more to say about this common sense fact than that our background knowledge includes knowledge of the fact that if a match is wet, it will not light. The role of the inference rules is to encode how we use such background knowledge in our inferential practice, including using it under suppositions and to introduce conditional speech acts. Someone might not know that if a match is wet, it will not light, but they might nonetheless fully grasp the conditions under which conditionals and conditional speech acts may be introduced.

Thus, general background knowledge can be represented in terms of conditional background assumptions such as *If a match is wet, will it light? No!* or *If I am outside, then would I be outside if it were raining? Yes!*. Once background knowledge is represented in this way, our coordination principles for supposition and the restrictions on the Conditional Proof rules ensure that the compatibility facts are respected.

Consider again the case of someone who is outside while it is sunny and concludes *If it were raining, would I get wet? Yes!*. We can validate their reasoning as follows. If one is outside and it is sunny, one may use the fact that one is outside and the background knowledge *If I am outside, then would I be outside if it were raining? Yes!* to infer the subjunctive conditional assertion *If it were raining, would I be outside? Yes!*. This conditional assertion tells us that the subjunctive supposition of *It is raining* entails the subjunctive supposition of *I am outside*. Thus, compatibly with the restrictions on the subjunctive Conditional Proof rule, the premiss *I am outside* is available when reasoning from the assumption that it is raining, allowing one to infer *I would get wet*. By discharging the assumption, one may infer *If it were raining, would I get wet? Yes!*. This reconstruction makes it clear that there is no circularity involved in the use of conditionals as background assumptions. Although many practical uses of the Conditional Proof rules do require making use of conditionals as background assumptions, the conditions under which conditional speech acts may be inferred are specified without appealing to any background assumption whatsoever. It is just that in our practices of inferring and evaluating conditionals and conditional speech acts, we frequently rely on background knowledge expressed in conditional terms.

A comparison with model-theoretic accounts will shed further light on what can and cannot be expected of a characterization of compatibility. Model-theoretic accounts of conditionals such as Stalnaker's (1968) and Lewis's (1973) abstract from compatibility by appealing to a selection function or closeness measure. Model-theoretic accounts of supposition such as Yalcin's (2007) do so by appealing to a set of worlds compatible with the active suppositions (see fn. 8 in Chapter 4). Carter's (2021) suppositional theory of conditionals appeals to a revision function to compute what background knowledge is available under a supposition. Model theorists can say something more about the nature of selection, closeness, compatibility, or revision by laying down constraints on these abstractions. But, similarly to the restrictions on the Conditional Proof rules, such constraints cannot tell us why *I am outside* is factually compatible with *It is raining*. They can only specify some of the structural properties of compatibility. This is how it should be: facts about what is compatible with what are not for semantics to settle. When considering our practices of evaluating and inferring conditionals, therefore, everyone must treat facts about what is compatible with what as tacit background premisses, available in virtue of our general world knowledge.

We have motivated the restrictions on the Conditional Proof rules on the basis of the Ramsey test, and have explained how, despite these restrictions, the Conditional Proof rules can be used to validate pieces of reasoning involving conditionals. Indeed, it turns out that, far from being an impediment, the restrictions allow us to meet a famous challenge to formal accounts of the indicative conditional: the Gibbard Collapse Argument.

8.4 The Gibbard Collapse Argument

Gibbard (1980) proved that, given some seemingly plausible assumptions, the indicative conditional is equivalent to the material conditional. This is known as the *Gibbard Collapse Argument*. The assumptions are as follows.[10]

- *Modus Ponens*: $A \rightarrow B$ and A jointly entail B.
- *Import-Export*: $A \rightarrow (B \rightarrow C)$ is equivalent to $(A \wedge B) \rightarrow C$.
- *Identity*: $A \rightarrow A$ is a logical truth.
- *Conjunctive Consequents*: $A \rightarrow (B \wedge C)$ entails $A \rightarrow B$.

And the argument goes as follows. First, we show that the material conditional entails the indicative conditional, that is that $A \supset B$ entails $A \rightarrow B$. By Modus Ponens, it suffices to show that $(A \supset B) \rightarrow (A \rightarrow B)$. By Import-Export, this is equivalent to $((A \supset B) \wedge A) \rightarrow B$, which can be derived as follows. By Identity, we have that $((A \supset B) \wedge A) \rightarrow ((A \supset B) \wedge A)$. By substitution of material equivalents, this entails that $((A \supset B) \wedge A) \rightarrow (A \wedge B)$ and so, by Conjunctive Consequents, it follows that $((A \supset B) \wedge A) \rightarrow B$. Next, we show that the indicative conditional entails the material conditional. This follows from Modus Ponens and the conditional proof rule for the material conditional.

That the indicative conditional entails the material conditional is generally considered unproblematic. But that the material conditional should entail the indicative conditional is troublesome. A material conditional is true if its antecedent is false, but the indicative *If Cotton is a rabbit, he is a bird* appears to be false even if Cotton is not a rabbit. While some theories of the indicative attempt to explain away such examples by means of some pragmatic (Jackson 1987) or heuristic (Williamson 2020) mechanism, most

[10] There are different reconstructions of this argument, with its assumptions spelled out in different ways. Here we follow Mandelkern's (2021) reconstruction.

theories take an indicative conditional such as *If Cotton is a rabbit, he is a bird* to be false. They must therefore hold that the Gibbard Collapse Argument fails. However, all of the argument's assumptions are prima facie plausible for the indicative conditional, which means that they are all on trial. And indeed proposals for how to deal with the Gibbard Collapse Argument include rejecting Modus Ponens (McGee 1985), Identity (Gillies 2010), and Import-Export. Now, Modus Ponens, Identity, and Conjunctive Consequents are trivial consequences of our inference rules for the indicative conditional and indicative conditional assertion. So, on our account, the culprit must be the Import-Export Principle.

Proposals that reject Import-Export fall into two camps. The first camp (see, e.g., McGee 1989; Edgington 1995) holds that conditionals cannot be embedded in conditional consequents, which makes Import-Export a non-well-formed principle. The second camp (see, e.g., Khoo and Mandelkern 2018) holds that conditionals can embed in conditional consequents but nonetheless do not validate Import-Export.

Our account of conditionality agrees with the first camp as far as conditional speech acts are concerned: conditional speech acts do not embed and hence cannot validate anything akin to Import-Export. However, in our discussion of the Frege–Geach Problem for conditionals, we have argued that, *pace* the first camp, conditionals can embed. As far as embeddable conditional operators are concerned, we agree with the second camp that Import-Export is a well-formed but invalid principle for indicative conditionals. We arrive at this conclusion by considering how we may derive Import-Export from the inference rules for conditionals and conditional assertion.

The target of the Gibbard Collapse Argument is the indicative conditional, since the Import-Export Principle does not enjoy the same prima facie plausibility for subjunctive conditionals. But to fix ideas, it is helpful to first consider the subjunctive case. The conditionals in the following example, discussed by Matthew Mandelkern (2021) who credits it to David Etlin, do not appear to be equivalent: the first seems false, whereas the second appears to be a trivial truth.

(20) a. If the match had lit, then it would have lit if it had been wet.
b. If the match had lit and it had been wet, then it would have lit.

Standard possible-world semantics for the subjunctive predict such judgements. To evaluate the conditional (20a), one first considers the worlds closest to the actual world in which the match lights. Then one considers the worlds closest to *these* in which the match is wet. In these worlds, the match plausibly does not light, since wet matches do not typically light and so worlds in which a wet match lights are further away. Thus, the consequent of (20a) is false at those worlds, and so the whole conditional is false. By contrast, to evaluate the conditional (20b), one only needs to consider the worlds closest to the actual world in which the match lights and it is wet, however implausible and far away they might be. Since these are in particular worlds in which the match lights, the consequent of (20b) is true at these worlds, and so the whole conditional is true.

The discussion reveals a general recipe for generating counterexamples to the Import-Export Principle. Take two sentences *A* and *B* whose conjunction is highly implausible and entails some *C* such that *B and C* is also highly implausible. Then the intuitive judgement is to reject *If it were the case that A, then if it were the case that B, then it would be the case that C* and accept *If it were the case that A and B, then it would be the case that C*. The simplest case, as in Etlin's example, is when *C* is just *A*.

From the proof-theoretic perspective, however, the situation is prima facie puzzling, since the Import-Export Principle appears to be *derivable* from the Conditional Proof rule. We present here only the putative derivation of the right-to-left direction, stating that $(A \wedge B) \,\Box\!\!\rightarrow C$ entails $A \,\Box\!\!\rightarrow (B \,\Box\!\!\rightarrow C)$, since this is the direction used in the Gibbard Collapse Argument and the direction which, when applied to Etlin's example, takes us from a seemingly true subjunctive to a seemingly false one.

$$\dfrac{\dfrac{\dfrac{\dfrac{\dfrac{\dfrac{+(A\wedge B)\,\Box\!\!\rightarrow C}{\Box\!\!\Rightarrow_+ (A\wedge B, C)}\,(+\,\Box\!\!\rightarrow\text{E.}) \qquad \dfrac{[+A]^1 \qquad [+B]^2}{+(A\wedge B)}\,(+\wedge\text{I.})}{+C}\,(\Box\!\!\Rightarrow_+\text{MP})}{\Box\!\!\Rightarrow_+ (B,C)}\,(\Box\!\!\Rightarrow_+\text{CP})^2}{+B\,\Box\!\!\rightarrow C}\,(+\,\Box\!\!\rightarrow\text{I.})}{\Box\!\!\Rightarrow_+ (A, B\,\Box\!\!\rightarrow C)}\,(\Box\!\!\Rightarrow_+\text{CP})^1}{+A\,\Box\!\!\rightarrow (B\,\Box\!\!\rightarrow C)}\,(+\,\Box\!\!\rightarrow\text{I.})$$

In spite of appearances, this derivation is invalid. When considered in full generality, it violates the restrictions on the subjunctive conditional proof rule. For when we apply the rule for the first time, discharging $+B$ to infer $\Box\!\!\Rightarrow_+ (B, C)$, the derivation leading up to $+C$ requires the additional undischarged premiss $+A$. And according to the restrictions on the subjunctive

Conditional Proof rule, this is only licit when A is factually compatible with B. But in Etlin's example, this is not the case, since A is *The match lights* and B is *The match is wet*, and it is part of our world knowledge that wet matches do not light. The same goes for all counterexamples to Import-Export generated using the recipe extracted from Etlin's pair of subjunctives.

Thus, Import-Export is not generally valid on our account of subjunctive conditionals. In particular, one cannot use the principle to derive the seemingly false subjunctive in Etlin's pair from the seemingly true one. And the Gibbard Collapse Argument cannot be applied to subjunctive conditionals, which is as it should be. But such blanket conclusions understate the complexity of the matter. Very many instances of Import-Export are valid for the subjunctive conditional, namely those where the two sentences in the antecedent are factually compatible. Can we therefore not run the Gibbard Collapse Argument for subjunctives using those instances of Import-Export?

When applied to subjunctive conditionals, the argument uses the instance of Import-Export stating that $((A \supset B) \wedge A) \mathrel{\Box\!\!\rightarrow} B$ entails $(A \supset B) \mathrel{\Box\!\!\rightarrow} (A \mathrel{\Box\!\!\rightarrow} B)$. The putative derivation of this instance, using the conditional proof rule for subjunctives, goes as follows.

$$
\dfrac{\dfrac{\dfrac{\dfrac{\dfrac{\dfrac{+((A \supset B)\wedge A) \mathrel{\Box\!\!\rightarrow} B}{\Box\!\!\Rightarrow_{+}((A\supset B)\wedge A, B)}\,(+\Box\!\!\rightarrow\text{E.}) \qquad \dfrac{[+(A\supset B)]^1 \quad [+A]^2}{+(A\supset B)\wedge A}\,(+\wedge\text{I.})}{+B}\,(\Box\!\!\Rightarrow_{+}\text{MP.})}{\Box\!\!\Rightarrow_{+}(A,B)}\,(\Box\!\!\Rightarrow_{+}\text{CP})^2}{+A \mathrel{\Box\!\!\rightarrow} B}\,(+\Box\!\!\rightarrow\text{I.})}{\Box\!\!\Rightarrow_{+}(A\supset B, A\mathrel{\Box\!\!\rightarrow} B)}\,(\Box\!\!\Rightarrow_{+}\text{CP})^1}{+(A\supset B)\mathrel{\Box\!\!\rightarrow}(A\mathrel{\Box\!\!\rightarrow} B)}\,(+\Box\!\!\rightarrow\text{I.})
$$

According to the restrictions on conditional proof, this derivation is valid only if A is factually compatible with $A \supset B$. But this is not the case for every choice of A and B. For instance, if A and B are themselves factually incompatible, then so are A and $A \supset B$. To see this, again let A be *The match lights* and B be *The match is wet*. Then $A \supset B$ is equivalent to *The match does not light or the match is wet*, and A is incompatible with $A \supset B$, since it is incompatible with either disjunct. And indeed, *The match does not light or the match is wet* should not be equivalent to *If the match had lit, it would be wet*, since the latter is clearly false, whereas the former is true if, for example, the match is wet.

Nonetheless, some valid instances of Import-Export do suffice to derive some instances of the Gibbard Collapse Argument, which means that some subjunctive conditionals are indeed equivalent to the corresponding material conditionals. But those seem to be precisely the cases in which subjunctive conditionals *should* be equivalent to the corresponding material conditionals. For instance, *The match is wet or it is dry* is equivalent to the subjunctive conditional *If the match were not wet, it would be dry* simply because both are true.[11]

We can now address the Gibbard Collapse Argument for its intended target, namely analyses of the indicative conditional. Similarly to the case of the subjunctive, Import-Export is not generally valid on our account of the indicative because of the restrictions on the indicative conditional proof rule. The putative derivation of the right-to-left direction of Import-Export for the indicative proceeds in an analogous manner as the one for the subjunctive.

$$\dfrac{\dfrac{\dfrac{\dfrac{\dfrac{\dfrac{+(A \wedge B) \to C}{\Rightarrow_+ (A \wedge B, C)}\,(+\to\text{E.}) \qquad \dfrac{[+A]^1 \qquad [+B]^2}{+(A \wedge B)}\,(+\wedge\text{I.})}{+C}\,(\Rightarrow_+\text{MP})}{\Rightarrow_+ (B, C)}\,(\Rightarrow_+\text{CP})^2}{+B \to C}\,(+\to\text{I.})}{\Rightarrow_+ (A, B \to C)}\,(\Rightarrow_+\text{CP})^1}{+A \to (B \to C)}\,(+\to\text{I.})$$

This derivation is valid if and only if (i) $\mathbb{S}^I B \vdash \mathbb{S}^I A$, (ii) $\mathbb{S}^I B \vdash \mathbb{S}^I (A \wedge B) \to C$, and (iii) $\mathbb{S}^I A \vdash \mathbb{S}^I (A \wedge B) \to C$. The first two conditions must be satisfied for the first application of Conditional Proof to be legitimate; the third condition must be satisfied for the second application of Conditional Proof to be legitimate. However, all three conditions can fail. We present here a case in which the first condition fails, since it is the easiest to describe. So let A be *The match might be wet* and B be *The match lights*. To infer $\mathbb{S}^I A$ from $\mathbb{S}^I B$—that is, that A is epistemically compatible with B—one must either rely on background knowledge that can be represented as the conditional assertion $\Rightarrow_+ (B, A)$ or infer $+A$ from $+B$ by using only Boolean or supposed premisses. It is clearly not possible to infer $+A$ from $+B$ by using only Boolean or supposed premisses, since the above derivation does

[11] Of course, they are equivalent *modulo* the presupposition of subjunctive conditionals that their antecedents are false.

not contain further premisses at all. Relying on background knowledge is not promising either, since it is implausible that the following conditional assertion realizing $\Rightarrow_+ (B, A)$ should be part of the background knowledge.

(21) If the match lights, then might it be wet? Yes!

Indeed, one would intuitively *reject* such conditionals. We can arrive at this judgement by considering a hypothetical context in which the match lights and conclude that it is not wet by using some background knowledge about what is the case when matches light.

Even if, in the context in which the conditional assertion (21) is performed, *The match might be wet* is actually the case, we may not import it into the hypothetical context to conclude the consequent of the conditional assertion. If it is actually the case that the match might be wet, we would rather reason as follows.

(22) a. The match might be wet.
b. Suppose it lights. Might it then be wet? No!

This further supports our claim that *The match might be wet* is epistemically incompatible with *The match lights*, since it cannot be imported into a hypothetical context in which it is indicatively supposed that the match lights.

Thus, the above derivation of Import-Export is not generally valid, since it uses Conditional Proof in ways which are illegitimate for certain choices of A and B. We have considered the case in which A is *The match might be wet* and B is *The match lights*. By letting C be the same as A, we can use this case to generate an explicit counterexample to Import-Export for the indicative.

(23) a. If the match might be wet, then it might be wet if it lights.
b. If the match lights and might be wet, then the match might be wet.

Our assessment of these conditionals is analogous to our assessment in the counterfactual case: the first seems false, whereas the second appears to be a trivial truth. But if Import-Export were valid, the trivially true conditional would immediately entail the seemingly false one.

Nevertheless, as in the case of subjunctive conditionals, our account validates many instances of Import-Export for indicative conditionals as well as many instances of the Gibbard Collapse Argument. Thus, many indicatives are in fact equivalent to the corresponding material conditionals. But again, those seem to be precisely the cases in which indicative conditionals should be equivalent to the corresponding material conditionals. The cases in which the equivalence should fail are ruled out because the relevant instance of Import-Export is not available.

Recall that the Gibbard Collapse Argument uses the instance of Import-Export stating that $((A \supset B) \wedge A) \rightarrow B$ entails $(A \supset B) \rightarrow (A \rightarrow B)$. The putative derivation of this instance, using the conditional proof rule for indicatives, proceeds in exactly the same manner as the putative derivation for subjunctives we gave above. Consider, for instance, what are perhaps the most prominent problematic cases, namely those involving conditionals with a false antecedent, such as *If Cotton is a rabbit, then he is a bird* where Cotton is in fact not a rabbit. So let A be *Cotton is a rabbit* and B be *Cotton is a bird*. Then it follows from the restrictions on the conditional proof rule for indicatives that the derivation of the relevant instance of Import-Export is valid only if *Cotton is a rabbit* is epistemically compatible with the material conditional equivalent to *Cotton is not a rabbit or Cotton is a bird*. But the two sentences are not epistemically compatible, since indicatively supposing that Cotton is not a rabbit or a bird does not entail indicatively supposing that Cotton is a rabbit. It is neither the case that *if Cotton is not a rabbit or Cotton is a bird, then Cotton is a rabbit* is a common sense conditional, nor that the assertion of Cotton being a rabbit can be derived from a Boolean combination of atoms in the global proof context, since, by stipulation, it is the case in the global context that Cotton is not a rabbit.

To sum up, the indicative conditional does not validate the Import-Export Principle and hence does not collapse into the material conditional. The reason why this is the case is the same as the reason why the subjunctive conditional does not validate a version of Import-Export and hence does not collapse into the material conditional. The conditional proof rules governing the conditional speech acts corresponding to both indicative and subjunctive conditionals disallow arbitrary side premisses which are used in the derivation of the relevant version of Import-Export, thereby blocking the derivation of the Gibbard Collapse Argument and its counterpart for subjunctives.

In Chapter 2, we argued that one of the advantages of the inferentialist approach lies in its methodology. Our strategy for diagnosing where the Gibbard Collapse Argument fails illustrates this advantage. Referentialist semantics proceeds in a bottom-up fashion by assigning referents to expressions so that truth conditions are generated that are compatible with the linguistic data. However, the available data vastly underdetermine what these truth conditions should be. The case of indicative conditionals makes this point especially vivid: despite the wide range of available linguistic data on the indicative, extant theories offer different diagnoses as to where the Gibbard Collapse Argument fails, in a way which appears to be compatible with the linguistic data. By contrast, our diagnosis of where the argument goes wrong has been reached in a top-down fashion, by formulating independently motivated inference rules satisfying a number of theoretical constraints and considering how one might derive the Import-Export Principle from these rules. We tentatively conclude that this provides evidence in favour of the view that the culprit is the Import-Export Principle.

The discussion in this section illustrates another methodological advantage of the inferentialist approach, namely that we can precisely determine which instances of Import-Export are valid by inspecting putative derivations of this principle and establishing under which conditions they are legitimate. Using this strategy, we have determined above that the right-to-left direction of Import-Export holds if and only if three conditions about compatibility are satisfied. Similarly, proof analysis of the putative derivation of the left-to-right direction reveals that this direction holds if and only if $\mathbb{S}^I A \wedge B \vdash \mathbb{S}^I A \rightarrow (B \rightarrow C)$. Like the three conditions on the right-to-left direction, this condition can fail, but will be satisfied in many cases. Thus, this methodological advantage of the inferentialist approach allows us to explain the intuitive appeal of the Import-Export Principle: despite not being generally valid, it has many valid instances. We now turn to some issues that arise when we combine our account of conditionals with the accounts of epistemic modality and truth developed in previous chapters.

8.5 Generalized Yalcinean sentences

Paolo Santorio (2017) observed that sentences such as (24)—call them *Santorian sentences*—sound bad in much the same way in which Yalcinean sentences such as *It is raining and it might not be* sound bad.

(24) If it's dry on the pavement, then it is not raining, and if it's dry on the pavement, then it might be raining.

In particular, like Yalcinean sentences, Santorian sentences continue to sound bad when embedded under *suppose*. In Chapter 4, we argued that one should treat Yalcinean sentences as semantically contradictory. But then, Santorio argues, one should also treat Santorian sentences as semantically contradictory. However, on our account, the strong assertion of the natural formalization of (24), namely $(p \to \neg q) \wedge (p \to \Diamond q)$, does not entail absurdity. This is because, on our account, an indicative conditional can be vacuously true, so the absurdity of $+\neg q \wedge \Diamond q$ does not force us to conclude that $+(p \to \neg q) \wedge (p \to \Diamond q)$ is itself absurd, but only that p satisfies the vacuity condition for our indicative conditional.

Santorio defines a conditional and a consequence relation which make Santorian sentences semantically contradictory, but we want to offer a different diagnosis of the infelicity of Santorian sentences, one which does not require any departure from our account of conditionals or any revision of our notion of consequence. Note that an utterance of the following sentence, which does not contain epistemic modals, already sounds odd.

(25) If it's dry on the pavement, then it is not raining, and if it's dry on the pavement, then it is raining.

This sentence and the Santorian sentence above seem to sound odd for similar reasons. Uttering either of these sentences appears to prompt one to suppose that it is dry on the pavement—that is, to consider what follows from it being dry on the pavement. But to suppose that it is dry on the pavement is absurd, given the information the sentences contain. Thus, these sentences seem to sound bad for the same reason that the following sentences would seem to sound bad: their antecedents cannot be supposed.

(26) a. If it's dry on the pavement and it's not dry on the pavement, then it's raining.
b. If it's dry on the pavement and it might not be dry on the pavement, then it's raining.

We therefore propose to explain the infelicity of all of these sentences by using a notion of *supposability* based on our characterization of supposition.

Following Stalnaker, we have taken the essential effect of supposing A to be a proposal to temporarily add A to the common ground. The supposability of A in a given context can then be characterized as the possibility of supposing A in that context. Formally, let A and C be sentences in $\mathcal{L}_{\Diamond}$. We say that A is *indicatively supposable* in a context C if $\mathbb{S}^I A \wedge C$ does not entail absurdity in EML together with the $\mathbb{S}^I$-Inference rule. Our suggestion is then that a strong assertion of an indicative conditional pragmatically presupposes the indicative supposability of the conditional's antecedent in a context that corresponds to the current common ground updated with the strong assertion's content.

Why should the indicative conditional have such a presupposition? Following Stalnaker (1978), the pragmatic presuppositions of a strong assertion include all the information that can be inferred from the performance of the strong assertion itself. In particular, a strong assertion presupposes that the context is such that its essential effect changes the context in a non-trivial and well-defined way and everyone in the conversation can compute this change. Now, when one proposes to update the common ground with an indicative conditional, one proposes to change it in such a way that if the antecedent is added to the common ground, its consequent is added too. Everyone must be able to compute what this change amounts to, since it is on this that they base their decision to accept or reject the update proposal. Thus, everyone must be able to consider the common ground updated with the conditional and then temporarily add the conditional's antecedent and arrive at a well-defined result. This may be seen as a discoursive analogue of the Ramsey test on which we based our semantics for conditionals. Many have claimed that indicative conditionals presuppose, in some sense, that their antecedents are possible (Gillies 2010; Mandelkern and Romoli 2017; Crespo et al. 2018). Our argument shows that supposability is the right way to specify what kind of possibility should be meant here, at least within a broadly Stalnakerian framework.

Hence, strong assertions of indicative conditionals pragmatically presuppose that the indicative conditional's antecedent is supposable in the context of the current common ground updated with the assertion's content. But this presupposition cannot be met for the conditionals above, including the Santorian sentence *If it's dry on the pavement, then it is not raining, and if it's dry on the pavement, then it might be raining.* Recall that we are formalizing this Santorian sentence as $(p \rightarrow \neg q) \wedge (p \rightarrow \Diamond q)$, so we need to check the indicative supposability of p with respect to a context containing at least

$(p \rightarrow \neg q) \wedge (p \rightarrow \Diamond q)$. But $\mathbb{S}^I p \wedge (p \rightarrow \neg q) \wedge (p \rightarrow \Diamond q)$ entails an absurdity in EML together with the $\mathbb{S}^I$-Inference rule. So p is not supposable.

One might reply that our pragmatic explanation is not general enough. For the Santorian sentence above also sounds bad when the conditionals it contains are replaced by the corresponding *subjunctives*.

(27) If it were dry on the pavement, then it would not be raining, and if it were dry on the pavement, then it would be the case that it might be raining.

And, the reply goes, the assertion of a subjunctive conditional does not have the supposability presupposition associated with the assertion of its indicative counterpart. While indicative conditionals change the common ground in a way that can be evaluated by provisionally *updating* the common ground with their antecedents, subjunctive conditionals change the common ground in a way that can be evaluated by provisionally *revising* the common ground with their antecedent (Stalnaker 1968). Thus, the strong assertion of a subjunctive conditional does not presuppose that its antecedent be supposable in the common ground updated with the strong assertion's content, since some of this content might be revised in order to suppose the antecedent.

The reply is unsuccessful. Although the assertion of a subjunctive does not have the same supposability presupposition as the assertion of the corresponding indicative, it does have *a* supposability presupposition. In particular, say that A is *subjunctively supposable* in a context C if $\mathbb{S}^{\Box} A \wedge C$ does not entail absurdity in EML together with the $\mathbb{S}^{\Box}$-Inference rule. Then, since not all information in the common ground is up for revision when considering a subjunctive antecedent, the assertion of a subjunctive conditional presupposes that its antecedent be subjunctively supposable in the context consisting of the non-revisable part of the common ground updated with the conditional. This presupposition cannot be met in the case of the subjunctive Santorian sentence above.

If we formalize the subjunctive Santorian sentence above as $(p \mathrel{\Box\!\!\Rightarrow} \neg q) \wedge (p \mathrel{\Box\!\!\Rightarrow} \Diamond q)$, this can be seen as follows. We need to check the subjunctive supposability of p with respect to the context c that results from updating the non-revisable part of the common ground with $(p \mathrel{\Box\!\!\Rightarrow} \neg q) \wedge (p \mathrel{\Box\!\!\Rightarrow} \Diamond q)$. But $\mathbb{S}^{\Box} q \wedge c$ entails $\mathbb{S}^{\Box} q \wedge \Diamond \neg q$ in EML together with the $\mathbb{S}^{\Box}$-Inference rule. And $\mathbb{S}^{\Box} q \wedge \Diamond \neg q$ entails absurdity. So p is not subjunctively supposable in c.

One may wonder how this pragmatic explanation of the infelicity of Santorian sentences can account for their infelicity when embedded under *suppose*. After all, the special problem raised by Yalcinean sentences is that, unlike Moorean sentences, they continue to sound bad under *suppose* and similar environments. This, the usual story goes, precludes a pragmatic explanation of their infelicity similar to the one given for Moorean sentences, since pragmatic inferences are suspended under *suppose*.

This story overplays the power of *suppose* to suspend pragmatic inferences: although the pragmatic inferences used to explain the infelicity of Moorean sentences are suspended under *suppose*, it does not follow that *all* pragmatic inferences are suspended. The pragmatic inference used to explain the infelicity of Moorean sentences is suspended under supposition: in asserting a Moorean sentence, the speaker undertakes an immediate commitment to displaying behaviour suitable for inferring the strong assertion of a belief ascription to themselves and to strongly rejecting this ascription. No such commitment is undertaken when one utters a Moorean sentence under *suppose*. By contrast, the pragmatic inference we outlined in the previous paragraphs clearly goes through under supposition: just as the actual update of the common ground with a conditional is only well defined if its antecedent is supposable in the relevant context, so is the *temporary* update with the same conditional only well defined if its antecedent is supposable in that context. Thus, the supposability requirement for conditionals applies to both strong assertions and suppositions.

There are some related observations that we can explain using the presuppositional strategy we have used to explain the infelicity of Santorian sentences. Mandelkern (2019) observed that sentences such as the following—call them *Mandelkernian sentences*—sound bad in the same way in which Yalcinean sentences do.

(28) a. It is raining and it might not be raining, or it is windy and it might not be windy.
b. It might be that (it is raining and it might not be raining).

On our account Mandelkernian sentences are not absurd: if A and B are classically consistent, there are models of EML in which $+(A \wedge \Diamond\neg A) \vee (B \wedge \Diamond\neg B)$ and $+\Diamond(A \wedge \Diamond\neg A)$ hold. Nevertheless, the presuppositional strategy we presented above allows us to account for the infelicity of Mandelkernian sentences in pragmatic terms. We can explain the infelicity of *It is raining and it might not be raining, or it is windy and it might not be windy* by taking the strong assertion of *A or B* to presuppose the indicative supposability of *A* and of *B*. And we can explain the infelicity of *It might be that (it is raining and it might not be raining)* by taking the strong assertion of *might A* to pragmatically presuppose the supposability of *A*. The existence of these pragmatic presuppositions is supported by the felicity of sequences such as the following.

(29) a. It is raining or it is windy. So suppose that it is in fact raining.
b. It might be raining. So suppose that it is in fact raining.

One might suggest that these sequences support a stronger conclusion, namely that supposability should be part of the meaning of *or* and *might*, not simply a presupposition of the sentences containing these expressions. In particular, one might identify the meaning of *might* with supposability and the meaning of *A or B* with *might A and might B*.[12] However, having asserted *not A*, one may go on to suppose that *A*, possibly just for the sake of argument, since supposition can be counterepistemic and counterfactual. By contrast, having asserted *not A*, it is a mistake to go on to assert *might A*. Thus the meaning of *might* cannot be identified with supposability.

Our pragmatic explanation of the infelicity of Mandelkernian sentences also has an empirical advantage over Mandelkern's own semantic approach. He claims that the infelicity of disjunctive Mandelkernian sentences is explained by the fact that Yalcinean sentences are semantic contradictions and that disjunctions of semantic contradictions are themselves semantic contradictions. But now consider the following sentence.

(30) It is raining and it might not be raining, or it is windy.

If *It is windy* is true, then according to the usual truth-functional meaning of disjunction, which Mandelkern does not dispute, the whole sentence is true, since it has a true disjunct. However, the sentence sounds odd. Thus, Mandelkern would seem to need some further mechanism to explain its oddity, such as our pragmatic presupposition or another principle to the effect that disjunctions one of whose disjuncts is a classical contradiction sound generally bad. But then, any such mechanism would also account for

[12] For a similar proposal concerning the meaning of *or*, see Zimmermann (2000).

the infelicity of the original disjunction considered by Mandelkern. Hence, the more parsimonious approach is to stick with a consequence relation that validates all classical arguments and explain the infelicity of Mandelkernian sentences pragmatically, as we have done.

8.6 Curry's Paradox

In addition to the paradoxes discussed in the previous chapter, any theory of truth must face Curry's Paradox, which arises from the interaction between the truth predicate and conditionals. The paradox is based on Curry sentences, which say of themselves that if they are true, then some other sentence holds. The paradox purports to derive this other sentence from no assumptions. Since this sentence can be any sentence whatsoever, the result is paradoxical. For present purposes, let m be the sentence *The moon is made of cheese* and consider the Curry sentence c such that $c \subset\supset (Tr\ulcorner c\urcorner \supset m)$. We can give a purported formal derivation of the moon's dairy status as follows. The material conditional $Tr\ulcorner c\urcorner \supset Tr\ulcorner c\urcorner$ is a classical logical truth. By eliminating the truth predicate in the consequent, we obtain another material conditional, $Tr\ulcorner c\urcorner \supset c$. But, by the definition of c, the consequent of this material conditional is equivalent to $Tr\ulcorner c\urcorner \supset m$, so we have $Tr\ulcorner c\urcorner \supset (Tr\ulcorner c\urcorner \supset m)$. By Contraction—the classically valid principle that $(A \supset (A \supset B)) \supset (A \supset B)$ is a logical truth—this entails that $Tr\ulcorner c\urcorner \supset m$. But, again by the definition of c, $Tr\ulcorner c\urcorner \supset m$ is equivalent to c, which, by an application of the introduction rule for the truth predicate, implies $Tr\ulcorner c\urcorner$. So we have derived both $Tr\ulcorner c\urcorner \supset m$ and $Tr\ulcorner c\urcorner$, which jointly entail m by *modus ponens*.

The argument is invalid according to our semantics for the truth predicate. As we stressed in the previous chapter, the inferential expressivist account of truth is not fully transparent. One cannot simply eliminate the truth predicate in the consequent of a material conditional. Rather, one must first eliminate the conditional, then eliminate the truth predicate, and finally introduce the conditional again. This means that the first step in the above argument, concluding $Tr\ulcorner c\urcorner \supset c$ from $Tr\ulcorner c\urcorner \supset Tr\ulcorner c\urcorner$, must be formally reconstructed as follows.

$$\dfrac{\dfrac{\dfrac{+Tr\ulcorner c\urcorner \supset Tr\ulcorner c\urcorner \qquad [+Tr\ulcorner c\urcorner]^1}{+Tr\ulcorner c\urcorner}\,(+\supset\text{E.})}{+c}\,(+Tr\text{E.})}{+Tr\ulcorner c\urcorner \supset c}\,(+\supset\text{I.})^1$$

However, this derivation is invalid. For the rule (+*Tr*E.) does not preserve evidence and its application is therefore disallowed under the rule of conditional proof for the material conditional.

But what about indicative and subjunctive conditionals? Consider the following indicative version of the Curry sentence *c*.

$$c_I \leftrightarrow (Tr\ulcorner c_I \urcorner \to m)$$

As before, Curry's argument would begin with the conditional $Tr\ulcorner c_I \urcorner \to Tr\ulcorner c_I \urcorner$, which is a logical truth on our account of indicatives. But nothing prevents one from eliminating the truth predicate in the consequent of this conditional. Nevertheless, Curry's argument fails, since the instance of Contraction required for the argument is not available. To see this, consider how one might attempt to derive the relevant instance.

$$\dfrac{\dfrac{\dfrac{\dfrac{\dfrac{\dfrac{+Tr\ulcorner c_I \urcorner \to (Tr\ulcorner c_I \urcorner \to m)}{\Rightarrow_+ (Tr\ulcorner c_I \urcorner, Tr\ulcorner c_I \urcorner \to m)}\,(+\to\text{E.}) \qquad [+Tr\ulcorner c_I \urcorner]^1}{+Tr\ulcorner c_I \urcorner \to m}\,(\Rightarrow_+\text{MP})}{\Rightarrow_+ (Tr\ulcorner c_I \urcorner, m)}\,(+\to\text{E.}) \qquad [+Tr\ulcorner c_I \urcorner]^1}{+m}}{\Rightarrow_+ (Tr\ulcorner c_I \urcorner, m)}\,(\Rightarrow_+\text{CP})^1}{+Tr\ulcorner c_I \urcorner \to m}\,(+\to\text{I.})$$

This is a valid derivation only if $Tr\ulcorner c_I \urcorner \to (Tr\ulcorner c_I \urcorner \to m)$ is epistemically compatible with $Tr\ulcorner c_I \urcorner$, that is if $\mathbb{S}^I Tr\ulcorner c_I \urcorner \vdash \mathbb{S}^I Tr\ulcorner c_I \urcorner \to (Tr\ulcorner c_I \urcorner \to m)$. By a simple application of conditional proof, this condition simplifies to $\mathbb{S}^I Tr\ulcorner c_I \urcorner \vdash \mathbb{S}^I m$. But in the cases in which m ought not to be derivable—that is, in which Curry's Paradox *is* paradoxical—we do not have that $\mathbb{S}^I Tr\ulcorner c_I \urcorner \vdash \mathbb{S}^I m$. It is implausible that, for instance, the following conditional should be part of our background knowledge.

(31) If 'If this sentence is true, then the moon is made of cheese' is true, then the moon is made of cheese.

Alternatively, one might attempt to show that $\mathbb{S}^I Tr\ulcorner c_I \urcorner \vdash \mathbb{S}^I m$ by deriving $+m$ from $+Tr\ulcorner c_I \urcorner$. But there is no reason to think that such a derivation is available, unless the very reasoning involved in Curry's Paradox goes through, which is what we were trying to establish. The situation for subjunctive conditional versions of the paradox is the same. The compatibility constraints on the subjunctive conditional proof rule prevent the derivation of the instance of Contraction needed for the argument.

8.7 Content conditionals and inferential conditionals

There remains a problematic consequence of our treatment of conditionals. Since our logic validates epistemic strengthening, that is the inference from *A* to *It must be that A*, we can derive the indicative conditional *If A, then it must be that A*. However, since we also validate a version of *modus tollens* for the indicative conditional, it would seem to follow that we have again trivialized the epistemic modal, deriving the indicative conditional *If it might be that not A, then not A*.

To address this problem, we need to revisit the fourth step in the multilateral methodology and take a closer look at which inferences preserve evidence and which inferences need to be restricted. In the subderivation of the conditional proof rules for conditional speech acts, we allowed rules that do not preserve evidence. This presents us with two options.

The first option is to restrict the conditional proof rules for conditional speech acts to only allow evidence-preserving rules in their subderivations. This would make these rules more similar to the conditional proof rule for the material conditional and, in particular, prohibit the derivation of *If A, then it must be A*. The second option is to leave the conditional proof rules as they are, which means that we must count the *modus ponens*-esque rules for conditional speech acts among the rules that do not preserve evidence. For if we do not modify the conditional proof rules, then a conditional speech act may be justified on the basis of an inference that does not preserve evidence. This means that when we use such a conditional speech act to infer its consequent from its antecedent by a *modus ponens*-esque rule, the resulting inference does not preserve evidence. On the second option, *If A, then it must be that A* is derivable, but *modus tollens* for indicative conditionals is not.

Thus, either option blocks the trivialization of the epistemic modal. At the same time, either option seems to have drawbacks. Taking the first option blocks the derivation of the intuitively acceptable conditional *If A, then it must be that A*. Taking the second option blocks the derivation of the intuitively valid *modus tollens*-esque inferences.

We are now going to argue that the two options track a distinction made in the linguistics literature and that, once this distinction is properly taken into account, the drawbacks turn out to be only apparent. Some linguists have argued that, in addition to the distinction between indicative and subjunctive conditionals, there is another typological distinction to be made among conditionals, namely that between *content* and *inferential* conditionals (Dancygier 1999; Dancygier and Sweetser 2005; Krzyżanowska et al. 2013). Following the linguistic literature, we focus on the distinction as applied to indicative conditionals; we return to the subjunctive case below.

A content conditional, roughly, describes a relation between states of affairs: that for some specific reason (such as a causal, nomological, or formal reason), the obtaining of the state of affairs described in the antecedent is a sufficient condition for the obtaining of the state of affairs described in the consequent. A paradigm example is the following conditional, adapted from Dancygier 1999: 63.

(32) If I drink milk, I get a rash.

This says that my drinking milk is a sufficient condition for my getting a rash. An inferential conditional, by contrast, indicates that the speaker subscribes to an *inference* from the antecedent to the consequent. Consider, for instance, the following, also adapted from Dancygier 1999: 87.

(33) If they left at nine, they have arrived at home by now.

In uttering this conditional, the speaker subscribes to the inference from *They left at nine* to *They have arrived at home by now*. Inferential conditionals can indicate that the speaker is subscribing to various types of inferences: not just deductive ones, but also abductive, inductive, and other inferences (Sweetser 1990; Krzyżanowska et al. 2013). Barbara Dancygier (1999: 86–88) describes a linguistic test for inferential conditionals: inferential conditionals typically occur (and sound better) with *must* in the consequent. Indeed, sometimes the addition of *must* turns a content conditional into an inferential conditional. Consider the following examples.

(34) If Alicia is teaching at the local university, she is a professor.

(35) If Alicia is teaching at the local university, she must be a professor.

The assertibility conditions of (34) and (35) differ: one should only assert (34) if the local university has a policy that only professors can teach (and not, say, postdoctoral researchers). If there is no policy of this sort, but one has some background knowledge about Alicia's research expertise that, when combined with the information that she teaches at the local university, leads one to infer that she is a professor, then one may still assert (35). By contrast, whenever the assertion of (34) is licensed, then so is the assertion of (35). These judgements can be explained by taking the conditional in (34) to be a content conditional and the conditional in (35) to be an inferential conditional. For this means that if (34) is assertible, then so is (35), since the truth of a content conditional licenses the assertion of the corresponding inferential conditional: if one accepts a content conditional, one can reason from its antecedent to its consequent. Thus, the fact that Alicia is teaching at the local university is sufficient condition for her being a professor licenses the inference from *Alicia is teaching at the local university* to *Alica must be a professor*. The converse, however, need not be the case. For not every kind of reasoning licensing this inference (for instance, heuristic reasoning based on what one knows about Alicia and her scholarly achievements) also establishes that Alicia is teaching at the local university is sufficient for her being a professor.

The foregoing considerations reveal that content conditionals sanction the existence of a *specific* inferential relation between the antecedent and the consequent: the antecedent holding is a sufficient condition for the consequent holding. This immediately implies that content conditionals must entail the corresponding material conditional, as per the seemingly unproblematic direction of the Gibbard Collapse Argument. Moreover, the existence of a specific inferential relation holding between the antecedent and the consequent of content conditionals supports *modus tollens*-like reasoning: if the specific inferential relation holds, and the consequent fails, then so does the antecedent. For example, if one asserts *If Alicia teaches at the local university, she is a professor* on the basis of evidence provided by university policy, then one may assert *If she is not a professor, Alicia does not teach at the local university* on the basis of the same evidence. The evidence justifying the assertion of content conditionals may be appealed to in evidence-preserving arguments using such conditionals as premisses. Therefore, *modus ponens* for content conditionals preserves evidence. Finally, since *modus ponens* for content conditionals does preserve evidence, we must restrict the subderivations of the conditional proof rules for content conditional speech acts to evidence-preserving rules, analogously to how we restricted the introduction rule for the material conditional.

Unlike content conditionals, inferential conditionals do not sanction the existence of a specific inference from the antecedent to the consequent. Rather, they seem to be correctly assertible simply in virtue of evidence for the existence of any inferential relation between the antecedent and the consequent. Among those, there are inferences that do not preserve evidence, which must therefore be allowed in the corresponding conditional proof rules. For similar reasons, inferential conditionals ought not to imply the corresponding material conditionals: the material conditional asserts the existence of a specific inferential relation between the antecedent and the consequent, not just of any inferential relation. Finally, inferential conditionals do not support *modus tollens*: from *If Alicia is teaching at the local university, she must be a professor* and *It is not the case that Alicia must be a professor* it does not follow that she is not teaching at the local university. From it not being the case that Alicia must be a professor, it follows that she *might* not be a professor. But this is insufficient to infer that she is not teaching. Thus, when dealing with inferential conditionals, we must ban the application of *modus ponens* in evidence-preserving inferences, thereby preventing the derivation of *modus tollens*.

Thus, content conditionals correspond in their formal properties to our first option to spell out the coordination principles for conditional speech acts: they are associated with a version of conditional proof restricted to evidence-preserving rules and a version of *modus ponens* that preserves evidence. Inferential conditionals correspond in their formal properties to our second option: they are associated with a version of conditional proof that is not so restricted, but with a version of *modus ponens* that does not preserve evidence. Inferential conditional readings explain why inferences such as the one from *If it is raining, then it is true that the streets are wet* to *If it is raining, then the streets are wet* sound acceptable.

We therefore introduce a further distinction between indicative content conditionals and indicative inferential conditionals and a corresponding distinction between indicative conditional speech acts. Once this distinction is incorporated into our account of conditionals, we see that *modus tollens* for indicatives and the seemingly good direction of the Gibbard Collapse Argument fail exactly in those cases in which they should fail.

Similar considerations apply to the subjunctive case. We distinguish between subjunctive content conditionals and subjunctive inferential conditionals and draw a corresponding distinction between subjunctive conditional speech acts. In uttering a subjunctive inferential conditional, the speaker indicates that she subscribes to an inference from the antecedent to the consequent, which need not be specific. In uttering a subjunctive content conditional, by contrast, the speaker subscribes to the existence of a certain relation holding between the antecedent and the consequent, a relation which immediately entails the existence of a specific inferential relation between them. The difference between indicatives and subjunctives lies in the background assumptions that are allowed in performing the inference—whether specific or not—from the antecedent to the consequent: as noted earlier in the chapter, indicatives only allow epistemically compatible assumptions, whereas subjunctives only allow factually compatible assumptions.

Having introduced the distinction between content and inferential conditionals, we can also address McGee's (1985) purported counterexample to *modus ponens*, which we already encountered in Chapter 2. For inferential readings of indicative conditionals allow us to account for McGee's counterexample. McGee constructs a scenario in which one has strong evidence for *If a Republican wins the election, then if it's not Reagan who wins it will be Anderson* and *A Republican will win the election* but no evidence for *If it's not Reagan who wins, it will be Anderson.* Under an inferential conditional reading of the embedded conditional in *If a Republican wins the election, then if it's not Reagan who wins it will be Anderson*, the inference does not preserve evidence, as noted by McGee. Nonetheless, *pace* McGee, the inference is valid in that it preserves commitment. Someone asserting *If a Republican wins the election, then if it's not Reagan who wins it will be Anderson* commits to being prepared to infer *It will be Anderson* from *It's not Reagan* should they also assert *A Republican wins the election.* If they do assert *A Republican wins the election* and *It's not Reagan*, they must, when suitably challenged, assert *It will be Anderson* or admit to a mistake.

Mandelkern (2020) has recently presented a related counterexample for subjunctive conditionals that cannot be dealt with by reading it as an inferential conditional. His example is the complex, conditional-embedding subjunctive *If Jones had had bronchitis, then if he had been in genotype A, he would be showing the symptoms he in fact is showing*, which we may suppose to be true even though people with genotype A are highly unlikely to get bronchitis. Now assume it is established by incontrovertible medical evidence, such as a bacterial assay, that Jones has in fact bronchitis. By *modus ponens*, we appear to be able to derive the embedded conditional *If Jones had genotype A, he would be showing the symptoms he in fact is showing*. Mandelkern argues that this is problematic: since people with genotype A are unlikely to show symptoms of bronchitis, we do not accept the embedded conditional. Whether deriving the embedded conditional is indeed problematic depends on whether we treat Jones's having bronchitis to be factually compatible with him having genotype A. If it *is* compatible, the embedded conditional *If Jones had genotype A, he would be showing the symptoms he in fact is showing* is true and it is not problematic to be able to derive it. If Jones's having bronchitis is not factually compatible with him having genotype A, the complex conditional *If Jones had had bronchitis, then if he had been in genotype A, he would be showing the symptoms he in fact is showing* is false, so we cannot derive the embedded conditional. Mandelkern's counterexample rests on treating two claims as both compatible and incompatible.

8.8 On the plurality of conditionals

We argued in this chapter that conditionality in natural language can manifest itself both in the form of embeddable conditionals and in the form of conditional speech acts. One immediate benefit of our approach is that it makes sense of both the data motivating conditional speech acts and the data showing that conditionals can embed. Moreover, by using the multilateral framework, in which we can distinguish conditional assertion from conditional rejection, we can vindicate intuitively compelling patterns of inference such as the inferential equivalence between the negative assertion of *If A, then B* and the assertion of *If A, then not B*.

We then observed a plurality of conditionals. Each conditional speech act can be performed indicatively and subjunctively, giving rise to a corresponding division within conditionals in natural language. Indicatives and subjunctives are all subject to a compatibility constraint, which suffices to block the derivation of the Gibbard Collapse Argument and of Curry's Paradox. The characteristic mark of indicatives is that they are subject to an epistemic compatibility constraint; the characteristic mark of subjunctives is that they are subject to a factual compatibility constraint. Understanding the distinction between indicatives and subjunctives in this way allows us to implement the distinction within the multilateral framework in terms of restrictions on the conditional proof rule for the corresponding conditional speech acts.

Within conditionals, we can further distinguish between content and inferential conditionals. Content conditionals sanction the existence of a specific inferential relation between the antecedent and the consequent, whereas inferential conditionals sanction the existence of an inference from the antecedent to the consequent, where this inference need not be specific. The distinction between content and inferential conditionals too can be implemented within the multilateral framework in terms of restrictions on the conditional proof rule for the corresponding conditional speech acts.

We have, therefore, the following situation. The conditional proof rule for indicative conditional speech acts is restricted to epistemically compatible premisses, whereas the conditional proof rule for subjunctive conditional speech acts is restricted to factually compatible premisses. The conditional proof rule for content conditional speech acts is restricted to evidence-preserving rules, whereas the conditional proof rule for inferential conditional speech acts carries no such restrictions, although their *modus ponens*-esque rules do not preserve evidence. Finally, the conditional proof rule for the material conditional is restricted to evidence-preserving rules, but is not subject to any compatibility constraints. Thus, the material conditional is a content conditional in that it asserts the existence of a specific inference from the antecedent to the consequent, but it is neither an indicative nor a subjunctive conditional, in that it is neither counterfactual nor counterepistemic.

Among the conditional phenomena we have considered, the indicative and subjunctive content conditional speech acts are therefore the most demanding ones, with their conditional proof rules combining the restrictions on their less demanding brethren. As we have argued, conditionality in natural language appears to be characterized by the presence of some compatibility constraint. By contrast, the restriction to evidence-preserving rules appears to be the hallmark of a specific class of natural language conditionals, namely that of content conditionals. All conditionals seem to sanction an inference from the antecedent to the consequent, but some of them are more demanding in that they sanction a *specific* inference. We are inclined to tentatively conclude that all conditionals are, *au fond*, inferential. This further supports the centrality we have ascribed to the Ramsey test and conditional proof in the study of conditionals.

9
Probability

In Chapter 4 we applied the multilateral methodology to develop an inferential expressivist account of epistemic modals. In this chapter we apply the methodology to develop an inferential expressivist account of epistemic probability operators. As we shall see, similar arguments to those typically given in favour of an expressivist account of epistemic modals can be given in favour of an expressivist account of probability operators. However, and again similarly to the case of epistemic modals, probability operators can embed and are therefore subject to a version of the Frege–Geach Problem. Once again, inferential expressivism allows one to deal with the Frege–Geach Problem whilst retaining the advantages of an expressivist approach to probability operators.

9.1 Traditional expressivism about *probable*

In English, one can use expressions such as *probable* and *likely* to construct the probability operators *it is probable that* and *it is likely that*. As in the case of epistemic modals, the orthodox approach to probability operators is a contextualist one and is due to Kratzer (1991). Recall that the orthodox approach takes an epistemic modal possibility claim to be true just in case the prejacent is compatible with some contextually determined body of knowledge, represented by a set of epistemically accessible possible worlds. On Kratzer's approach, there is another contextual element to the semantics of modal expressions: a contextually supplied ordering source, which induces an ordering of the contextually determined set of possible worlds. In the case of epistemic modals, when the ordering source is non-empty, this ordering is given by stereotypicality or normality, and the epistemically possible worlds that are quantified over are those that are closest according to this ordering. The orthodox approach to probability operators similarly takes their meaning to be given by quantification over epistemically closest possible worlds ordered according to a contextually given stereotypicality or

Reasoning with Attitude: Foundations and Applications of Inferential Expressivism. Luca Incurvati and Julian J. Schlöder, Oxford University Press. © Oxford University Press 2023.
DOI: 10.1093/oso/9780197620984.003.0009

normality ordering. The approach is thus very much along the lines of the orthodox approach to epistemic modals except that the ordering source is now exploited to account for the gradability of probability operators: one uses the ordering to define the meaning of expressions such as *A is at least as probable as B* and *A is more probable than B*. One then says that *A is probable* is true just in case *A is more probable than not A*.

The descriptivist understanding of the constitutive function of epistemic modal talk typically associated with the orthodox approach carries over to probabilistic talk. That is, probabilistic talk serves to describe a special aspect of the world: in uttering a sentence having an epistemic modal as its main operator, the speaker describes an epistemic position. Thus, *It is probable that it is raining* serves to communicate the information that the speaker's own evidence or some contextually determined body of evidence makes raining more likely than not raining.

The same challenges that arise for the orthodox approach to epistemic modals can be raised to its application to probability operators. First, probability operators give rise to sentences whose behaviour is similar to those of Yalcinean sentences. As early as 1950, Stephen Toulmin pointed out the existence of Moore-like sentences involving probability operators.

> No one person is permitted, in one and the same breath, to call the same thing both 'improbable' and 'true' ...: the form of words 'Improbable but true' is therefore ruled out. (Whether or no we are to say that it is a 'contradiction', is another question, and one that might get us into deep water, though I think a strong case could be made out for calling it one.)
>
> (Toulmin 1950: 37)

Toulmin is correct that there is a strong case to be made for calling Moore-like sentences involving probability operators contradictions, namely their behaviour in embedded contexts. As Yalcin (2007: 1015) himself notes, it is infelicitous to utter either (1) or (2) but the latter, unlike the former, remains infelicitous in suppositional contexts or conditional antecedents.

(1) # It is raining and given my evidence it is likely that it isn't.

(2) # It is raining and it is probable that it isn't.

(3) Suppose it's raining and given my evidence it is likely that it isn't.

(4) # Suppose it's raining and it is probable that it isn't.

Second, probability operators give rise to disagreement phenomena that are difficult to account for on the standard contextualist approach. Consider the following dialogue.

(5) *Alice*: I can't find the keys.
Bob: It is probable that they are in the car.
Alice: No, it's not probable. They are almost certainly in the drawer.

Similarly to the case of epistemic modals, it is difficult to locate a proposition about a single body of evidence that Bob is warranted in asserting and Alice is warranted in rejecting. Indeed, that this is so was noted by Price (1983) well before cases of modal disagreement were invoked against contextualist accounts of epistemic modality. Price considers the idea of analysing an utterance of *It is probable that A* as elliptical for *Given the existing evidence, it is probable that A*. Interpreting *the existing evidence* as the evidence the speaker possesses, Price notes, makes it impossible to account for disagreement phenomena involving probabilistic operators.

> If I disagree with your claim that it is probably going to snow, I am not disagreeing that given *your* evidence it is likely that this is so, but indicating what follows from *my* evidence. Indeed, I might *agree* that it is probably going to snow and yet think it false that this follows from your evidence. (Price 1983: 404)

Interpreting *the existing evidence* as the evidence available in principle to the speaker, Price continues, forces one to regard many prima facie legitimate probabilistic utterances as unjustified.

> consider the surgeon who says, 'Your operation has probably been successful. We could find out for sure, but since the tests are painful and expensive, it is best to avoid them.' The accessibility, in principle, of evidence which would override that on which the [single case probabilistic] judgement is based, is here explicitly acknowledged. (Price 1983: 405)

Price goes on to argue that no interpretation of the phrase *the existing evidence* will result in an account capable of both accounting for disagreement phenomena involving probabilistic operators and vindicating our ordinary probabilistic talk as justified. He concludes that utterances of the form

It is probable that A are non-assertoric: *it is probable that* modifies the force with which a speech act is made, rather than its content.

An expressivist approach promises to have the resources to account for epistemic contradictions and disagreement phenomena involving probability operators. Predictably, however, the fact that expressions such as *probable* and *likely* may feature in conditional antecedents gives rise to a version of the Frege–Geach Problem.

(6) a. If it is probable that it is raining, we'd better take an umbrella.
b. It is probable that it is raining.
———
c. We'd better take an umbrella.

This inference appears to be valid and so the Frege–Geach argument can be used to establish that *probable* in the second premiss must modify content, on pain of the inference not being an instance of *modus ponens*. We are now going to show that an inferential expressivist approach can deal with the Frege–Geach Problem whilst retaining the core of the expressivist solution to the difficulties faced by the orthodox approach to probability operators.

9.2 *Probable* and *probably*

In Chapter 4, we argued that while the Frege–Geach argument does show that *might* modifies content, the same is not true of *perhaps*, which should instead be treated as a force modifier. We now present linguistic evidence to the effect that, analogously, while *probable* modifies content, *probably* modifies force. Similar evidence could be provided for the claim that adjectival uses of *likely* modify content but adverbial uses modify force.

For obvious reasons, the similarities between *probable* and *probably* are even more apparent than in the case of *might* and *perhaps*. It should come as no surprise, then, that *it is probable that* and *probably* are often treated as interchangeable. For instance, Price titles his 1983 article 'Does "probably" modify sense?' but discusses the proper semantic analysis of expressions of the form *It is probable that A*, only to then use utterances involving *probably* in the examples, as we saw in the previous section. As a matter of fact, unembedded occurrences of *it is probable that* and *probably* give rise to inferentially equivalent sentences. For instance, the following sentences can be uttered in exactly the same circumstances.[1]

(7) a. Probably it is raining.
    b. It is probable that it is raining.

Nonetheless, *probable* and *probably* have different embedding behaviour. To begin with, *probable* can be felicitously used in conditional antecedents, whereas the use of *probably* in the same context results in odd-sounding sentences.

(8) # If probably it is raining, we'd better take an umbrella.

(9) If it is probable that it is raining, we'd better take an umbrella.

Moreover, *probably* does not embed under negation and *never*, whereas *probable* does.

(10) # It is not probably raining.

(11) It is not probable that it is raining.

(12) # It never probably rains in the summer.

(13) It is never probable that it rains in the summer.

These data have suggested to some linguists that *probably* is a positive polarity item, that is an expression which may only occur in positive contexts. (Similarly, a negative polarity item is an expression which may only occur in negative contexts.) The problem, however, is to explain why *it is probable that* seems capable of appearing in the very same contexts in which *probably* cannot. Standard accounts of what makes a linguistic expression a positive or negative polarity item are semantic, the reason for this being that the

[1] We are setting aside cases such as *I should probably leave*. Although in such cases *probably* does seem to work as a force modifier—the speaker is hedging their assertion—*I should probably leave* is not inferentially equivalent to *It is probable that I should leave*. These uses of *probably* appear to be additional uses of the expression resulting from a conventionalized form of politeness. Indeed, cross-linguistic evidence shows that equivalent expressions in other languages cannot always be employed in this way but only have a probabilistic use.

distribution of these items is hard to predict on the basis of their syntactic category alone (Giannakidou 2011). But what could the semantic difference between *probably* and *probable* be?

Øystein Nilsen (2004) attempts to account for the difference between *probably* and *probable* by arguing that the former is more restrictive than the latter: the semantic value of *probably* is given in the meta-language by a quantification over a smaller set of possible worlds. He offers a similar explanation for the difference in embedding behaviour between *possible* and *possibly*. As Thomas Ernst (2009) notes, however, Nilsen's proposal is not supported by the linguistic data: if *probable* and *probably* were on a scale, we would expect the following utterance to sound felicitous, but it doesn't.

(14) # It's probable that it's raining—in fact, it's probably raining.

Ernst suggests distinguishing between epistemic modal adjectives and epistemic modal adverbs in terms of the subjective/objective distinction. Thus, *probably* expresses likelihood given the evidence available to the subject, while *it is probable that* expresses likelihood given the evidence available in principle. Similarly, *possibly* expresses compatibility with the evidence available to the subject while *it is possible that* expresses compatibility with the evidence available in principle. Our discussion in the previous section should make it clear that Ernst's proposal about the meaning of *probable* makes it hard to account for epistemic contradictions involving probability operators and forces one to regard much probabilistic talk, as in Price's surgeon example, as unjustified.

We submit that the source of the problem is the view that the embedding behaviour of *probably* is due to its status as a positive polarity item. This leads both Nilsen and Ernst to drive a semantic wedge between *probably* and *probable*, thereby preventing them from accounting for the linguistic data. What is more, on closer inspection *probably* exhibits certain occurrence patterns that are hard to reconcile with the view that it is a positive polarity item. In particular, *probably* appears to resist embedding not only in conditional antecedents but also under supposition—or, at the very least, its use in suppositional contexts is no more felicitous than its use in conditional antecedents.

(15) # Suppose probably it is raining.

(16) Suppose it is probable that it is raining.

However, *suppose*, unlike *if*, is not typically taken to generate a negative context, since it does not license the use of prototypical negative polarity items such as *any* and *ever*.

(17) # Suppose you have any questions.

(18) If you have any questions, let me know.

In addition, speakers tend to find narrow scope readings of *probably* hard to obtain, whereas *probable* may be used adjectivally to easily obtain such readings. For instance, (19) is a true description of a standard lottery, whereas (20) is not (compare Swanson 2010 on *possible*).

(19) All tickets are probable losers.

(20) All tickets will probably lose.

By taking *probably* to modify force and *probable* to modify content, one can explain their different embedding behaviour. At the same time, postulating this difference between the two expressions does not force one to drive a semantic wedge between them, which was the problem with Nilsen's and Ernst's proposals.

We have been taking assertions to be realizable by positive answers to polar questions. *Probably* may be used to modify such answers, as in (21a). The result is the speech act we call *moderate assertion*. Typically, the *yes* is elided, as in (21b). Finally, in (21c), *probably* turns into a moderate assertion an otherwise strong assertion of *it is raining*.

(21) a. Is it raining? Probably yes.
b. Is it raining? Probably.
c. Probably it is raining.

We take these three utterances to be linguistic variants of each other: all three utterances may be used to realize the moderate assertion of *It is raining*. This is a different speech act than the strong assertion of *It is probable that it is raining*. Nonetheless, there is ample linguistic evidence that the moderate assertion of *It is raining* and the strong assertion of *It is probable that it is raining* are inferentially equivalent. Using our multilateral methodology, this inferential equivalence will enable us to explain the meaning of *probable* in terms of the speech act of moderate assertion.

Alongside the speech act of moderate assertion, it is natural to also consider a speech act of moderate rejection, realized by the following utterances.

(22) a. Is it raining? Probably not.
b. Is it raining? Improbably.

Having identified the linguistic realizations of the speech acts of moderate assertion and rejection, our next step is to explain what these speech acts are.

9.3 Moderate assertion and moderate rejection

In Chapter 4, we presented the multilateral square of opposition. As we noted there, the square tells us that the speech act of weak assertion is subaltern to the speech act of strong assertion. Thus, strong assertion and weak assertion are related in the same way in which, according to the traditional square of opposition, the determiners *all* and *some* are related. Now there is a host of determiners in English which are located in strength between *all* and *some*, such as *most*, *many*, and *few*, and which can be located within the square of opposition (see Peterson 1979). Of particular interest for current purposes is *most*. This determiner resembles *all* in that *Most As are Bs* and *Most As are not Bs*, like the corresponding sentences involving *all*, are contraries. At the same time, *most* resembles *some* in that *Most As are Bs* is compatible with *Some As are not Bs*. Moderate assertion appears to stand to strong and weak assertion as *most* stands to *all* and *some*. In particular, while *Probably A* and *Probably not A* are contraries, it is plainly not absurd to utter *Probably A* and *Perhaps not A*. Similarly, moderate rejection stands to strong and weak rejection as *Most As are not Bs* stands to *No As are Bs* and *Some As are not Bs* in the traditional square of opposition. We can use the parallel between the case of determiners and the case of speech acts to guide our account of moderate assertion and moderate rejection.[2]

Recall that when explaining the nature of a speech act, we have described the attitude it conventionally expresses and its essential effect on the

[2] The analogy between assertoric speech acts of various strength and determiners raises the question of whether there are expressions in English that serve to realize the speech act that stands to strong, moderate, and weak assertion as *few* stands to *all*, *most*, and *some*. It might well be that the adverb *plausibly* fits the bill, but we will focus on moderate assertion and rejection in this chapter.

conversation. Let us start from the first aspect of the speech act of moderate assertion. Once one takes *probably* to be a force modifier, it is natural to take an utterance of the form *Probably A* to express a strong partial belief towards *A*. The attitude of partial belief sits in strength between the attitude of belief and the attitude of refraining from disbelieving—as it should, given what we have said. However, we need to say more about what this attitude is. Those that have suggested that probability operators are force modifiers or, at any rate, serve to express a strong partial belief, have taken a strong partial belief to consist, in effect, in a high degree of belief or credence (Forrest 1981; Price 1983). However, as Richard Holton (2008: 33) notes, the psychological findings suggest, unsurprisingly, that we are much happier making qualitative probability judgements, rather than quantitative ones. In general, we normally talk of an outcome being likely or unlikely, or being more likely than another. We do not assign numerical values to possibilities. Although no doubt something must be said about our probability talk involving numerical values—and we shall return to the issue below—a more economical approach would seem to be desirable.

Holton (2008: 39) suggests that the notion of live possibility may be of help here, and we concur. Holton argues that a subject believes that *A* just in case they take *A* to be a live possibility but do not take *not A* to be a live possibility. Without endorsing Holton's account of belief, we can rephrase Holton's insight here by saying that by expressing belief towards *A*, one commits to regarding *A* as a live possibility and ruling out *not A* as a live possibility. This is very much in keeping with our account of epistemic modality: expressing the belief that *A* commits one to believing that it might be that *A* and to refraining from believing that it might be that it is not the case that *A*. Holton goes on to argue that using the notion of live possibility one can also give an account of *partial* belief: a subject partially believes that *A* just in case they take both *A* and its negation to be live possibilities. In our framework, we can capture this insight by saying that expressing the partial belief that *A* commits one to believing that it might be that *A* but not to ruling out that it might be that it is not the case that *A*. Note that we do not say that expressing a partial belief towards *A* commits one to also believing that it might be that it is not the case that *A* because someone might express a partial belief while in fact having a full belief that *A*.

Now this correctly classifies the attitude expressed by moderate assertion as a case of partial belief: as noted above, it is not absurd to utter *Probably A* and *Perhaps not A*. However, a partial belief may also be expressed by a weak assertion: weakly asserting *A* obviously commits one to believing that it might be that *A* but does not commit one to ruling out that it might be that it is not the case that *A*. What is expressed by weakly asserting a sentence is a *weak* partial belief: a subject may (and often will) weakly assert a sentence because they are completely undecided about what the sentence says. What further commitments are incurred by expressing a *strong* partial belief, the attitude expressed by a moderate assertion?

The parallel with determiners might help us to find an answer. As noted above, moderate assertion seems to stand to strong and weak assertion as *most* stands to *all* and *some*. Now in the study of generalized quantifiers (Barwise and Cooper 1981), it is usually assumed that in saying that most *A*s are *B*s one is committing to there being at least one *A* which is *B* but also to there being more *A*s that are *B*s than *A*s that are not. We can say something similar in the case of strong partial belief. In expressing a strong partial belief that *A* one is not only committing to believing that it might be that *A* but also to believing that the live possibilities in which *A* is true outnumber those in which it is not. In effect, this follows from the fact that it is absurd to moderately assert both a sentence and its negation (whereas it is not absurd to weakly assert them).

The commitments incurred by expressing a strong partial belief help to explain the commitments Toulmin (1950) takes to be engendered by utterances of the form *Probably A*. According to Toulmin,

> [w]hen I say 'S is probably P' I *commit myself guardedly* (tentatively, with reservations) to the view that S is P, and (likewise guardedly) *lend my authority* to that view. (Toulmin 1950: 35)

By qualifying one's utterance with *probably*, Toulmin (1950: 31) says, one expressly avoids *unreservedly* committing oneself. This follows from the fact that an utterance of *Probably A* does not commit one to refraining from believing that it might be that it is not the case that *A*. Nonetheless, the commitment to *A* is stronger than in the case in which one simply utters *Perhaps A*: by committing oneself to there being more live possibilities in which *A* holds than live possibilities in which it doesn't, one is reasonably taken as, for instance, licensing others to rely on *A* being the case (rather than its negation).

What goes for moderate assertion goes, *mutatis mutandis*, for moderate rejection. Thus, moderate rejection expresses strong partial disbelief. If someone has a strong partial disbelief towards *A*, then they have an attitude towards *A* which is compatible with refraining from fully disbelieving *A* but incompatible with strong partial belief that *A*. The latter means, in particular, that they are committed to regarding the possibilities in which *not A* holds to outnumber those in which *A* holds.

We now turn to the second aspect of the speech act of moderate assertion, that is its essential effect on conversation. Our starting point is again the observation that moderate assertion is similar to strong assertion in that *Probably A* and *Probably not A* are contraries, but it is dissimilar from strong assertion in that it is not absurd to utter *Probably A* and *Perhaps not A*. The fact that *Probably A* and *Probably not A* are contraries suggests, at the level of essential effects on the conversation, that in moderately asserting one does make some kind of proposal to add a sentence to the common ground. It is absurd to simultaneously propose the addition of a sentence to the common ground and to the negative common ground. At the same time, the proposal being effected by a moderate assertion must be such that, should it be accepted, the possibility that *It might be that not A* be added to the common ground is not ruled out. We take this to suggest that the proposal being made is a *tentative* one. Thus, we take the essential effect of moderately asserting *A* to be that of tentatively proposing the addition of *A* to the common ground.

The effect of accepting a tentative proposal may be modelled in terms of possible worlds. Recall that the context set is the collection of possible worlds compatible with the common ground. One can then take the effect of accepting a tentative proposal to add *A* to the common ground to make it so that most worlds in the context set are *A*-worlds. But, to stress, this is just a formal model of the effect of tentative assertion on the common ground. One need not think of the common ground as a set of possible worlds.

Again, what goes for moderate assertion goes, *mutatis mutandis*, for moderate rejection. So we take the effect of moderately rejecting a sentence to be a tentative proposal to add that sentence to the negative common ground.

9.4 Coordination principles

Our next task is to provide coordination principles to integrate the speech acts we have introduced into the multilateral framework. We let $\boxplus$ stand for moderate assertion and $\boxminus$ stand for moderate rejection. We begin with principles governing the relation between these two speech acts. As noted above, *Most As are Bs* and *Most As are not Bs* are contraries, and moderate assertion and rejection stand in an analogous relation. We can therefore lay down analogues of the coordination principles* governing the interaction of strong assertion and strong rejection, where, as usual, $\vdots *$ denotes evidence-preserving inference.

$$\text{(Moderate Contrariety)}\ \frac{\boxplus A \qquad \boxminus A}{\bot} \qquad (SR_3{}^*)\ \frac{\begin{array}{c}[\boxplus A]\\ \vdots *\\ \bot\end{array}}{\boxminus A} \qquad (SR_4{}^*)\ \frac{\begin{array}{c}[\boxminus A]\\ \vdots *\\ \bot\end{array}}{\boxplus A}$$

The Moderate Contrariety rule says that it is absurd to moderately assert and reject the same sentence. The third and fourth Smileian *reductio** rules say that if it is absurd for a speaker to moderately assert a sentence, then the speaker is committed to moderately rejecting that sentence, and that if it is absurd for a speaker to moderately assert a sentence, then the speaker is committed to moderately rejecting that sentence.

We now turn to principles governing the interaction of moderate assertion and moderate rejection with the other speech acts of the multilateral square of opposition. We have been arguing that moderate assertion stands to strong and weak assertion as *most* stands to *all* and *some*. That is, weak assertion is subaltern to moderate assertion, which in turn is subaltern to strong assertion. The following coordination principles codify these relations between the three assertoric speech acts.

$$(SA_3)\ \frac{+A}{\boxplus A} \qquad (SA_3{}^*)\ \frac{\boxplus A \qquad \begin{array}{c}[+A]\\ \vdots *\\ +B/\bot\end{array}}{\boxplus B/\bot}$$

$$(SA_4)\ \frac{\boxplus A}{\oplus A} \qquad (SA_4{}^*)\ \frac{\oplus A \qquad \begin{array}{c}[\boxplus A]\\ \vdots *\\ \boxplus B/\bot\end{array}}{\oplus B/\bot}$$

The third subalternity rule says that if the speaker strongly asserts A, then they are committed to expressing a strong partial belief towards A. The fourth subalternity rule says that if the speaker moderately asserts A, then they are committed to expressing refraining from disbelieving A. This is in line with our analysis of the attitude expressed by moderate assertion and its essential effect on the conversation. Belief is a stronger attitude than strong partial belief, which is in turn a stronger attitude than refraining from disbelieving. Similarly, proposing to add a sentence to the common ground is stronger than tentatively doing the same thing, which is in turn stronger than preventing the sentence from being added to the common ground. The starred subalternity rules ensure that weak and moderate assertions are preserved under evidence-preserving inference.

Using the appropriate versions of the Smileian *reductio** rules to be discussed below, analogous coordination principles can be derived governing the interaction of moderate rejection with its strong and weak counterparts.

$$(\mathrm{SA}_5)\ \frac{-A}{\boxminus A} \qquad (\mathrm{SA}_5{}^*)\ \frac{\boxminus A \qquad \begin{matrix}[-A]\\ \vdots{*}\\ -B/\bot\end{matrix}}{\boxminus B/\bot}$$

$$(\mathrm{SA}_6)\ \frac{\boxminus A}{\ominus A} \qquad (\mathrm{SA}_6{}^*)\ \frac{\ominus A \qquad \begin{matrix}[\boxminus A]\\ \vdots{*}\\ \boxminus B/\bot\end{matrix}}{\ominus B/\bot}$$

Finally, we can derive principles stating that strong assertion and moderate assertion are incompatible, and that the same is the case for moderate assertion and strong rejection.

$$\text{(Moderate Rejection)}\ \frac{+A \qquad \boxminus A}{\bot}$$

$$\text{(Moderate Assertion)}\ \frac{\boxplus A \qquad -A}{\bot}$$

This is how it should be: it is absurd to both express belief towards A and a strong partial disbelief towards A, or to both express disbelief towards A and a strong partial belief towards A. Similarly, it is absurd to propose to add some sentence to the common ground while tentatively proposing to add it to the negative common ground, or to tentatively propose to add some sentence to the common ground while proposing to add it to the negative common ground.

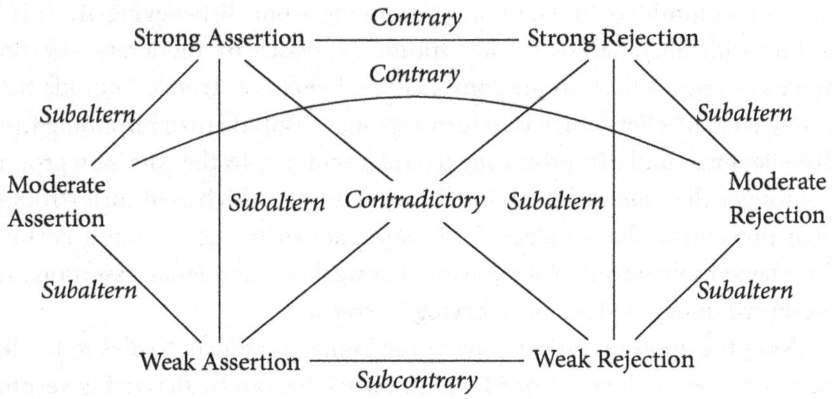

Fig. 9.1. Opposition in multilateral logic with probability.

The coordination between the six speech acts is summarized in Figure 9.1, which depicts the relations between strong, moderate, and weak assertion and rejection by means of what one may call the *hexagon of opposition* in multilateral logic with probability.

9.5 Operational rules

The next stage in our multilateral methodology is to explain the meaning of *it is probable that* in terms of the speech act of moderate assertion. As we noted in Section 9.2, *Probably, it is raining* and *It is probable that it is raining* appear to be inferentially equivalent. We can codify this inferential equivalence in the following rules, where Δ formalizes *it is probable that.*

$$(+\Delta\text{I.})\ \frac{\boxplus A}{+\Delta A} \qquad (+\Delta\text{E.})\ \frac{+\Delta A}{\boxplus A}$$

Similarly to the case of *might* and *perhaps*, we also have rules ensuring the inferential equivalence of *Probably, it is probable that it is raining* and *Probably, it is raining.*

$$(\boxplus\Delta\text{I.})\ \frac{\boxplus A}{\boxplus\Delta A} \qquad (\boxplus\Delta\text{E.})\ \frac{\boxplus\Delta A}{\boxplus A}$$

We also have rules governing the behaviour of *probable* under weak assertion and of *might* under weak assertion.

$$(\boxplus\Diamond\text{I.})\ \frac{\oplus A}{\boxplus\Diamond A} \qquad (\boxplus\Diamond\text{E.})\ \frac{\boxplus\Diamond A}{\oplus A}$$

$$(\oplus\triangle\text{I.})\ \frac{\boxplus A}{\oplus\triangle A} \qquad (\oplus\triangle\text{E.})\ \frac{\oplus\triangle A}{\boxplus A}$$

These four rules together imply that the compositional value of a sentence is affected neither by iterations of *might* or *probable* alone nor by iterations of the two combined. This would seem to be in line with the observation that iterations of epistemic and probabilistic operators often do not sound right (compare Nilsen 2004: 819).

(23) ?? Probably, it might be raining.

At the same time, the result might seem to be problematic. For Sarah Moss (2015: §§1.1–1.2) has offered a number of examples which purport to show that nesting epistemic vocabulary changes the meaning of an utterance. All examples involve interactions of possibility modals such as *might* and *possible* with probabilistic expressions such as *probable* and *likely*. The examples, therefore, cannot undermine the view that iterations of *might* or *probable* alone do not affect the compositional content of an utterance. Nonetheless, the examples do threaten the analogous point concerning iterations of the two combined. So it is worth explaining how we can deal with the examples in a similar way in which we dealt with purported examples of non-redundant iterations of *might* in Chapter 4. We only need to focus on one of the examples, since the strategy generalizes in the obvious way.

Moss asks us to consider a situation in which Alice and Bob are both being considered for certain job positions. In such a situation, says Moss, one might observe the following.

(24) Alice is a likely hire, and Bob might be a likely hire.

(25) Alice is a possible hire, and Bob is probably also a possible hire.

The content of these utterances, Moss continues, could then be spelled out as follows.

(26) It is likely that we will hire Alice, and we might also be likely to hire Bob.

(27) We might hire Alice, and it is probably the case that we might hire Bob too.

As Moss notes, if one utters (24), there seems to be a difference between one's expressed opinion with regards to Alice and Bob being likely hires. This difference can be captured on our account. In saying that Alice is a likely hire, one is tentatively proposing to add to the common ground *Alice will be hired*, whereas in saying that Bob might be a likely hire one is preventing from making it common ground that one will not tentatively propose to add to the common ground *Bob will be hired* once, say, the candidate's file has been looked at more closely or interviews have been held. As in the case of epistemic modals, this is compatible with the $\oplus\vartriangle$-rules being valid, since they apply when the common ground is held fixed. Similar considerations apply to (25). In saying that Alice is a possible hire, one is preventing from making it common ground that Alice will not be hired, whereas in saying that Bob is probably a possible hire one is tentatively proposing to making it common ground that, once more information about the candidates is in, one will not prevent from making it common ground that Bob will be hired. Again, this is compatible with the $\oplus\vartriangle$-rules being valid when the common ground is held fixed.

We now turn to the rules for negation under moderate assertion and rejection, which are analogous to their counterparts under strong and weak assertion and rejection.

$$(\boxplus\neg\text{I.})\ \frac{\boxminus A}{\boxplus\neg A}\qquad\qquad(\boxplus\neg\text{E.})\ \frac{\boxplus\neg A}{\boxminus A}$$

$$(\boxminus\neg\text{I.})\ \frac{\boxplus A}{\boxminus\neg A}\qquad\qquad(\boxminus\neg\text{E.})\ \frac{\boxminus\neg A}{\boxplus A}$$

The rules tell us how to pass from tentative proposals to add to the common ground to tentative proposals to add to the negative common ground and vice versa by introducing and eliminating negation. Thus, for instance, the

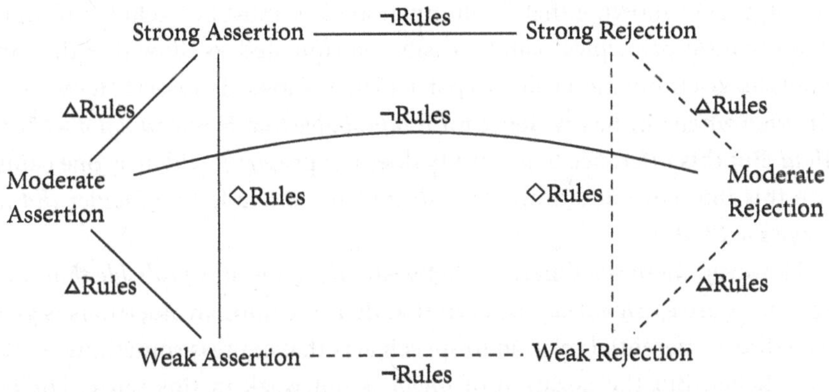

Fig. 9.2. Operations in multilateral logic with probability. Dashed lines are derivable.

(⊞¬I.)-rule says that if one tentatively proposes to add *A* to the negative common ground one is thereby committed to accepting a tentative proposal to add *not A* to the common ground.

Given the rules we have presented so far, one can derive negative counterparts to the positive rules for △. We omit them here. The situation is summarized in Figure 9.2, which depicts the hexagon of operations in multilateral logic using negation, epistemic modals, and probabilistic operators.

9.6 Evidence and probability

We now come to the final stage of our multilateral methodology. That is, we must attend once again to the restrictions on the coordination principles*, which now include not only rules coordinating strong assertion and strong rejection but also rules coordinating moderate assertion and moderate rejection.

Now, coordination principles* are restricted so that their subderivations must preserve evidence. We have seen that in EML the inferences that preserve evidence are those that do not use premisses signed with markers for weak assertion or weak rejection and that do not employ rules eliminating a ◇. This remains to be the case, but the language now includes two force markers for speech acts that are not strong and an additional epistemic operator, namely △. The arguments showing that ⊖ and ⊕ must be excluded as premisses from the coordination principles* carry over to ⊟ and ⊞. And the argument showing that $\Diamond$-elimination rules must be excluded from the coordination principles* can be easily reformulated to show that the same holds for Δ-elimination rules. In particular, it is absurd to assert *Homer wrote the Iliad* when one has evidence for *It is probable that Homer did not write the Iliad.* But this inference to absurdity does not preserve evidence: one cannot use this inference and Smileian *reductio** to conclude that *Homer did not write the Iliad.*

However, there is a difference between *might* and *it is probable that.* Both operators are *epistemically weak* in that their meaning in assertions is given by a direct inferential relation to speech acts that express something weaker than belief. But the negation of *might* is not weak in this sense. For *It is not the case that it might be that not A* is equivalent to *It must be that A*, and *must* can be introduced or eliminated in assertions without moving to a weaker speech act, as per the rules of (+$\Box$I.) and (+$\Box$E.) we discussed in Chapter 4. From this point of view, *it is probable that* differs from *might*, since *It is not probable that not A* (and, similarly, *It is not probable that A*) is also epistemically weak. The point can be appreciated formally by comparing the following two derivations.

$$\dfrac{\dfrac{\dfrac{+\neg\Diamond A}{-\Diamond A}\,(+\neg\text{E.}) \qquad \dfrac{\dfrac{[+A]^1}{\oplus A}\,(\text{SA}_1)}{+\Diamond A}\,(+\Diamond\text{I.})}{\bot}\,(\text{Strong Rejection})}{-A}\,(\text{SR}_1{}^*)^1$$

$$\dfrac{\dfrac{\dfrac{+\neg\Delta A}{-\Delta A}\,(+\neg\text{E.}) \qquad \dfrac{\dfrac{[+A]^1}{\boxplus A}\,(\text{SA}_3)}{+\Delta A}\,(+\Delta\text{I.})}{\bot}\,(\text{Strong Rejection})}{-A}\,(\text{SR}_1{}^*)^1$$

The derivations appear to be entirely analogous. In particular, they both seem to show that a negated modal disappears. Yet, the first derivation, involving $\Diamond$, is not a modal collapse. Rather, it is a valid derivation, which simply serves to derive a version of epistemic weakening. By contrast, the second derivation, involving Δ, is indeed a modal collapse. It purports to show that *It is not probable that A* entails *not A* and is therefore invalid. The difference is that the negation of *probable* is epistemically weak, so its disappearance is a mistake, whereas the negation of *might* is not epistemically weak, so its disappearance is harmless. This means that it is not only the elimination of Δ that fails to preserve evidence, but also its introduction. This is because—through a minor detour exhibited in the second derivation above—the introduction rule for Δ is what allows one to eliminate $\neg\Delta$. And the elimination of an epistemically weak operator does not preserve evidence.

We therefore arrive at the following formulation of the Smileian *reductio** principles. The same restrictions apply to the subderivations in the subalternity rules (SA_3*) and (SA_4*).

$$(SR_1{}^*)\ \frac{\begin{array}{c}[+A]\\ \vdots\\ \bot\end{array}}{-A}$$ if the derivation of $\bot$ does not use $\Diamond$-elimination or any Δ rules, or premisses signed with $\ominus$, $\oplus$, $\boxminus$, or $\boxplus$

$$(SR_2{}^*)\ \frac{\begin{array}{c}[-A]\\ \vdots\\ \bot\end{array}}{+A}$$ if the derivation of $\bot$ does not use $\Diamond$-elimination or any Δ rules, or premisses signed with $\ominus$, $\oplus$, $\boxminus$, or $\boxplus$

$$(SR_3{}^*)\ \frac{\begin{array}{c}[\boxplus A]\\ \vdots *\\ \bot\end{array}}{\boxminus A}$$ if the derivation of $\bot$ does not use $\Diamond$-elimination or any Δ rules, or premisses signed with $\ominus$, $\oplus$, $\boxminus$, or $\boxplus$

$$(SR_4{}^*)\ \frac{\begin{array}{c}[\boxminus A]\\ \vdots *\\ \bot\end{array}}{\boxplus A}$$ if the derivation of $\bot$ does not use $\Diamond$-elimination or any Δ rules, or premisses signed with $\ominus$, $\oplus$, $\boxminus$, or $\boxplus$

Everything is now in place to provide a multilateral logic of epistemic modality and probabilistic operators. Thus, we let Probabilistic Epistemic Multilateral Logic (PEML for short) be the result of adding to the coordination principles and operational rules of EML the operational rules discussed in the previous section, the subalternity principles for moderate speech acts, and Moderate Contrariety. Given these coordination principles, the third and fourth Smileian *reductio** principles are derivable.

It is straightforward to obtain a sound model theory for PEML along similar lines as the model theory for EML. First, we extend the modal logic **S5** with an operator Δ governed by a clause stating that, for finite models, ΔA is true at a world just in case it is true at more than half of the worlds. This is in line with the analogy we have been drawing between *probable* and *most*, since in the literature of generalized quantifiers, *most* is typically analysed so that its truth conditions are equivalent to those of *more than half* (Barwise and Cooper 1981). Next, we take the embedding of EML into **S5** and turn it into an embedding of PEML into the extension of **S5** with Δ by translating $\boxplus A$ as ΔA and $\boxminus A$ as $\Delta\neg A$. The soundness argument from Chapter 4 can be straightforwardly extended to the current case.

PEML immediately validates a number of compelling principles about the logical behaviour of *probable* (see Yalcin 2010). First, it follows immediately from the coordination principles and the interaction of *probable* with *probably* and of *might* with *perhaps* that from *It must be that A* it follows that *It is probable that A*, from which, in turn, it follows that *It might be that A*. Moreover, the logic immediately validates that from *It is probable that A and B*, it follows that it is probable that *A* and that it is probable that *B*. Finally, we have that from *It is probable that A* it follows that *It is probable that A or B*. In the next section, we explain how the inferentialist expressivist account of *probable* retains the advantages of its traditional expressivist counterpart whilst avoiding the Frege–Geach Problem.

9.7 Inferential expressivism about probability

According to inferential expressivism about probability, the meaning of *probable* is given by the inference rules for Δ, which tell us to which attitude expression the speaker is committed in strongly, moderately or weakly asserting or rejecting ΔA, and when they are committed to these speech acts. The account immediately explains why *It is raining and it is probable that it isn't* sounds infelicitious and remains so under supposition. The following derivation shows that $+A \wedge \Delta\neg A$ is absurd.

$$\dfrac{\dfrac{\dfrac{+A \wedge \Delta\neg A}{+A}\,(+\wedge\mathrm{E.})}{\boxplus A}\,(\mathrm{SA}_3) \qquad \dfrac{\dfrac{\dfrac{\dfrac{+A \wedge \Delta\neg A}{+\Delta\neg A}\,(+\wedge\mathrm{E.})}{\boxplus\neg A}\,(+\Delta\mathrm{E.})}{\boxminus A}\,(\boxplus\neg\mathrm{E.})}{\bot}\,(\text{Moderate Contrariety})$$

Given our account of supposition, it immediately follows that $\mathbb{S}(A \wedge \Delta\neg A)$ is also absurd, regardless of whether one takes $\mathbb{S}$ to denote indicative or subjunctive supposition. The above derivation tells us that in uttering *It is raining and it is probable that it isn't* one displays incompatible attitudes:

it is absurd to express a strong partial belief and a strong partial disbelief towards the same sentence. Similarly, it is absurd to tentatively propose to add a sentence and its negation to the common ground. This remains the case under supposition: it is absurd to temporarily tentatively propose to add a sentence and its negation to the common ground.

Turning to cases of probabilistic disagreement, consider again the following dialogue.

(28) *Alice*: I can't find the keys.
Bob: It is probable that they are in the car.
Alice: No, it's not probable. They are almost certainly in the drawer.

Similarly to the case of the inferential expressivist account of epistemic modals, there is indeed a sentence that Bob is warranted in asserting and Alice is warranted in rejecting, namely the sentence *It is probable that the keys are in the car*. In particular, Bob asserts *It is probable that the keys are in the car*. By the inferential equivalence of *It is probable that* A and *Probably* A, this means that he is tentatively proposing to add *The keys are in the car* to the common ground. Alice, on the other hand, thinks that the keys are probably in the drawer. Thus, she thinks that *The keys are in the drawer* should be tentatively added to the common ground. Since it is absurd to tentatively propose to add to the common ground both *The keys are in the car* and *The keys are in the drawer*, she disagrees with Bob and rejects *It is probable that the keys are in the car*.

Price (1983) offered another argument for endorsing traditional expressivism about probability talk, namely that it solves the *Confidence Problem*.

> Why should the full belief that it is probable that *q* give rise to such behaviour (in addition, note, to the type of behaviour standardly associated with a full belief, in contexts in which the outcome of a course of action depends on whether it is probable that *q*, rather than just on whether *q*)? Without loss of generality, this may be taken to be the question as to why the full belief that it is probable that *q* should be accompanied by a strong partial belief that *q*. (Price 1983: 397)

Traditional expressivist views such as Price's or Forrest's (1981), which take *it is probable that* to function as a force modifier, straightforwardly solve the Confidence Problem: to utter *It is probable that A* just *is* to express a strong partial belief that *A*. However, these views fall prey to the Frege–Geach Problem. The inferential expressivist account of probability operators can solve the Confidence Problem: the reason why the full belief that it is probable that *A* is accompanied by a strong partial belief that *A* is that the meaning of *probable* is given in terms of the speech act of moderate assertion, which expresses strong partial belief. Thus, someone asserting that it is probable that *A* but refusing to moderately assert that *A* would display incompatible commitments, in virtue of the meaning of the expressions involved. The Commitment-to-Incoherence principle we defended in the context of moral motivation in Chapter 5 then entails that, in the relevant circumstances, someone who has a full belief that it is probable that *A* also has a strong partial belief that *A*, on pain of incoherence.

At the same time, and unlike traditional expressivist views, our account has the resources to address the Frege–Geach Problem. Recall the Frege–Geach inference involving probability operators:

(29) a. If it is probable that it is raining, we'd better take an umbrella.
b. It is probable that it is raining.
c. We'd better take an umbrella.

According to the inferential expressivist account of *probable*, *it is probable that* modifies content in both premisses of the inference. Thus, the argument is validated simply by the validity of *modus ponens* for asserted sentences. Nonetheless, the meaning of *probable* is explained in terms of the speech act of moderate assertion, which may be realized by using *probably*. As usual, this gives rise to a revenge version of the Frege–Geach Problem.

(30) a. If it is probable that it is raining, we'd better take an umbrella.
b. Probably it is raining.
c. We'd better take an umbrella.

Given that *Probably A* serves to perform a moderate assertion, the argument cannot be validated by simple *modus ponens* for asserted sentences. But the argument is validated by exploiting the inferential equivalence of *It is probable that A* and *Probably A*.

9.8 Gradability

So far, our account of *it is probable that* has been very much along the lines of our account of *might*. On the face of it, however, there is a striking difference between the two expressions, namely that *probable* is gradable. One feature of gradable adjectives is that they can be combined with a range of modifiers such as *highly* or *very*. From an inferential expressivist viewpoint, it is natural to account for this aspect of *probable* by saying that, for instance, the meaning of *It is very probable that A* is inferentially explained in terms of a speech act whose strength lies in between that of strong assertion and moderate assertion and which can be realized by *Very probably A*. This does not mean that *very probably* must be treated as lexically simple. Rather, the inferential expressivist can take *very* to be a force intensifier, so that an utterance of *Very probably it is raining* serves to perform a speech act of strength intermediate between that of moderate and strong assertion, expressing an attitude intermediate between full belief and strong partial belief.

Another feature of gradable adjectives is that they can be used in comparative language. We can say *It is more probable that it will rain than that it will snow* or *That it will hail is at least as likely as that it will sleet*. An adequate treatment of probabilistic expressions should be able to account for their use in comparatives. In this connection, Daniel Lassiter (2015, 2017: §3.4.7) has criticized Kratzer's (1991) treatment of modality on the grounds that it fails to validate the following inference pattern.

(31) a. It must be that *A*.
b. *B* is at least as probable as *A*.
———————————
c. It must be that *B*.

We agree with Lassiter that it is a serious problem for Kratzer's semantics that it does not validate this inference pattern. A satisfying account of comparative expressions involving probabilistic adjectives should validate it. However, we disagree with Lassiter on the prospects for validating the pattern using a non-scalar semantics being dim. In fact, paying close attention to this pattern can serve as the basis for an inferentialist account of the meaning of comparative constructions involving probabilistic vocabulary. From an inferential expressivist perspective, one of the consequences of asserting that it is at least as probable that it will rain as it is that it will snow is that one is committing oneself to believing that it will rain should one strongly assert that it will snow. Another consequence is that one is committing oneself to refraining from disbelieving that it will rain should one weakly assert that it will snow.

To account for these consequences, we can follow our treatment of conditionals in the previous chapter. We begin by defining a binary speech act for comparative probability, realized by *A no more probably than B* and formalized as $\precsim(A, B)$.[3] By now familiar arguments show that *no more probably*, just like *probably*, does not embed. For instance, *It is snowing no more probably than it is raining* can be felicitously uttered unembedded, but its use in conditional antecedents, *If it is snowing no more probably than it is raining ...*, is infelicitous. It is more idiomatic to say *If that it is snowing is no more probable than that it is raining, ...*

The coordination of $\precsim(A, B)$ with the other speech acts is given by the following principles. We only lay down a set of *modus ponens*-like principles for the assertoric speech acts, since the corresponding set of *modus tollens*-like principles for the rejective speech acts is derivable. We also give a principle akin to conditional proof.

$$(\precsim\mathrm{MP}^{+})\ \frac{+A \qquad \precsim(A, B)}{+B} \qquad (\precsim\mathrm{MP}^{\boxplus})\ \frac{\boxplus A \qquad \precsim(A, B)}{\boxplus B}$$

$$(\precsim\mathrm{MP}^{\oplus})\ \frac{\oplus A \qquad \precsim(A, B)}{\oplus B} \qquad (\precsim\mathrm{CP})\ \frac{\begin{matrix}[+A]\\ \vdots *\\ +B\end{matrix}}{\precsim(A, B)}$$

We can then inferentially explain the meaning of *A is at least as probable as B*, formalized as $A\vartriangle_{\geq}B$, in terms of *B no more probably than A* via the following rules.

$$(+\vartriangle_{\geq}\mathrm{I.})\ \frac{\precsim(B, A)}{+A\vartriangle_{\geq}B} \qquad (+\vartriangle_{\geq}\mathrm{E.})\ \frac{+A\vartriangle_{\geq}B}{\precsim(B, A)}$$

[3] As in the case of *very probably*, this does not mean that we need to treat *A no more probably than B* as lexically simple. Rather, we could treat *more* and *no more* as *force modifiers* which turn a unary speech act such as moderate assertion into a binary, comparative speech act.

These rules and the coordination principles jointly imply that when one asserts that it is at least as probable that it will rain as it is that it will snow, one is committing oneself to disbelieving that it will snow should one strongly reject that it will rain. Analogous results can be obtained concerning the interaction of *at least as probable as* with the other rejective speech acts.

The inferential expressivist account of *A is at least as probable as B* immediately validates the inference isolated by Lassiter which failed on Kratzer's semantics, namely the inference from *It must be that A* and *B is at least as probable as A* to *It must be that B.*

$$\dfrac{\dfrac{\dfrac{+\Box B}{+B}\,(+\Box\text{E.}) \qquad \dfrac{+A\Delta_{\geq}B}{\precsim(B,A)}\,(+\Delta_{\geq}\text{E.})}{+A}\,(\precsim\text{MP}^{+})}{+\Box A}\,(+\Box\text{I.})$$

Similarly, the inferential expressivist account of *A is at least as probable as B* validates the following intuitively valid inference pattern, which again fails on Kratzer's account (Yalcin 2010: 921):

(32) a. Probably *A*.
b. *B* is at least as probable as *A*.
c. Probably *B*.

At the same time, the inferential expressivist account does not validate the following inference pattern—call it the *Disjunctive Inference*—which, as Lassiter and Yalcin (2010) have independently noticed, is validated by Kratzer's semantics.

(33) a. *A* is at least as probable as *B*.
b. *A* is at least as probable as *C*.
c. *A* is at least as probable as $B \vee C$.

This is another problematic consequence of Kratzer's account, since the pattern has intuitively invalid instances, such as the following (see Halpern 1997, 2003).

(34) a. That the coin lands heads is at least as probable as it landing tails.
b. That the coin lands heads is at least as probable as it landing heads.
c. That the coin lands heads is at least as probable as it landing heads or tails.

At first sight, it might seem that our account faces the same problem, since the following derivation would seem to establish the general validity of the Disjunctive Inference.

$$\dfrac{\dfrac{\dfrac{[+B\vee C]^3 \qquad \dfrac{[+B]^1 \quad \dfrac{+A\triangle_{\geq}B}{\precsim(B,A)}\,(+\triangle_{\geq}\text{E.})}{+A}\,(\precsim\text{MP}^+) \qquad \dfrac{[+C]^2 \quad \dfrac{+A\triangle_{\geq}C}{\precsim(C,A)}\,(+\triangle_{\geq}\text{E.})}{+A}\,(\precsim\text{MP}^+)}{+A}\,(+\vee\text{E.})^{1,2}}{\precsim(B\vee C,A)}\,(\precsim\text{CP})^3}{+A\triangle_{\geq}B\vee C}\,(+\triangle_{\geq}\text{I.})$$

However, this derivation fails. For, as we argued above, rules that eliminate a $\triangle$ do not preserve evidence. This applies to rules that eliminate comparative occurrences of $\triangle$ just as much as non-comparative ones. This means that the $(+\triangle_{\geq}\text{E.})$-rule must be excluded from the coordination principles* and, consequently, from rules whose subderivations are restricted to evidence-preserving inferences. This includes both the disjunction elimination rule and the *no more probably than* introduction rule, both of which are used in the derivation.

Wesley Holliday and Thomas Icard (2013) have proposed a sophisticated modification of Kratzer's semantics which invalidates the Disjunctive Inference. However, as Lassiter (2015, 2017) points out, this modification also entails that, in the presence of Kratzer's semantics for epistemic modals, the following intuitively valid inference pattern fails.

(35) a. It must be that *A*.
―――――――――――――
b. *A* is more probable than not *A*.

If, as usual, we define *A is more probable than B* as *A is at least as probable as B but B is not at least as probable as A*, it is easy to see that the pattern is validated by our account of comparative probability.

Holliday and Icard, for their part, adopt an account of *might* and *must* according to which these expressions are treated as quantifiers over epistemically accessible possible worlds, as in epistemic logic (see also, e.g., von Fintel and Gillies 2010). The result is that *must* is *strong*: it satisfies the epistemic weakening inference from *It must be that A* to *A*. Given this account of epistemic modality, the inference from *It must be that A* to *A is more probable than not A* is indeed validated, but then so is the following inference.

(36) a. *A* is more probable than *B*.
b. It might be that *A*.

Although plausible at first sight, however, this inference is subject to counter-examples, as Lassiter (2015: 678) observes. In a year in which Watford has a much better team than Crystal Palace, it would be reasonable to assert *It is at least as probable that Watford will win the Premier League as it is that Crystal Palace will do*. Nonetheless, it would seem coherent, at the same time, to reject *Watford might win the Premier League* on the grounds that the chances of them winning it are minuscule.

While we have been taking *must* to be strong, our account of comparative probabilistic expressions does not have the untoward consequence pointed out by Lassiter. On our account, it follows that Watford might win the Premier League from *It is at least as probable that Watford will win the Premier League as it is that Crystal Palace will do* only in the presence of the additional premiss that Crystal Palace might win.

Bringing the discussion together, our account has the resources to account for the validity and invalidity of a number of inference patterns that are problematic for several accounts of probability operators, notably the orthodox one developed by Kratzer. At the same time, our account does not need the full machinery of probability spaces or other quantitative models to account for the validity or invalidity of these inference patterns, as has been suggested by Lassiter and Yalcin among others. We now briefly turn to another proposal that advocates the use of probability spaces to provide the semantics for probability operators and indeed to account for a vast range of phenomena. This will also allow us to say something about what our account has to say about probability talk involving numerical values.

9.9 Moss on probabilistic belief

Moss (2016) has launched a sustained defence of the thesis that we can have probabilistic knowledge and of its importance in a number of domains. Probabilistic knowledge requires probabilistic belief, where a probabilistic belief is not simply a full belief in a proposition (understood as a set of possible worlds) about the objective probability of some event occurring or the subjective probability of some event occurring given certain evidence. Rather, a probabilistic belief is the sort of belief one has when one believes that it is 0.6 likely that it will rain.

Moss assumes that we have probabilistic beliefs—this is her starting point. However, this assumption is compatible, she says, with two accounts of what it is to have probabilistic beliefs (Moss 2016: §§1.2–1.3). The first, and the one she favours, is that to have a probabilistic belief is to have a belief towards a probabilistic content, which is just a set of probability spaces. The second is that to have a probabilistic belief is to have some degree of belief towards a non-probabilistic content, such as a proposition. On the first account, one has a simple attitude towards a complex content, so we may call it the *complex content account*; on the second account, one has a complex attitude towards a simple content, so we may call it the *complex attitude account*.

Let us set aside for the moment the case of believing that it is 0.6 likely that it will rain and focus on the simple case in which one believes that it is likely that it will rain. On our account, one can express such beliefs by asserting *It is probable that it will rain*. One expresses a simple attitude, namely belief, towards a complex content. Thus, in a sense, our account can be considered a version of the complex content account. However, on our proposal, what it is to assert the complex content *It is probable that it will rain* is explained in terms of its inferential equivalence with moderately asserting—that is, expressing a strong partial belief towards—the simple content *it will rain*. Thus, although one expresses a simple belief towards a complex content, what it is to do so is explained in terms of its inferential relation to expressing another attitude towards a simple content. In the case at hand, this may be done by uttering *Probably, it will rain*. Thus, our proposal also includes elements of the complex attitude account. Indeed, on our account the reason why *It is probable that it will rain* and *It is probable that it will not rain* are inconsistent is that they inferentially commit the speaker to expressing a strong partial belief and a strong partial disbelief towards *it will rain*. Thus, the inconsistency of *It is probable that it will rain* and *It is probable that it will not rain* is explained in terms of a *B-type* inconsistency, to use Schroeder's terminology introduced in Chapter 5.

Moss (2016: 10ff.) argues that although the complex content account and the complex attitude account are empirically equivalent, there are theoretical reasons to favour the complex content account. These theoretical reasons are in effect a version of Schroeder's explanatory challenge to B-type expressivism which we discussed in Chapter 5. Moss claims that the complex attitude account must take the absurdity of, for instance, having a strong partial belief and a strong partial disbelief towards the same content as primitive. By contrast, she continues, the simple belief attitude account can take the belief that it is likely that it will rain and the belief that it is likely that it will not rain to be incompatible in virtue of the fact that the intersection of the probability spaces associated with *It is likely that it will rain* and *It is likely that it will not rain* is empty. The reply we gave to Schroeder's explanatory challenge in Chapter 5 carries over to the present context: defenders of the complex content account too must take some inconsistency as primitive. In this case, they must assume that the axioms of the probability calculus hold of probabilistic contents and insist that the incompatibilities they entail are transmitted by belief. Of course, there are good reasons to make these assumptions. For instance, it might be argued that these assumptions provide the best explanation of probabilistic beliefs. But defenders of the complex attitude account can say something similar about the structure of relations between the attitudes they take as basic.

So the explanatory challenge raised by the negation version of the Frege–Geach Problem does not affect our account of probability operators. But what does Moss have to say about the original, conditional version of the problem? Moss takes sentences such as *It is raining* to denote regular contents and sentences such as *It is probably raining* to denote probabilistic contents. Logical operators are then polymorphic: when they operate on regular contents, they have their traditional intensional semantic values; when they operate on probabilistic contents, they have probabilistic semantic values. But what happens when we are dealing with mixed discourses? This is the question made vivid by the Frege–Geach Problem. On Moss's account, the following inference would seem to mix probabilistic and regular contents.

(37) a. If it is probable that it is raining, I'll take an umbrella.
b. It is probable that it is raining.
____
c. I'll take an umbrella.

Moss's (2016: 55) strategy is to introduce a covert operator 𝔠 (for *certain*) that shifts regular contents (sets of worlds) to probabilistic ones (sets of probability spaces). Thus, the logical form of *If it is probable that it is raining, I'll take an umbrella* is *If it is probable that* 𝔠 *[it is raining],* 𝔠 *[I'll take an umbrella]*, and the conditional operates on probabilistic contents. Mixed inferences are then only so on the face of it; in reality, they only involve probabilistic contents.

The strategy is reminiscent of Blackburn's (1984) early approach to the Frege–Geach Problem, which takes the conditional to operate on attitudes rather than on (standard) contents. The problem with Blackburn's approach, pointed out by Hale (1993), was that it makes refusing to accept the conclusion of the original Frege–Geach inference, having accepted the premisses, a moral mistake, rather than a logical one. Moss's strategy makes refusing to accept the conclusion of a probabilistic Frege–Geach inference a probabilistic mistake. This might be an acceptable consequence on its own, but it becomes more problematic once one notices that Moss must also say that when we are dealing with inferences involving ordinary contents such as *If it is raining, I'll take an umbrella,* the mistake of refusing to accept the conclusion having accepted the premisses is not a probabilistic one. Thus, despite their apparent similarity, the first of the two following inferences is, in Moss's view, an application of probabilistic reasoning, the second of standard logical reasoning.

(38) a. If it is probable that it is raining, I'll take an umbrella.
b. It is probable that it is raining.
____
c. I'll take an umbrella.

(39) a. If it is raining, I'll take an umbrella.
b. It is raining.
____
c. I'll take an umbrella.

It follows that Moss's approach faces a version of the logicality problem we encountered in connection with the Negation Problem in Chapter 5. The logicality problem is that the inconsistency of *Lying is wrong* and *Lying is not wrong* ought to be recognizable simply by grasping the meaning of *not*, without understanding the non-logical vocabulary featured in these sentences. Similarly, it is plausible to require that the validity of the two inferences above ought to be recognizable simply by grasping the meaning of *if.*

Moss (2016: 239) argues that postulating the covert operator c has advantages of its own. In particular, it allows her to account for the invalidity of the following inference (in a context in which we are throwing a standard six-faced die).

(40) a. If it is low, it is probable that it is odd.
b. It is not probable that it is odd.
____
c. It is not low.

On her account, the premisses and conclusion of this inference have the following logical form.[4]

(41) a. If [c it is low], [it is probable that c it is odd].
b. Not [it is probable that c it is odd].

---

c. c not [it is low].

Since the negation operators in the second premiss and in the conclusion have two different semantic types—the first operates on probabilistic contents, the second operates on regular contents—the argument is not an instance of *modus tollens*. However, by the same token, the following seemingly valid inference is not an instance of *modus tollens* either.

(42) a. If it is probable that it is sunny, we are going to the beach.
b. We are not going to the beach.

---

c. It is not probable that it is sunny.

For, according to Moss's proposal it has the following logical form and so the negation operators in the second premiss and the conclusion have different semantic types.

(43) a. If [it is probable that c it is sunny], [c we are going to the beach].
b. c not [we are going to the beach].

---

c. Not [it is probable that c it is sunny].

Thus, in addition to facing a version of the logicality problem, Moss's account is also confronted with a version of the New Negation Problem we discussed in Chapter 5. The initial challenge raised by the Negation Problem is to explain how different kinds of contents can be negated: for the traditional expressivist, these are cognitive and non-cognitive contents; for Moss, they are regular and probabilistic contents. Moss's covert operator c makes room for additional places to insert a negation to account for this, but this results in some negations being inserted in the wrong place.

[4] In her formalization, Moss also uses indices to indicate the partition of the probability space that pieces of logical and probabilistic vocabulary operate on. We omit the indices, since they do not affect the discussion to follow.

It might be argued that the advantages of Moss's approach emerge when one considers probabilistic talk involving numerical values and corresponding probabilistic beliefs, such as the belief that it is 0.6 likely that it will rain. As mentioned earlier, however, the psychological findings do suggest that, in general, we are much happier in making qualitative probabilistic judgements than quantitative ones. What to say, though, about probability talk involving numerical values?

One option would be to say that such talk is technical talk imported into everyday language. Yalcin (2010: 923) replies that giving quantitative talk an autonomous semantics might make it difficult to account for its connection to ordinary probabilistic talk. For instance, we want our semantics to account for the validity of the inference from *It's 0.6 likely that it will rain* to *It's not likely that it will not rain*. However, to say that numerical probability talk is technical talk imported into everyday language is not necessarily to say that it should receive a completely autonomous semantics. The idea might rather be that we borrow such talk to convey mental attitudes that are not as fine-grained as the numerical talk would seem to suggest. Thus, for instance, when one says that *It is 90% likely that it will rain*, the meaning of this utterance is to be explained in terms of the expression of a very strong partial belief, something which can be already captured by our semantics, once we account for intensifiers like *very*. But the precision of the numerical talk is not necessary. Indeed, to voice one's disagreement with the utterance by saying that it is only 89% likely that it will rain would come across as facetious outside technical contexts. Technical talk might not be adopted with an autonomous, quantitative semantics, but merely as an alternative way of expressing non-quantitative attitudes. More problematic might be the worry that numerical talk such as *It is twice as likely that it will rain than that it will snow* is harder to rule out as simply technical talk. But again, it is not clear whether such talk really expresses a quantitative comparison or is instead a different way of making a qualitative comparative judgement. Outside technical contexts, it also seems facetious to voice one's disagreement with *It is twice as likely that it will rain than that it will snow* by responding that it is only 1.9 times as likely.

Having said this, there is a second option for dealing with probability talk involving numerical values. Having shown how one can explain the meaning of *it is probable that* in terms of *probably*, we have gone on to provide an inferential expressivist account of comparative uses of probabilistic talk. But once this account is in place, one can appeal to *comparativism*—or some qualified version thereof, as we will shortly see—to obtain the full spectrum of partial beliefs. Comparativism is the view that our partial beliefs are grounded in comparative beliefs, such as believing that *A* is more probable than *B*.[5] Comparativists appeal to representation theorems to show that if the relation of comparative probability *being at least as probable as* satisfies certain (rather minimal) conditions, then one can represent the full quantitative spectrum of partial beliefs corresponding to the whole of the probability space in terms of comparative beliefs. For present purposes, one would need to subscribe only to a qualified version of comparativism: while qualitatively characterized partial belief would be accounted for directly in terms of what is expressed by utterances of the form *Probably A*, the full quantitative spectrum of partial beliefs is accounted for in terms of comparative probability judgements. Thus, quantitative partial beliefs such as the belief that it is 71% likely that it will rain are grounded in comparative beliefs.

Historically, probability talk was one of the first areas of application in the project of extending expressivism beyond the realm of ethical discourse. However, as we have been stressing throughout the book, expressivism must contend with the Frege–Geach Problem wherever it is applied. As one might have expected, traditional expressivism about probability talk is undermined by the original, conditional version of the Frege–Geach Problem. But Moss's recent account of probabilistic beliefs faces versions of the problem too. Inferential expressivism about probability, by contrast, has the resources to deal with the Frege–Geach Problem in its various forms. At the same time, it provides satisfying accounts of both absolute and comparative probability talk. In line with the psychological findings, moreover, the account does not explain our everyday qualitative probability talk in terms of fine-grained distinctions between many different belief states. Nonetheless, there are attractive options for extending it to cover quantitative probability talk as well.

[5] Precursors of comparativism include Keynes 1921, De Finetti 1931, and Ramsey 1931b. For modern defences of comparativism, see Zynda 2000 and Stefánsson 2017 among others.

10
The Road Ahead

Peter Geach (1965: 455) once remarked that a sign for rejection 'cometh of evil'. Indeed 'it almost takes an apology to earn toleration [for negative judgements], let alone favor and esteem', as Immanuel Kant lamented in the *Critique of Pure Reason* (A708–709/B736–737). We hope to have won favour and esteem not just for rejection, the expression of disbelief, but for the many attitude expressions that can be fruitfully treated in the multilateral, inferential expressivist approach to semantics. Indeed, we hope to have won favour and esteem for an ongoing plan to keep extending the multilateral framework with further attitude expressions, in order to account for further semantic phenomena.

Our starting point was the pragmatist insight, shared by expressivism and inferentialism, that semantic explanations should not go beyond what is needed to explain the role of language in our practices. This pragmatist insight lies behind the Pragmatist Razor, which urges us to avoid semantic explanations that are not needed to account for an expression's or sentence's functional role. Inferential expressivism retains the pragmatist insight and complies with the Pragmatist Razor. Indeed, as we have stressed throughout, inferential expressivist semantic explanations remain neutral with respect to questions on which referentialism must take a stand. The Razor does not demand avoiding such questions altogether, but only that they not be answered *by semantics*. Extra-semantic reasons might lead one to endorse substantial metaphysical claims about minds and attitudes, moral properties, or truth. Indeed, as witnessed by the model theories we have developed for our proof-theoretic semantics, inferential expressivism can be reconciled with substantial metaphysical claims by insisting that the meanings of certain expressions determine denotations that, the realist claims, must be mentioned when describing the etiology of those meanings. But, to stress, such etiological claims are not within the remit of semantics.

Our overall pragmatist outlook does not merely provide us with constraints for selecting a semantic framework. It also equips us with a *methodology* for semantic theorizing: when searching for the meaning of an

Reasoning with Attitude: Foundations and Applications of Inferential Expressivism. Luca Incurvati and Julian J. Schlöder, Oxford University Press. © Oxford University Press 2023.
DOI: 10.1093/oso/9780197620984.003.0010

expression, consider its functional role and assign a meaning to it which is fit for this role. When this role is connected to the expression of an attitude, this gives rise to the multilateral methodology for semantic explanation which we have adopted in the book. According to this methodology, an attitude expression and a corresponding embeddable piece of language are treated by identifying a linguistic realization of the attitude expression, specifying how the attitude expression interacts with other attitude expressions, laying down rules relating the expression to the embeddable piece of language, and determining whether these new inferences preserve evidence.

Both expressivism and inferentialism are confronted with challenges not faced by the referentialist view on semantics. On the expressivist side of the framework, we focused on the Frege–Geach Problem and showed how the combination with inferentialism avoids the various versions of the problem in a uniform manner. The rich literature on expressivism includes more problems than just Frege–Geach, of course, and we addressed the most prominent ones along the way. On the inferentialist side, our focus has been on the Problem of Limited Applicability. Inferentialism has the ill repute of being applicable to nothing more than the core logical constants. The combination with expressivism, we have shown, paves the way for applications to epistemic modals, moral vocabulary, attitude talk, the truth predicate, conditionals, and probability talk. We think this is far from limited and should give inferentialists the impetus to press on.

Besides enabling one to address their key challenges, combining expressivism and inferentialism presents some distinct advantages over referentialism. We have already mentioned that inferential expressivism provides semantic explanations that remain adequately neutral on matters of metaphysics. But, additionally, inferential expressivist theorizing is subject to certain constraints, such as the harmony constraint and the need to assign semantic values by means of inference rules, that cut down the space of permissible meanings. In many cases throughout this book, the semantics we have given for an expression seems to be the only one that fits with our methodology. This, we submit, helps to limit the phenomenon of underdetermination of theory by data, in sharp contrast with the embarrassment of riches that referentialism typically faces. This is not to say that inferential expressivism makes semantic theorizing *easy*. It is sometimes far from obvious how to carry out certain other parts of the multilateral methodology, such as the coordination of speech acts and the identification of evidence-preserving rules. These modelling tasks might indeed appear to make semantic theorizing in the inferential expressivist style more arduous than referentialist theorizing. We submit that this is due to the relative unfamiliarity of our methods compared to the well-entrenched referentialist tools. Finally, some of the specific applications we have explored seem to have their own advantages over received referentialist treatments of the same expressions, for instance in how they deal with purported paradoxes and collapse results. Thus, we contend, inferential expressivism is not only supported by general meta-theoretical considerations but also by domain-specific semantic considerations.

The various philosophical, logical, and linguistic themes running through this book converge therefore into a manifesto for a novel approach to semantics. The current dominance of the referentialist paradigm may just be the outcome of contingent historical circumstances, rather than being due to it being intrinsically more applicable. Indeed, we have barely scratched the surface of the potential applications of inferential expressivism, and of the range of theoretical questions that are opened up by embracing this approach to semantics. In closing, we highlight a fraction of the issues that stand before us.

One question we have not addressed despite its salience concerns the probability of conditionals. *Stalnaker's Thesis* is the prima facie appealing thesis that the probability of a conditional is equal to the probability of its consequent conditional on the antecedent, unless the antecedent has probability 0 (Stalnaker 1970). However, Lewis (1976) proved a result purporting to show that Stalnaker's Thesis entails that the probability of the conditional is just the probability of its consequent. Since we have provided a semantics for conditionals in Chapter 8 and a semantics for probability operators in Chapter 9, one may reasonably ask where their combination leaves Lewis's triviality result.

We cannot provide an answer here, since our semantics for probability talk does not yet provide a definitive account of conditional probabilities as applied to indicative conditionals. Nonetheless, we have reason to believe that Lewis's triviality result may be avoided analogously to how we avoided Gibbard's collapse problem. For Lewis's triviality result relies on Factorization, a probabilistic analogue of the Import-Export Principle used in the Gibbard Collapse Argument, and we strongly suspect that, analogously to the case of Import-Export, the inferential expressivist account of conditional probability will validate the instances of Factorization that motivate it, but not those required for the triviality result. Rejecting Factorization is not a new response to Lewis's triviality result (see, e.g., McGee 1989), but a proof-theoretic approach might once again be advantageous, by allowing us to precisely circumscribe the class of valid instances of Factorization and hence of Stalnaker's Thesis. So, if an inferential expressivist account along these lines can be made out, it will not only explain why Factorization and Stalnaker's Thesis are not generally valid, but also why they are intuitively appealing and can be properly used in many cases.

Another salient open question is how to treat phenomenological self-reports such as *I am in pain*. From an inferential expressivist standpoint, it is natural to inferentially explain the meaning of *I am in pain* in terms of the expression of pain, *Ouch!*. As usual, this semantic explanation stands in contrast with the traditional expressivist view that takes *I am in pain* to replace *Ouch!* (Wittgenstein 1953). As Bar-On (2004: 233ff.) notes, traditional expressivism about phenomenological self-reports is undermined by the Frege–Geach Problem. Once again, inferential expressivism promises to deal with the problem without having to confront the challenges faced by the hybrid expressivist view that *I am in pain* expresses both the state of being in pain and the attitude of belief towards being in pain (Finkelstein 2003; Bar-On 2004), or concluding that *I am in pain* is ambiguous (Freitag 2018).

Phenomenological self-reports also raise more specific challenges. An example by Harman (1986) provides a case in point. As Harman notes, valid arguments containing *I am in pain* as their conclusion are unsuitable to get someone who believes their premisses to believe that they are in pain. Harman points out that the validity of such arguments has nevertheless a social function: if we can convince someone of the validity of an argument that would lead them to conclude that they are in pain when they are not, we can force them to reject a premiss of the argument. This observation—that some inferences cannot do the epistemic work of establishing belief—is familiar from the literature on failures of warrant transmission (see, e.g., Wright 2002) and seems related to our distinction between evidence-preserving and commitment-preserving inferences. While evidence-preserving inferences are suitable to gain new beliefs, merely commitment-preserving ones need not be. A plausible hypothesis is thus that first-person phenomenological assertions are associated with particular evidential standards that make them unsuitable to appear in conclusions of evidence-preserving arguments. For instance, it may be that evidence for first-person phenomenological assertions must be somehow direct and cannot be obtained inferentially. The matter requires further investigation, but its connection with the notion of evidence preservation suggests that inferential expressivism can make its own distinct contribution here as well.

Another class of expressions that appear to interact with evidence in an interesting way are predicates of personal taste such as *tasty* and *funny*. The evidential profile of utterances such as *This is tasty* seems to share important features with the evidential profile of first-person phenomenological assertions, in that when someone claims something to be tasty, there seems to be a tendency to assume they have direct experience of the item (Ninan 2014). At the same time, the behaviour of assertions concerning matters of taste seems to be interestingly distinct from that of phenomenological self-reports, in that they can give rise to disagreement phenomena. In this respect, utterances of *This is tasty* seem analogous to moral assertions, although the status of disagreements concerning matters of taste vis-à-vis moral disagreements is controversial (Eriksson 2016; Pietroiusti 2021). We leave it open here how predicates of personal taste are to be treated. But it seems reasonable to think that an explanation in terms of attitude expressions and evidence preservation might again prove fruitful.

Many more linguistic phenomena remain in need of inferential expressivist treatment. We have not discussed important objects of semantic study such as tense and aspect, pronouns and anaphora, and generalized quantifiers. Inferential expressivism must furthermore be extended to the coherence structure of multiple speech acts in sequence (Asher and Lascarides 2003) and to further non-assertoric speech acts such as questions and directives. By incorporating questions and directives in the multilateral framework, one could then extend the account of conditional speech acts from Chapter 8 to speech acts performed by means of sentences such as *If it is raining, will you take an umbrella?* or *If it is raining, take an umbrella!*. In addition, one could apply the familiar inferential expressivist mode of semantic explanation by, for instance, inferentially explaining the meaning of deontic modals such as *must* in terms of directives.

There also remain a host of open questions concerning the meta-theory of inferential expressivism and its relation to more abstract questions in the study of meaning. Chief among them is the question of compositionality. The referentialist assigns meanings to subsentential expressions and then proceeds to explain how they compose to give rise to sentential meanings. Thus, the referentialist has a general story about how the meaning of expressions contribute to the meaning of the more complex expressions in which they occur. The inferential expressivist may appear not to have given such a story. However, this appearance is misleading. The inferential expressivist has not proceeded by assigning meanings to subsentential expressions in isolation, but it is mistaken to think that this is required for one's semantics to be compositional. Indeed, Frege (1884) already stressed that it is only in the context of a sentence that words have meaning, thus warning against the possibility of assigning meaning to subsentential expressions in isolation (see also Dummett 1956). In keeping with Frege's admonitions, the inferential expressivist lays down, for each linguistic item, the rules specifying the inferences that complete attitude expressions, in virtue of containing the item, can feature in. Compositionality is then achieved because the totality of rules governing the behaviour of the linguistic items in a sentence wholly determines the meaning of the sentence itself. We take it that the treatment of compositionality—indeed, the very understanding of compositionality—is another significant difference between referentialist and inferential expressivist semantics.

Finally, there are many open technical questions concerning the multilateral framework within which inferential expressivist semantics is carried out. One prominent such question concerns the expressive power of force markers. That bilateral systems enjoy more expressive power than their unilateral cousins has been known since Smiley (1996) showed that they rule out the non-standard interpretations of the propositional calculus given by Carnap (1943). But recent work of ours (Incurvati and Schlöder 2022a,b, Forthcoming) seems to show that the increase in expressive power afforded by the addition of force markers goes well beyond this. The presence of force markers allows us to provide natural deduction calculi which are sound and complete with respect to model-theoretic frameworks whose axiomatization had before seemed to require sophisticated technical machinery or even to be beyond reach. Further investigation is needed to determine the exact additional expressive power bestowed by force markers and where it hails from. The best we can say at this time is that this expressive power seems to have to do with the fact that force markers represent model-theoretically global information, whereas embeddable operators typically represent local information.

We have given a glimpse of the extent and variety of questions that await further investigation. Our hope, in so doing, is that this will inspire others to address these questions, perhaps by joining forces with us. But the range of open questions should not lead one to underestimate what has already been achieved by inferential expressivism. Indeed, we hope that this manifesto has made clear the many and profound benefits of reasoning with attitude.

References

Adams, E. W. (1970), Subjunctive and indicative conditionals, *Foundations of Language* **6**, 89–94.

Aloni, M., Incurvati, L., and Schlöder, J. J. (Forthcoming), Epistemic modals in hypothetical reasoning, *Erkenntnis*.

Asher, N. and Lascarides, A. (2003), *Logics of Conversation*, Cambridge University Press, Cambridge.

Asher, N. and Lascarides, A. (2008), Commitments, beliefs and intentions in dialogue, *Proceedings of the 12th Workshop on the Semantics and Pragmatics of Dialogue*.

Ayer, A. J. (1936), *Language, Truth and Logic*, 2nd edn, Gollancz, London.

Bach, K. and Harnish, R. M. (1979), *Linguistic Communication and Speech Acts*, MIT Press, Cambridge, MA.

Bacon, A. (2015), Can the classical logician avoid the revenge paradoxes?, *Philosophical Review* **124**, 299–352.

Baker, D. (2021), Quasirealism as semantic dispensability, *Philosophical Studies* **178**, 2313–2333.

Baker, D. (2022), Deflating the many attitudes problem, *Philosophical Quarterly* **72**, 1–18.

Baker, D. and Woods, J. (2015), How expressivists can and should explain inconsistency, *Ethics* **125**, 391–424.

Bar-On, D. (2004), *Speaking My Mind: Expression and Self-Knowledge*, Oxford University Press, Oxford.

Bar-On, D. (2015), Transparency, expression, and self-knowledge, *Philosophical Explorations* **18**, 134–152.

Barker, S. J. (2000), Is value content a component of conventional implicature?, *Analysis* **60**, 268–279.

Bartsch, R. (1973), 'Negative transportation' gibt es nicht, *Linguistische Berichte* **27**, 1–7.

Barwise, J. and Cooper, R. (1981), Generalized quantifiers and natural language, *Linguistics and Philosophy* **4**, 159–219.

Beall, Jc. (2009), *Spandrels of Truth*, Oxford University Press, Oxford.

Beddor, B. (2020), A solution to the many attitudes problem, *Philosophical Studies* **177**, 2789–2813.

Beisecker, D. (2012), Normative functionalism and its pragmatist roots, *Social Epistemology Review and Reply Collective* **2**, 109–116.

Beisecker, D. (2019), Denial has its consequences: Peirce's bilateral semantics, *Transactions of the Charles S. Peirce Society* **55**, 361–386.

Bellert, I. (1977), On semantic and distributional properties of sentential adverbs, *Linguistic Inquiry* **8**, 337–351.

Belnap, N. D. (1962), Tonk, plonk and plink, *Analysis* **22**, 130–134.

Blackburn, S. (1984), *Spreading the Word: Groundings in the Philosophy of Language*, Clarendon Press, Oxford.

Blackburn, S. (1988), Attitudes and contents, *Ethics* **98**, 501–517.

Blackburn, S. (1993), *Essays in Quasi-Realism*, Oxford University Press, Oxford.
Blackburn, S. (1998), *Ruling Passions: A Theory of Practical Reasoning*, Oxford University Press, Oxford.
Bledin, J. (2015), Modus ponens defended, *Journal of Philosophy* **112**, 57–83.
Bledin, J. and Lando, T. (2018), Closure and epistemic modals, *Philosophy and Phenomenological Research* **97**, 3–22.
Bledin, J. and Rawlins, K. (2020), Resistance and resolution: Attentional dynamics in discourse, *Journal of Semantics* **37**, 43–82.
Boghossian, P. (1996), Analyticity reconsidered, *Noûs* **30**, 360–391.
Boghossian, P. (2011), Williamson on the *a priori* and the analytic, *Philosophy and Phenomenological Research* **82**, 488–497.
Boisvert, D. R. (2008), Expressive-assertivism, *Pacific Philosophical Quarterly* **89**, 169–203.
Bonnay, D. and Westerståhl, D. (2016), Compositionality solves Carnap's problem, *Erkenntnis* **81**, 721–739.
Boolos, G. (1990), The standard of equality of numbers, *in* G. Boolos (ed.), *Meaning and Method: Essays in Honor of Hilary Putnam*, Cambridge University Press, Cambridge, pp. 261–277.
Boyd, R. N. (1983), On the current status of the issue of scientific realism, *Erkenntnis* **19**, 45–90.
Brandom, R. (1983), Asserting, *Noûs* **17**, 637–650.
Brandom, R. (1994), *Making it Explicit*, Harvard University Press, Cambridge, MA.
Brandom, R. (2000), *Articulating Reasons*, Harvard University Press, Cambridge, MA.
Brandom, R. (2007), Inferentialism and some of its challenges, *Philosophy and Phenomenological Research* **74**, 651–676.
Brink, D. O. (1989), *Moral Realism and the Foundations of Ethics*, Cambridge University Press, Cambridge.
Cantini, A. (1990), A theory of formal truth arithmetically equivalent to ID_1, *Journal of Symbolic Logic* **56**, 244–259.
Carnap, R. (1943), *Formalization of Logic*, Harvard University Press, Cambridge, MA.
Carter, S. (2021), A suppositional theory of conditionals, *Mind* **130**, 1059–1086.
Chrisman, M. (2008), Expressivism, inferentialism, and saving the debate, *Philosophy and Phenomenological Research* **77**, 334–358.
Churchland, P. M. (1981), Eliminative materialism and propositional attitudes, *Journal of Philosophy* **78**, 67–90.
Churchland, P. M. (1989a), Folk psychology and the explanation of human behavior, *Philosophical Perspectives* **3**, 225–241.
Churchland, P. S. (1989b), *Neurophilosophy: Toward a Unified Science of the Mind-Brain*, MIT press, Cambridge, MA.
Ciardelli, I. (2021), Restriction without quantification: Embedding and probability for indicative conditionals, *Ergo* **8**, 598–636.
Collins, C. and Postal, P. M. (2014), *Classical NEG Raising: An Essay on the Syntax of Negation*, MIT Press, Cambridge, MA.
Collins, C. and Postal, P. M. (2018), Disentangling two distinct notions of NEG raising, *Semantics and Pragmatics* **11**, 1–22.
Cook, R. T. (2005), What's wrong with tonk, *Journal of Philosophical Logic* **34**, 217–226.
Copp, D. (2001), Realist-expressivism: A neglected option for moral realism, *Social Philosophy and Policy* **18**, 1–43.

Crabill, J. D. (2013), Suppose Yalcin is wrong about epistemic modals, *Philosophical Studies* **162**, 625–635.

Crespo, I., Karawani, H., and Veltman, F. (2018), Expressing expectations, *in* D. Ball and B. Rabern (eds), *The Science of Meaning: Essays on the Metatheory of Natural Language Semantics*, Oxford University Press, Oxford, pp. 253–275.

Dancygier, B. (1999), *Conditionals and Prediction: Time, Knowledge and Causation in Conditional Constructions*, Cambridge University Press, Cambridge.

Dancygier, B. and Sweetser, E. (2005), *Mental Spaces in Grammar: Conditional Constructions*, Cambridge University Press, Cambridge.

Darwall, S. L. (1983), *Impartial Reason*, Cornell University Press, Ithaca, NY.

Davidson, D. (1974), Psychology as philosophy, *in* S. C. Brown (ed.), *Philosophy of Psychology*, Palgrave Macmillan, London, pp. 41–52.

Davis, W. A. (2003), *Meaning, Expression and Thought*, Cambridge University Press, Cambridge.

De Finetti, B. (1931), Sul significato soggettivo della probabilità, *Fundamenta Mathematicae* **17**, 298–329.

Dennett, D. C. (1987), *The Intentional Stance*, MIT press, Cambridge, MA.

Dennett, D. C. (1991), Real patterns, *Journal of Philosophy* **88**, 27–51.

DeRose, K. (1991), Epistemic possibilities, *Philosophical Review* **100**, 581–605.

Devitt, M. (1997), *Realism and Truth*, Princeton University Press, Princeton, NJ.

Dicher, B. (2016), Weak disharmony: Some lessons for proof-theoretic semantics, *Review of Symbolic Logic* **9**, 1–20.

Dickie, I. (2010), Negation, anti-realism, and the denial defence, *Philosophical Studies* **150**, 161–185.

Divers, J. and Miller, A. (1995), Platitudes and attitudes: A minimalist conception of belief, *Analysis* **55**, 37–44.

Dorr, C. (2002), Non-cognitivism and wishful thinking, *Noûs* **36**, 97–103.

Dreier, J. (1996), Expressivist embeddings and minimalist truth, *Philosophical Studies* **83**, 29–51.

Dreier, J. (2004), Meta-ethics and the problem of creeping minimalism, *Philosophical Perspectives* **18**, 23–44.

Dreier, J. (2006), Negation for expressivists: A collection of problems with a suggestion for their solution, *in* R. Shafer-Landau (ed.), *Oxford Studies in Metaethics*, Vol. 1, Oxford University Press, Oxford, pp. 217–233.

Dreier, J. (2018), The real and the quasi-real: Problems of distinction, *Canadian Journal of Philosophy* **48**, 532–547.

Dummett, M. (1956), Nominalism, *Philosophical Review* **65**, 491–505.

Dummett, M. (1973a), *Frege: Philosophy of Language*, Duckworth, London.

Dummett, M. (1973b), The justification of deduction, Britsh Academy Lecture. Reprinted in his *Truth and Other Enigmas*, pp. 290–318. Page references are to *Truth and Other Enigmas*.

Dummett, M. (1975a), The philosophical basis of intuitionistic logic, *in* H. E. Rose and J. C. Shepherdson (eds), *Studies in Logic and the Foundations of Mathematics*, Vol. 80, Elsevier, Amsterdam, pp. 5–40.

Dummett, M. (1975b), What is a theory of meaning?, *in* S. Guttenplan (ed.), *Mind and Language*, Clarendon Press, Oxford, pp. 97–138.

Dummett, M. (1976), What is a theory of meaning?(II), *in* G. Evans and J. McDowell (eds), *Truth and Meaning: Essays in Semantics*, Clarendon Press, Oxford, pp. 67–137.

Dummett, M. (1987), Reply to John McDowell, *in* B. Taylor (ed.), *Michael Dummett: Contributions to Philosophy*, Martinus Nijhoff Publishers, Dordrecht, pp. 253–268.

Dummett, M. (1991a), *Frege: Philosophy of Mathematics*, Harvard University Press, Cambridge, MA.

Dummett, M. (1991b), *The Logical Basis of Metaphysics*, Harvard University Press, Cambridge, MA.

Dummett, M. (2006), *Thought and Reality*, Oxford University Press, Oxford.

Dummett, M. (2007), Reply to John McDowell, *in* R. E. Auxier and L. E. Hahn (eds), *The Philosophy of Michael Dummett*, Vol. 31 of *The Library of Living Philosophers*, Open Court, Chicago, pp. 367–382.

Dutilh Novaes, C. (2015), A dialogical, multi-agent account of the normativity of logic, *Dialectica* **69**, 587–609.

Edgington, D. (1995), On conditionals, *Mind* **104**, 235–329.

Eriksson, J. (2016), Expressivism, attitudinal complexity and two senses of disagreement in attitude, *Erkenntnis* **81**, 775–794.

Ernst, T. (2001), *The Syntax of Adjuncts*, Cambridge University Press, Cambridge.

Ernst, T. (2009), Speaker-oriented adverbs, *Natural Language & Linguistic Theory* **27**, 497–544.

Eva, B., Shear, T., and Fitelson, B. (2022), Four approaches to supposition, *Ergo* **8**.

Ferrari, F. and Incurvati, L. (2022), The varieties of agnosticism, *Philosophical Quarterly* **72**, 365–380.

Ferrero, L. (2009), Conditional intentions, *Noûs* **43**, 700–741.

Field, H. (1994), Deflationist views of meaning and content, *Mind* **103**, 249–285.

Field, H. (1999), Deflating the conservativeness argument, *Journal of Philosophy* **96**, 533–540.

Field, H. H. (2008), *Saving Truth from Paradox*, Oxford University Press, Oxford.

Finkelstein, D. H. (2003), *Expression and the Inner*, Harvard University Press, Cambridge, MA.

Fitch, F. B. (1952), *Symbolic Logic: An Introduction*, Ronald Press Co., New York.

Fodor, J. A. (1975), *The Language of Thought*, Harvard University Press, Cambridge, MA.

Fodor, J. A. (1987), *Psychosemantics*, MIT Press, Cambridge, MA.

Forrest, P. (1981), Probabilistic modal inferences, *Australasian Journal of Philosophy* **59**, 38–53.

Francez, N. and Dyckhoff, R. (2012), A note on harmony, *Journal of Philosophical Logic* **41**, 613–628.

Frankena, W. K. (1958), Obligation and motivation in recent moral philosophy, *in* A. I. Melden (ed.), *Essays in Moral Philosophy*, University of Washington Press, Seattle, WA, pp. 40–81.

Franzén, N. and Soria-Ruiz, A. (Forthcoming), Moorean and moral incoherencies, *Ergo*.

Frege, G. (1884), *Die Grundlagen der Arithmetik: Eine logisch-mathematische Untersuchung über den Begriff der Zahl*, Verlag von Wilhelm Koebner, Breslau.

Frege, G. (1919), Die Verneinung: Eine logische Untersuchung, *Beiträge zur Philosophie des deutschen Idealismus* **1**, 143–157.

Frege, G. (1923), Gedankengefüge, *Beiträge zur Philosophie des deutschen Idealismus* **3**, 36–51.

Freitag, W. (2018), Wittgenstein on 'I believe', *Grazer Philosophische Studien* **95**, 54–69.

Gajewski, J. R. (2007), Neg-raising and polarity, *Linguistics and Philosophy* **30**, 289–328.

García-Carpintero, M. (2012), Foundational semantics I: Descriptive accounts, *Philosophy Compass* **7**, 397–409.

Geach, P. T. (1965), Assertion, *Philosophical Review* **74**, 449–465.

Gentzen, G. (1935), Untersuchungen über das logische Schließen I, *Mathematische Zeitschrift* **39**, 176–210.

Giannakidou, A. (2011), Positive polarity items and negative polarity items: Variation, licensing, and compositionality, *in* C. Maienborn, K. von Heusinger, and P. Portner (eds), *Semantics: An International Handbook of Natural Language Meaning*, De Gruyter Mouton, Berlin, pp. 1660–1712.

Gibbard, A. (1980), Two recent theories of conditionals, *in* W. L. Harper, R. Stalnaker, and G. Pearce (eds), *Ifs: Conditionals, Beliefs, Decision, Chance, and Time*, Reidel, Dordrecht, pp. 211–247.

Gibbard, A. (1990), *Wise Choices, Apt Feelings: A Theory of Normative Judgment*, Harvard University Press, Cambridge, MA.

Gibbard, A. (1993), Reply to Blackburn, *Philosophical Issues* **4**, 67–73.

Gibbard, A. (2003), *Thinking How to Live*, Harvard University Press, Cambridge, MA.

Gibbard, A. (2012), *Meaning and Normativity*, Oxford University Press, Oxford.

Gillies, A. S. (2010), Iffiness, *Semantics and Pragmatics* **3**, 4–1.

Goldstein, S. (2019), A theory of conditional assertion, *Journal of Philosophy* **116**, 293–318.

Graham, P. J. (2020), Assertions, handicaps, and social norms, *Episteme* **17**, 349–363.

Green, M. S. (2000), The status of supposition, *Noûs* **34**, 376–399.

Green, M. S. (2007), *Self-Expression*, Oxford University Press, Oxford.

Green, M. S. (2009), Speech acts, the handicap principle and the expression of psychological states, *Mind and Language* **24**, 139–163.

Grice, H. P. (1957), Meaning, *Philosophical Review* **66**, 377–388.

Grice, H. P. (1969), Utterer's meaning and intentions, *Philosophical Review* **78**, 147–177.

Grice, H. P. (1991), *Studies in the Way of Words*, Harvard University Press, Cambridge, MA.

Hale, B. (1993), Can there be a logic of attitudes?, *in* J. Haldane and C. Wright (eds), *Reality, Representation, and Projection*, Oxford University Press, Oxford, pp. 337–363.

Hale, B. (2002), Can arboreal knotwork help Blackburn out of Frege's abyss?, *Philosophy and Phenomenological Research* **65**, 144–149.

Halpern, J. Y. (1997), Defining relative likelihood in partially-ordered preferential structures, *Journal of Aritificial Intelligence Research* **7**, 1–24.

Halpern, J. Y. (2003), *Reasoning about Uncertainty*, MIT Press, Cambridge, MA.

Hare, R. M. (1952), *The Language of Morals*, Clarendon Press, Oxford.

Harman, G. (1986), *Change in View*, MIT Press, Cambridge, MA.

Harman, G. H. (1974), Meaning, *Journal of Philosophy* **71**, 224–229.

Hay, R. J. (2013), Hybrid expressivism and the analogy between pejoratives and moral language, *European Journal of Philosophy* **21**, 450–474.

Holliday, W. H. and Icard, T. F. (2013), Measure semantics and qualitative semantics for epistemic modals, *in* T. Snider (ed.), *Semantics and Linguistic Theory (SALT)*, Vol. 23, LSA and CLC Publications, Ithaca, NY, pp. 514–534.

Holton, R. (2008), Partial belief, partial intention, *Mind* **117**, 27–58.

Horgan, T. and Timmons, M. (2006), Cognitivist expressivism, *in* T. Horgan and M. Timmons (eds), *Metaethics after Moore*, Oxford University Press, Oxford, pp. 255–298.

Horn, L. (1978), Remarks on neg-raising, *in* P. Cole (ed.), *Syntax and Semantics*, Vol. 9, Academic Press, New York, pp. 129–220.
Horn, L. (1989), *A Natural History of Negation*, University of Chicago Press, Chicago.
Horn, L. (2020), Neg-raising, *in* V. Déprez and M. T. Espinal (eds), *Oxford Handbook of Negation*, Oxford University Press, Oxford, pp. 199–215.
Horsten, L. (2009), Levity, *Mind* **118**, 555–581.
Horwich, P. (1993), Gibbard's theory of norms, *Philosophy and Public Affairs* **22**, 67–78.
Horwich, P. (1998), *Truth*, Oxford University Press, Oxford.
Incurvati, L. and Schlöder, J. J. (2017), Weak rejection, *Australasian Journal of Philosophy* **95**, 741–760.
Incurvati, L. and Schlöder, J. J. (2019), Weak assertion, *Philosophical Quarterly* **69**, 741–770.
Incurvati, L. and Schlöder, J. J. (2021), Inferential expressivism and the negation problem, *in* R. Shafer-Landau (ed.), *Oxford Studies in Metaethics*, Vol. 16, Oxford University Press, Oxford, pp. 80–107.
Incurvati, L. and Schlöder, J. J. (2022a), Epistemic multilateral logic, *Review of Symbolic Logic* **15**, 505–536.
Incurvati, L. and Schlöder, J. J. (2022b), Meta-inferences and supervaluationism, *Journal of Philosophical Logic* **51**, 1549–1582.
Incurvati, L. and Schlöder, J. J. (Forthcoming), Inferential deflationism, *Philosophical Review*.
Incurvati, L. and Smith, P. (2009), Rejection and valuations, *Analysis* **70**, 3–10.
Jackson, F. (1987), *Conditionals*, Basil Blackwell, Oxford.
Jaśkowski, S. (1934), On the rules of suppositions in formal logic, *Studia Logica* **1**, 5–32.
Jeshion, R. (2013), Expressivism and the offensiveness of slurs, *Philosophical Perspectives* **27**, 231–259.
Kant, I. (1781/1787), *Critique of Pure Reason*, translated by P. Guyer and A. Wood, Cambridge University Press, Cambridge, 1998. References are to the standard pagination of the 1st and 2nd editions.
Kaplan, D. (1989a), Afterthoughts, *in* J. Almog, J. Perry, and H. Wettstein (eds), *Themes from Kaplan*, Oxford University Press, New York, pp. 565–614.
Kaplan, D. (1989b), Demonstratives: An essay on the semantics, logic, metaphysics and epistemology of demonstratives and other indexicals, *in* J. Almog, J. Perry, and H. Wettstein (eds), *Themes from Kaplan*, Oxford University Press, New York, pp. 481–563.
Keefe, R. (2000), *Theories of Vagueness*, Cambridge University Press, Cambridge.
Keynes, J. M. (1921), *A Treatise on Probability*, Dover Publications, London.
Khoo, J. (2015), Modal disagreements, *Inquiry* **58**, 511–534.
Khoo, J. and Mandelkern, M. (2018), Triviality results and the relationship between logical and natural languages, *Mind* **128**, 485–526.
Köhler, S. (2013), Do expressivists have an attitude problem?, *Ethics* **123**, 479–507.
Köhler, S. (2018), Expressivism, meaning, and all that, *Canadian Journal of Philosophy* **48**, 337–356.
Kölbel, M. (2000), Edgington on compounds of conditionals, *Mind* **109**, 97–108.
Kratzer, A. (1977), What 'must' and 'can' must and can mean, *Linguistics and Philosophy* **1**, 337–355.
Kratzer, A. (1981), The notional category of modality, *in* H.-J. Eikmeyer and H. Reiser (eds), *Words, Worlds, and Contexts*, de Gruyter, Berlin, pp. 38–74.

Kratzer, A. (1991), Modality, *in* A. von Stechow and D. Wunderlich (eds), *Semantics: An International Handbook of Contemporary Research*, De Gruyter, Berlin, pp. 639–650.
Kripke, S. (1976), Outline of a theory of truth, *Journal of Philosophy* **72**, 690–716.
Krzyżanowska, K., Wenmackers, S., and Douven, I. (2013), Inferential conditionals and evidentiality, *Journal of Logic, Language and Information* **22**, 315–334.
Künne, W. (2003), *Conceptions of Truth*, Oxford University Press, Oxford.
Lassiter, D. (2015), Epistemic comparison, models of uncertainty, and the disjunction puzzle, *Journal of Semantics* **32**, 649–684.
Lassiter, D. (2017), *Graded Modality: Qualitative and Quantitative Perspectives*, Oxford University Press, Oxford.
Leitgeb, H. (2008), Towards a logic of type-free modality and truth, *in* C. Dimitracopoulos (ed.), *Logic Colloquium 2005: Proceedings of the Annual European Summer Meeting of the Association for Symbolic Logic*, Cambridge University Press, Cambridge, pp. 28–68.
Lennertz, B. (2021), Quantificational attitudes, *Journal of Philosophy* **118**, 585–613.
Lewis, D. K. (1970), General semantics, *Synthese* **22**, 18–67.
Lewis, D. K. (1973), *Counterfactuals*, Blackwell, Oxford.
Lewis, D. K. (1976), Probabilities of conditionals and conditional probabilities, *in* W. L. Harper, R. Stalnaker, and G. Pearce (eds), *Ifs*, Springer, Dordrecht, pp. 129–147.
Lewis, D. K. (1980), Index, context, and content, *in* S. Kanger and S. Öhman (eds), *Philosophy and Grammar*, Reidel, Dordrecht, pp. 79–100.
Lycan, W. G. (2006), Conditional-assertion theories of conditionals, *in* J. Thomson and A. Byrne (eds), *Content and Modality: Themes from the Philosophy of Robert Stalnaker*, Oxford University Press, Oxford, pp. 148–164.
Lycan, W. G. (2022), A Rylean mental fictionalism, *in* T. Demeter, T. Parent, and A. Toon (eds), *Mental Fictionalism: Philosophical Explorations*, Routledge, London and New York, pp. 70–85.
MacFarlane, J. (2003), Future contingents and relative truth, *Philosophical Quarterly* **53**, 321–336.
MacFarlane, J. (2004), In what sense (if any) is logic normative for thought?, Unpublished manuscript.
MacFarlane, J. (2014), *Assessment Sensitivity: Relative Truth and its Applications*, Oxford University Press, New York.
Mackie, J. L. (1973), *Truth, Probability and Paradox: Studies in Philosophical Logic*, Oxford University Press, Oxford.
Mackie, J. L. (1977), *Ethics: Inventing Right and Wrong*, Penguin Books, London.
Maher, C. (2012), Normative functionalism, *Social Epistemology Review and Reply Collective* **2**, 13–15.
Mandelkern, M. (2019), Bounded modality, *Philosophical Review* **128**, 1–61.
Mandelkern, M. (2020), A counterexample to modus ponenses, *Journal of Philosophy* **117**, 315–331.
Mandelkern, M. (2021), If p, then p!, *Journal of Philosophy* **118**, 645–679.
Mandelkern, M. and Romoli, J. (2017), Parsing and presuppositions in the calculation of local contexts, *Semantics and Pragmatics* **10**, 1–40.
McCready, E. (2010), Varieties of conventional implicature, *Semantics and Pragmatics* **3**, 1–57.
McDowell, J. (1987), In defence of modesty, *in* B. Taylor (ed.), *Michael Dummett: Contributions to Philosophy*, Martinus Nijhoff Publishers, Dordrecht, pp. 59–80.
McDowell, J. (1996), *Mind and World*, Harvard University Press, Cambridge, MA.

McDowell, J. (1997), Another plea for modesty, *in* R. G. Heck (ed.), *Language, Thought, and Logic: Essays in Honour of Michael Dummett*, Oxford University Press, Oxford.

McDowell, J. (2007), Dummett on truth conditions and meaning, *in* R. E. Auxier and L. E. Hahn (eds), *The Philosophy of Michael Dummett*, Vol. 31 of *The Library of Living Philosophers*, Open Court, Chicago, pp. 351–366.

McGee, V. (1985), A counterexample to modus ponens, *Journal of Philosophy* **82**, 462–471.

McGee, V. (1989), Conditional probabilities and compounds of conditionals, *Philosophical Review* **98**, 485–541.

McGlynn, A. (2010), *Assertion and Belief without Knowledge*, PhD thesis, University of Texas at Austin.

Mele, A. R. (1996), Internalist moral cognitivism and listlessness, *Ethics* **106**, 727–753.

Misak, C. (2016), *Cambridge Pragmatism: From Peirce and James to Ramsey and Wittgenstein*, Oxford University Press, Oxford.

Mittwoch, A. (1977), How to refer to one's own words: Speech-act modifying adverbials and the performative analysis, *Journal of Linguistics* **13**, 177–189.

Montague, R. (1963), Syntactical treatments of modality, with corollaries on reflexion principles and finite axiomatizability, *Acta Philosophica Fennica* **16**, 153–157.

Montague, R. (1970), English as a formal language, *in* B. Visentini (ed.), *Linguaggi nella società e nella tecnica*, Edizioni di Comunità, Rome, pp. 188–221.

Moss, S. (2015), On the semantics and pragmatics of epistemic vocabulary, *Semantics and Pragmatics* **8**, 1–81.

Moss, S. (2016), *Probabilistic Knowledge*, Oxford University Press, Oxford.

Murzi, J. and Carrara, M. (2015), Denial and disagreement, *Topoi* **34**, 109–119.

Murzi, J. and Rossi, L. (2019), Generalized revenge, *Australasian Journal of Philosophy* **98**, 1–25.

Murzi, J. and Steinberger, F. (2017), Inferentialism, *in* B. Hale, C. Wright, and A. Miller (eds), *A Companion to the Philosophy of Language*, Wiley, Hoboken, NJ, pp. 197–224.

Nilsen, Ø. (2004), Domains for adverbs, *Lingua* **114**, 809–847.

Ninan, D. (2014), Taste predicates and the acquaintance inference, *in* M. W. Todd Snider and Sarah D'Antonio (ed.), *Semantics and Linguistic Theory (SALT)*, Vol. 24, LSA and CLC Publications, Ithaca, NY, pp. 290–309.

Oddie, G. (2016), Truthlikeness, *in* E. N. Zalta (ed.), *The Stanford Encyclopedia of Philosophy*, Winter 2016 edn, Metaphysics Research Lab, Stanford University.

Olson, J. (2014), *Moral Error Theory: History, Critique, Defence*, Oxford University Press, Oxford.

Peirce, C. S. (1905), What pragmatism is, *The Monist* **15**, 161–181.

Peterson, P. L. (1979), On the logic of 'few', 'many', and 'most', *Notre Dame Journal of Formal Logic* **20**, 155–179.

Pfenning, F. and Davies, R. (2001), A judgmental reconstruction of modal logic, *Mathematical Structures in Computer Science* **11**, 511–540.

Picollo, L. and Schindler, T. (2018), Deflationism and the function of truth, *Philosophical Perspectives* **32**, 326–351.

Pietroiusti, G. (2021), Disagreement and conflict: How moral and taste judgements do not differ, *Theoria* **87**, 837–846.

Poggiolesi, F. and Restall, G. (2012), Interpreting and applying proof theory for modal logic, *in* G. Restall and G. Russell (eds), *New Waves in Philosophical Logic*, Palgrave MacMillan, London, pp. 39–62.

Pollock, J. L. (1987), Defeasible reasoning, *Cognitive Science* **11**, 481–518.
Portner, P. (1995), Quantification, events, and gerunds, *in* E. Bach, E. Jelinek, A. Kratzer, and B. H. Partee (eds), *Quantification in Natural Languages*, Springer, Dordrecht, pp. 619–659.
Portner, P. (2007), Beyond the common ground: The semantics and pragmatics of epistemic modals, *in* J. Y. Yoon and K. A. Kim (eds), *The Perspectives of Linguistics in the 21st Century*, Hankook Publishing Company, Seoul, pp. 1–18.
Potts, C. (2007), The expressive dimension, *Theoretical Linguistics* **33**, 165–198.
Prawitz, D. (1965), *Natural Deduction. A Proof-Theoretical Study*, Almqvist & Wicksell, Uppsala.
Prawitz, D. (1974), On the idea of a general proof theory, *Synthese* **27**, 63–77.
Prawitz, D. (2015), Explaining deductive inference, *in* H. Wansing (ed.), *Dag Prawitz on Proofs and Meaning*, Springer, Dordrecht, pp. 65–100.
Price, H. (1983), Does 'probably' modify sense?, *Australasian Journal of Philosophy* **61**, 396–408.
Price, H. (1990), Why 'not'?, *Mind* **99**, 221–238.
Priest, G. (1979), The logic of paradox, *Journal of Philosophical Logic* **8**, 219–241.
Prior, A. N. (1960), The runabout inference ticket, *Analysis* **21**, 38–39.
Quine, W. V. (1950), *Methods of Logic*, Harvard University Press, Cambridge, MA.
Quine, W. V. (1951), Two dogmas of empiricism, *Philosophical Review* **60**, 20–43.
Quine, W. V. (1970), *Philosophy of Logic*, Harvard University Press, Cambridge, MA.
Rabern, B. (2012), Against the identification of assertoric content with compositional value, *Synthese* **189**, 75–96.
Ramsey, F. P. (1927), Facts and propositions, *Proceedings of the Aristotelian Society, Supplementary Volume* **7**, 153–206.
Ramsey, F. P. (1931a), General propositions and causality, *in* R. B. Braithwaite (ed.), *The Foundations of Mathematics and Other Logical Essays*, Kegan Paul, London, pp. 237–255.
Ramsey, F. P. (1931b), Truth and probability, *in* R. B. Braithwaite (ed.), *The Foundations of Mathematics and Other Logical Essays*, Kegan Paul, London, pp. 156–198.
Ramsey, W., Stich, S., and Garon, J. (1990), Connectionism, eliminativism and the future of folk psychology, *Philosophical Perspectives* **4**, 499–533.
Read, S. (2010), General-elimination harmony and the meaning of the logical constants, *Journal of Philosophical Logic* **39**, 557–576.
Read, S. (2015), Semantic pollution and syntactic purity, *Review of Symbolic Logic* **8**, 649–661.
Restall, G. (2005), Multiple conclusions, *in* P. Hájek, L. Valdés-Villanueva, and D. Westerståhl (eds), *Logic, Methodology and Philosophy of Science: Proceedings of the Twelfth International Congress*, Kings College Publications, London, pp. 189–205.
Restall, G. (2016), Review of *Advances in Proof-Theoretic Semantics*, *Notre Dame Philosophical Reviews*.
Ridge, M. (2006), Ecumenical expressivism: Finessing Frege, *Ethics* **116**, 302–336.
Ridge, M. (2014), *Impassioned Belief*, Oxford University Press, Oxford.
Ripley, D. (2015), Anything goes, *Topoi* **34**, 25–36.
Ripley, D. (2020), Denial, *in* M. T. E. Viviane Déprez (ed.), *The Oxford Handbook of Negation*, Oxford University Press, Oxford, pp. 47–57.
Rosenblatt, L. (2023), Paradoxicality without paradox, *Erkenntnis* **88**, 1347–1366.
Rumfitt, I. (1997), The categoricity problem and truth-value gaps, *Analysis* **57**, 223–236.

Rumfitt, I. (2000), 'Yes' and 'No', *Mind* **109**, 781–823.

Russell, J. S. and Hawthorne, J. (2016), General dynamic triviality theorems, *Philosophical Review* **125**, 307–339.

Santorio, P. (2017), Conditional excluded middle in informational semantics, *in* T. v. G. A. Cremers and F. Roelofsen (eds), *Proceedings of the 21st Amsterdam Colloquium*, ILLC, Amsterdam, pp. 385–394.

Santorio, P. (2019), Context-free semantics, *in* E. LePore and D. Sosa (eds), *Oxford Studies in Philosophy of Language*, Vol. 1, Oxford University Press, Oxford, pp. 208–239.

Scharp, K. (2013), *Replacing truth*, Oxford University Press, Oxford.

Schlöder, J. J. and Fernández, R. (2015), Pragmatic rejection, *Proceedings of IWCS 2015*, Association for Computational Linguistics, London, pp. 250–260.

Schlöder, J. J. and Fernández, R. (2019), How to reject what in dialogue, *in* C. Howes, J. Hough, and C. Kennington (eds), *Proceedings of the 23rd Workshop on the Semantics and Pragmatics of Dialogue*, SEMDIAL, London, pp. 1–10.

Schlöder, J. J., Venant, A., and Asher, N. (2018), Aligning intentions: Acceptance and rejection in dialogue, *in* R. Truswell, C. Cummins, C. H. B. Rabern, and H. Rohde (eds), *Proceedings of Sinn und Bedeutung 21*, University of Edinburgh, Edinburgh, pp. 1073–1089.

Schnieder, B. (2010), Expressivism concerning epistemic modals, *Philosophical Quarterly* **60**, 601–615.

Schroeder-Heister, P. (1985), Proof-theoretic validity and the completeness of intuitionistic logic, *in* G. Dorn and P. Weingartner (eds), *Foundations of Logic and Linguistics*, Plenum, New York, pp. 43–87.

Schroeder, M. (2008a), *Being For: Evaluating the Semantic Program of Expressivism*, Oxford University Press, Oxford.

Schroeder, M. (2008b), How expressivists can and should solve their problem with negation, *Noûs* **42**, 573–599.

Schroeder, M. (2009), Hybrid expressivism: Virtues and vices, *Ethics* **119**, 257–309.

Schroeder, M. (2013), Two roles for propositions: Cause for divorce?, *Noûs* **47**, 409–430.

Schultheis, G. (Forthcoming), Counterfactual probability, *Journal of Philosophy*, 1–36.

Schulz, M. (2010), Epistemic modals and informational consequence, *Synthese* **174**, 385–395.

Searle, J. (1962), Meaning and speech acts, *Philosophical Review* **71**, 423–432.

Searle, J. (1969), *Speech Acts: An Essay in the Philosophy of Language*, Cambridge University Press, Cambridge.

Searle, J. and Vanderveken, D. (1985), *Foundations of Illocutionary Logic*, Cambridge University Press, Cambridge.

Sellars, W. (1956), Empiricism and the philosophy of mind, *in* H. Feigl and M. Schriven (eds), *Minnesota Studies in the Philosophy of Science*, Vol. 1, University of Minnesota Press, Minneapolis, MN, pp. 253–329.

Sellars, W. (1969), Language as thought and as communication, *Philosophy and Phenomenological Research* **29**, 506–527.

Shapiro, S. (2003), The guru, the logician, and the deflationist: Truth and logical consequence, *Noûs* **37**, 113–132.

Sher, G. (2016), Substantivism about truth, *Philosophy Compass* **11**, 818–828.

Shiller, D. (2017), The problem of other attitudes, *American Philosophical Quarterly* **54**, 141–152.

Siebel, M. (2003), Illocutionary acts and attitude expression, *Linguistics and Philosophy* **26**, 351–366.
Simpson, M. (2018), Solving the problem of creeping minimalism, *Canadian Journal of Philosophy* **48**, 510–531.
Simpson, M. (2020), Creeping minimalism and subject matter, *Canadian Journal of Philosophy* **50**, 750–766.
Sinclair, N. (2006), The moral belief problem, *Ratio* **19**, 249–260.
Sinclair, N. (2011), Moral expressivism and sentential negation, *Philosophical Studies* **152**, 385–411.
Smiley, T. (1996), Rejection, *Analysis* **56**, 1–9.
Soames, S. (1999), *Understanding Truth*, Oxford University Press, Oxford.
Speaks, J. (2021), Theories of meaning, *in* E. N. Zalta (ed.), *The Stanford Encyclopedia of Philosophy*, Spring 2021 edn, Metaphysics Research Lab, Stanford University.
Stalnaker, R. C. (1968), A theory of conditionals, *in* W. L. Harper, R. Stalnaker, and G. Pearce (eds), *Ifs*, The University of Western Ontario Series in Philosophy of Science, D. Reidel, Dordrecht, pp. 41–55.
Stalnaker, R. C. (1970), Probability and conditionals, *Philosophy of Science* **37**, 64–80.
Stalnaker, R. C. (1978), Assertion, *in* P. Cole (ed.), *Syntax and Semantics*, Vol. 9, Academic Press, New York, pp. 315–332.
Stalnaker, R. C. (1997), Reference and necessity, *in* B. Hale and C. Wright (eds), *A Companion to the Philosophy of Language*, Blackwell, Oxford, pp. 534–554.
Stalnaker, R. C. (2002), Common ground, *Linguistics and Philosophy* **25**, 701–721.
Stalnaker, R. C. (2014), *Context*, Oxford University Press, Oxford.
Stefánsson, H. O. (2017), What is 'real' in probabilism?, *Australasian Journal of Philosophy* **95**, 573–587.
Steinberger, F. (2011), What harmony could and could not be, *Australasian Journal of Philosophy* **89**, 617–639.
Steinberger, F. (2013), On the equivalence conjecture for proof-theoretic harmony, *Notre Dame Journal of Formal Logic* **54**, 79–86.
Steinberger, F. (2019), Consequence and normative guidance, *Philosophy and Phenomenological Research* **98**, 306–328.
Stevenson, C. L. (1937), The emotive meaning of ethical terms, *Mind* **46**, 14–31.
Stojnić, U. (2017), One's modus ponens: Modality, coherence and logic, *Philosophy and Phenomenological Research* **95**, 167–214.
Strawson, P. F. (1949), Truth, *Analysis* **10**, 83–97.
Streumer, B. (2017), *Unbelievable Errors: An Error Theory about All Normative Judgments*, Oxford University Press, Oxford.
Svavarsdottir, S. (1999), Moral cognitivism and motivation, *Philosophical Review* **108**, 161–219.
Swanson, E. (2006), *Interactions with Context*, PhD thesis, Department of Linguistics and Philosophy, MIT.
Swanson, E. (2010), On scope relations between quantifiers and epistemic modals, *Journal of Semantics* **27**, 529–540.
Sweetser, E. (1990), *From Etymology to Pragmatics: Metaphorical and Cultural Aspects of Semantic Structure*, Cambridge University Press, Cambridge.
Szabo, Z. G. (2019), Semantic explanations, *in* E. Lepore and D. Sosa (eds), *Oxford Studies in Philosophy of Language*, Vol. 1, Oxford University Press, Oxford, pp. 240–275.

Tanter, K. (Forthcoming), Subatomic inferences: An inferentialist semantics for atomics, predicates, and names, *Review of Symbolic Logic*.

Tennant, N. (1997), *The Taming of the True*, Oxford University Press, Oxford.

Tennant, N. (1999), Negation, absurdity and contrariety, *in* D. Gabbay and H. Wansing (eds), *What Is Negation?*, Kluwer, Dordrecht, pp. 199–222.

Tennant, N. (2020), Inferentialism, logicism, harmony, and a counterpoint, *in* A. Miller (ed.), *Logic, Language, and Mathematics: Themes from the Philosophy of Crispin Wright*, Oxford University Press, Oxford, pp. 223–250.

Tiefensee, C. (2016), Inferentialist metaethics, bifurcations and ontological commitment, *Philosophical Studies* **173**, 2437–2459.

Tiefensee, C. (2019), Saving which differences? Creeping minimalism and disagreement, *Philosophical Studies* **176**, 1905–1921.

Tiefensee, C. (2021), Metasemantics for the relaxed, *in* R. Shafer-Landau (ed.), *Oxford Studies in Metaethics*, Vol. 16, Oxford University Press, Oxford, pp. 108–133.

Toppinen, T. (2014), Expressivism and Moore's paradox: A reply to Woods, *Journal of Ethics and Social Philosophy* **8**, 1–6.

Toulmin, S. E. (1950), Probability, *Proceedings of the Aristotelian Society, Supplementary Volume* **24**, 27–62.

Unwin, N. (1999), Quasi-realism, negation and the Frege-Geach problem, *Philosophical Quarterly* **49**, 337–352.

Unwin, N. (2001), Norms and negation: A problem for Gibbard's logic, *Philosophical Quarterly* **51**, 60–75.

van Fraassen, B. (1971), *Formal Semantics and Logic*, Macmillan, New York.

Veltman, F. (1996), Defaults in update semantics, *Journal of Philosophical Logic* **25**, 221–261.

Vendler, Z. (1967), Verbs and times, *in* Z. Vendler (ed.), *Linguistics in Philosophy*, Cornell University Press, Ithaca, NY, pp. 97–121.

von Fintel, K. and Gillies, A. (2007), An opinionated guide to epistemic modality, *in* T. S. Gendler and J. Hawthorne (eds), *Oxford Studies in Epistemology*, Vol. 2, Oxford University Press, pp. 32–62.

von Fintel, K. and Gillies, A. S. (2010), Must...stay...strong!, *Natural Language Semantics* **18**, 351–383.

von Plato, J. and Gentzen, G. (2008), Gentzen's proof of normalization for natural deduction, *Bulletin of Symbolic Logic* **14**, 240–257.

Wallace, M. (2022), Mental fictionalism, *in* T. Demeter, T. Parent, and A. Toon (eds), *Mental Fictionalism: Philosophical Explorations*, Routledge, London and New York, pp. 27–51.

Wansing, H. (2006), Connectives stranger than tonk, *Journal of Philosophical Logic* **35**, 653–660.

Warren, J. (2015), Talking with tonkers, *Philosopher's Imprint* **15**, 1–24.

Warren, J. (2021), Defending understanding-assent links, *Synthese* **199**, 9219–9236.

Wedgwood, R. (2007), *The Nature of Normativity*, Oxford University Press, Oxford.

Willer, M. (2013), Dynamics of epistemic modality, *Philosophical Review* **122**, 45–92.

Williams, J. R. G. (2020), *The Metaphysics of Representation*, Oxford University Press, New York.

Williams, M. (1988), Epistemological realism and the basis of scepticism, *Mind* **97**, 415–439.

Williamson, T. (1994), *Vagueness*, Routledge, Abingdon-on-Thames.

Williamson, T. (2000), *Knowledge and its Limits*, Oxford University Press, Oxford.
Williamson, T. (2003), Understanding and inference, *Proceedings of the Aristotelian Society, Supplementary Volume* 77, 249–293.
Williamson, T. (2008), *The Philosophy of Philosophy*, Blackwell, Oxford.
Williamson, T. (2010), Review of *Reason in Philosophy: Animating Ideas*, *Times Literary Supplement* **5579**, 22–23.
Williamson, T. (2020), *Suppose and Tell: The Semantics and Heuristics of Conditionals*, Oxford University Press, Oxford.
Wittgenstein, L. J. J. (1922), *Tractatus Logico-Philosophicus*, Routledge & Kegan Paul, London.
Wittgenstein, L. J. J. (1953), *Philosophical Investigations*, Wiley-Blackwell, New York.
Woods, J. (2014), Expressivism and Moore's paradox, *Philosophers' Imprint* **14**, 1–12.
Woods, J. (2018), A commitment-theoretic account of Moore's paradox, *in* K. P. Turner and L. Horn (eds), *Pragmatics, Truth and Underspecification*, Brill, Leiden, pp. 323–348.
Wright, C. (1983), *Frege's Conception of Numbers as Objects*, Aberdeen University Press, Aberdeen.
Wright, C. (1992), *Truth and Objectivity*, Harvard University Press, Cambridge, MA.
Wright, C. (1995), Critical notice of Michael Dummett's *Frege: Philosophy of Mathematics*, *Philosophical Books* **37**, 85–102.
Wright, C. (1998), On the (harmless) impredicativity of N=('Hume's Principle'), *in* M. Schirn (ed.), *Philosophy of Mathematics Today*, Clarendon Press, Oxford, pp. 393–368.
Wright, C. (2001), On being in a quandary, *Mind* **110**, 45–98.
Wright, C. (2002), (Anti-)sceptics simple and subtle: G. E. Moore and John McDowell, *Philosophy and Phenomenological Research* **65**, 330–348.
Wright, C. (2012), The pain of rejection, the sweetness of revenge, *Philosophical Studies* **160**, 465–476.
Yalcin, S. (2007), Epistemic modals, *Mind* **116**, 983–1026.
Yalcin, S. (2010), Probability operators, *Philosophy Compass* **5**, 916–937.
Yalcin, S. (2014), Semantics and metasemantics in the context of generative grammar, *in* A. Burgess and B. Sherman (eds), *Metasemantics: New Essays on the Foundations of Meaning*, Oxford University Press, Oxford, pp. 17–54.
Yalcin, S. (2018), Expressivism by force, *in* D. Fogal, D. W. Harris, and M. Moss (eds), *New Work on Speech Acts*, Oxford University Press, Oxford, pp. 400–430.
Zimmermann, T. E. (2000), Free choice disjunction and epistemic possibility, *Natural Language Semantics* **8**, 255–290.
Zynda, L. (2000), Representation theorems and realism about degrees of belief, *Philosophy of Science* **67**, 45–69.

Index